MEANS LEGAL REFERENCE FOR DESIGN & CONSTRUCTION

CHARLES R. HEUER, ESQ., AIA

MEANS LEGAL REFERENCE FOR DESIGN & CONSTRUCTION

CHARLES R. HEUER, ESQ., AIA

R. S. MEANS COMPANY, INC.
CONSTRUCTION CONSULTANTS & PUBLISHERS
100 CONSTRUCTION PLAZA
P.O. BOX 800
KINGSTON, MASSACHUSETTS 02364-0800
(617) 585-7880

10 9 8 7 6 5 4 3 2 1

This book was edited by Jill Farinelli and Ernest Williams. Typesetting was supervised by Joan Marshman. The book and jacket were designed by Norman Forgit.

Printed in the United States of America.

ISBN 0-87629-145-0
Library of Congress Cataloging in Publication Data

Dedication

This book is dedicated to my wife, Sheila, and daughter, Michelle. They support me in too many ways to mention. In addition to being resilient and good natured, they were able to forgive the disruptions to our lives which are inherent in a project such as this. I love them both.

TABLE OF CONTENTS

FOREWORD

The goal of this book is to enable its readers to recognize the legal implications of the situations in which they find themselves. They can then obtain additional information or specialized legal help. This book will not turn design and construction professionals into attorneys. Nor will it provide pat legal advice concerning every potential pitfall of a construction project. No book can do all that. Instead, this book will provide readers with a basic understanding of construction law so that they will be better able to plan their business operations and implement their business plans.

Design and construction law is a difficult topic to grasp. One look at the table of contents will explain why: The range of issues to be addressed is very broad indeed. What compounds the problem is that the material in this text is generally not covered in law schools, much less in educational programs focusing on real estate development, construction management, or construction contracting. Nor is this material taught as a separate topic in architectural and engineering programs. This is somewhat curious, given the myriad of legal issues that surround the typical construction project.

Furthermore, most people in the design and construction industry are *result-oriented*. They are interested in designing and building structures. Frequently, they perceive the law as so much paperwork, or procedural niceties that have little to do with the actual building process. All too often they describe the legal advice they have received as "100% correct and absolutely worthless."

This book will provide construction professionals with practical information on those legal issues commonly encountered in design and construction. It is intended, first, for nonlawyers working in the design and construction industry in the United States. This includes owners' and developers' representatives (field representatives, facilities managers, and project managers), design professionals (architects, engineers, and surveyors), constructors (general contractors and subcontractors), and suppliers. Although not primarily meant for use by attorneys, those legal professionals who are unfamiliar with the design and construction industry should also find it informative.

This book is organized into two parts. Part I, "The Foundations of Construction Law," is comprised of chapters on legal theories or areas of law that have particular relevance to the design and construction industry: basic business structures, contract law, agency law, tort law, real property law, intellectual property law, and labor law. It is important to understand that the coverage of the chapters in Part I is intentionally limited and focused. This part is NOT intended to be a legal treatise but, rather, the foundation, so to speak, on which Part II rests. Readers should be generally familiar with the substance of Part I before moving on to Part II.

Part II, "The Law and the Stages of Project Delivery," applies basic legal principles to situations commonly encountered by design and construction professionals during the course of project delivery. The chapters in this part are arranged to parallel the various stages of a typical construction project: financing, design, bidding, and construction. Licensing, insurance, and bonding are also examined in this part. Because disputes are often the final act on a construction project, this section concludes with chapters on common sources of dispute and methods of dispute resolution.

This book contains a comprehensive glossary of legal terms in a construction context. In addition, Appendix A contains a list of national trade associations, professional societies, regulatory agencies, and other construction-oriented organizations for reference purposes. Samples of some of the form contracts widely used in the industry are reprinted in Appendix B. They are the principal forms of contract between the owner and design professional, and the owner and construction contractor, as published by the American Institute of Architects (AIA) and the Engineers Joint Contract Documents Committee (EJCDC). The EJCDC is comprised of representatives from the National Society of Professional Engineers (NSPE), the American Consulting Engineers Council (ACEC), the American Society of Civil Engineers (ASCE), and the Construction Specifications Institute (CSI).

This book is designed to be used as a reference. It is quite possible that no one will read it from cover to cover but will, instead, *refer* to specific topics as the need and the interest arises. Those who are unfamiliar with legal concepts are best served by starting with some of the chapters in Part I. The chapters on contracts, torts, and real property will probably be of interest to most readers. The appeal of the other chapters will vary with the reader and the nature of his or her questions. Readers who already possess a working knowledge of legal principles, on the other hand, either from experience or from previous study, may want to skip directly to a specific chapter in Part II, mindful that he or she can refer to Part I as needed.

Finally, readers should understand that there is no shame in not knowing how to solve a particular problem. Rather, the shame is in not knowing that a problem exists. The key is to know enough to ask for help.

Caveat:
The material in this book is designed to inform readers about various legal and professional practice topics. Readers are cautioned to use this book as an information resource only. Determinations concerning particular legal matters or the application of the law to particular factual situations should not be made based solely on information contained in this book. Consult an attorney or other appropriate professional advisor before applying or acting on anything contained in or suggested by the materials in this book.

Acknowledgments

No major project is accomplished without help. In particular, I want to acknowledge Diane Ahearn, Esq., an associate attorney with The Heuer Law Group. Diane worked diligently, for an extended period of time, on this book. Her research and writing were of substantial assistance, and I am very grateful for her help and support.

I also want to acknowledge Jill Farinelli, the manuscript editor. Jill has a gift for sorting through complicated materials to make sure that they are organized and written in a clear and comprehensive manner. She was consistently helpful.

PART I
THE FOUNDATIONS OF CONSTRUCTION LAW

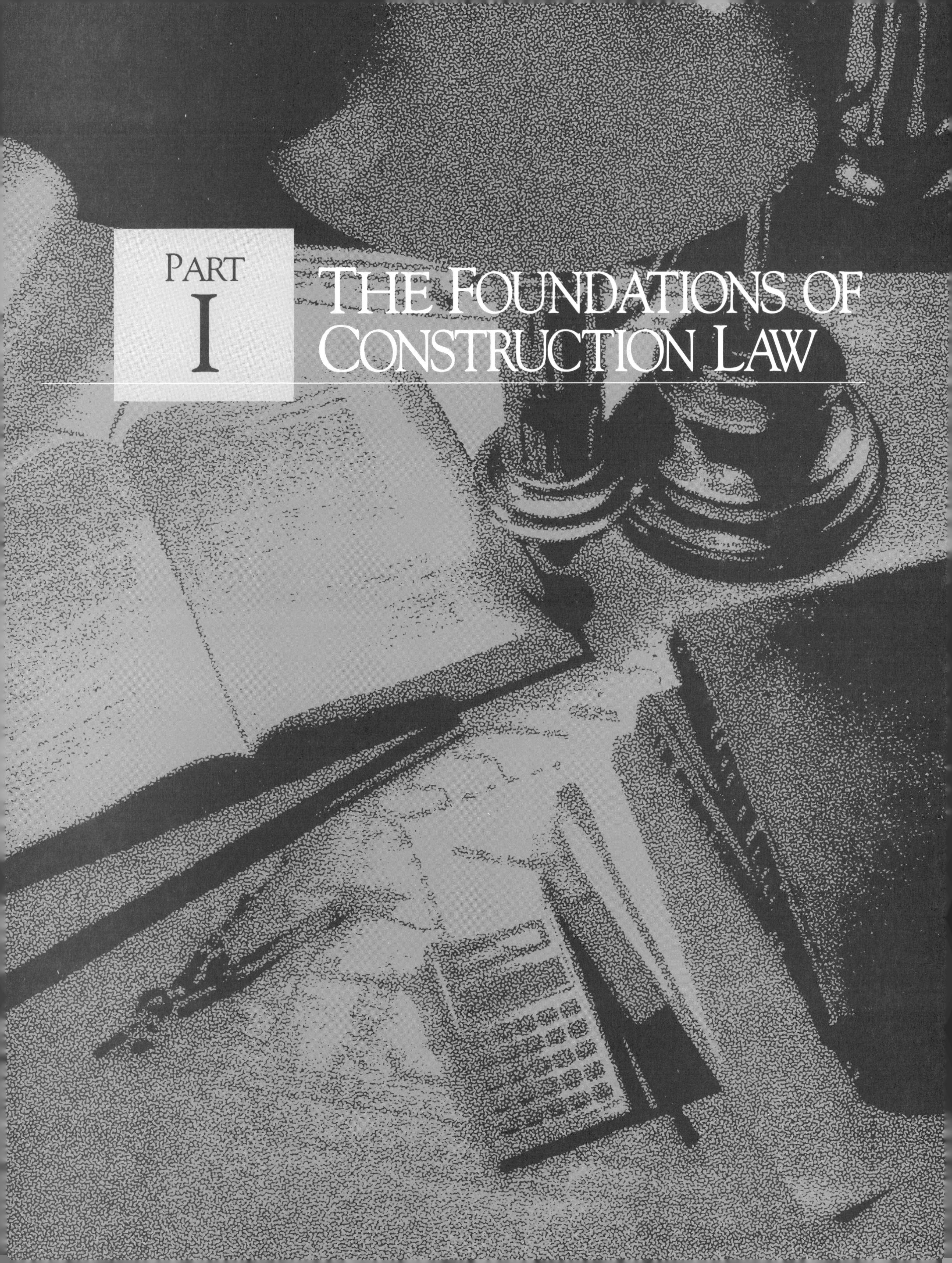

Disputes and other legal problems are, at best, expensive distractions from the primary business interests of design and construction professionals. Too frequently they result from misconceptions or lack of knowledge about the legal principles that govern professional and business relationships and transactions. Individuals sometimes adopt unrealistic positions in the mistaken belief that they are legally justified in doing so. To avoid this, design and construction professionals must understand the basic legal principles applicable to their business affairs. The following seven chapters provide a theoretical foundation for analyzing the legal implications of situations commonly encountered in design and construction.

CHAPTER 1
BASIC BUSINESS STRUCTURES

CHAPTER 1 BASIC BUSINESS STRUCTURES

Introduction

This chapter examines the basic business structures currently used in the design and construction industry. For those contemplating the establishment of a new business, the information that follows will help them choose a business structure that best suits their needs. For those who have already established businesses, this chapter will clarify aspects of their organization which perhaps are not well understood, as well as point out the opportunities and constraints of other types of organizations with which they conduct business or compete.

Types of Business Structures

Most businesses, including those in the construction industry, are structured in one of the following three ways:

- Sole proprietorship
- General partnership
- Corporation

Each of these business structures may vary in complexity depending on the business' particular needs. When organizing their businesses, owners should consider a number of factors, including the size of the business and its potential for growth, applicable tax laws, liability issues, capitalization and financing, and management. These factors determine how a business conducts its day-to-day affairs, and can affect the business' profitability as well as longevity. The material that follows discusses these factors with respect to each business structure.

Sole Proprietorship

A *sole proprietorship* is a business that is owned by one person. This person is referred to as the *proprietor*. Although there is only one owner in this type of business structure, there may be any number of employees providing production and management services. In fact, some sole propietorships are large operations. However, the owner of a sole proprietorship is ultimately responsible for the daily operations of the business as well as for fiscal matters and long-range planning.

Many businesses originate as sole propietorships, since this structure is the simplest form of business ownership. Because there is only one owner, there is no allocation of responsibility, and no one with which to share profits and losses. A large percentage of all architects in private practice are sole proprietors, for example. As a business grows, however, the owner will often restructure, usually to either a partnership or a corporation, to take advantage of the shared or limited liability achievable in business types with more than one owner. (Partnerships and corporations will be discussed later in this chapter.)

Establishing a Sole Proprietorship

A sole proprietorship is relatively easy to establish. When the owner has no employees, the process can be as simple as hanging out the proverbial shingle announcing that the proprietor is open for business. Unlike other types of business structures which must obtain a *federal employer identification number (EIN)* from the Internal Revenue Service (IRS) for tax purposes, the owner of a sole proprietorship with no employees is not required to obtain this number. Instead, the owner's personal Social Security Number (SSN) is used. If the sole proprietorship does have employees, however, the owner *must* obtain an EIN. The rules governing whether or not a business must have a *state employer identification number* are essentially the same.

Additionally, some states, counties, and municipalities require owners, depending on the nature of their business, to obtain professional and/or business licenses. For example, architects and engineers are required to be professionally licensed in the state(s) in which they practice. Often, contractors need to be licensed either in general, or to perform specific services such as asbestos removal.

Businesses that sell personal property items, such as building products and equipment, must obtain a *vendor's identification number* from the various states in which they do business. They must also collect and remit applicable state (and perhaps local) sales tax on the items they sell.

Finally, if the owner of the sole proprietorship is operating the business under a *fictitious name* or *trade name* (a name other than his or her own legal name), that name must be registered with the city or town in which he or she is conducting business. See Chapter 11 for more information on professional, trade, and business licensing.

Financing and Capitalization

The owner of a sole proprietorship must finance the business either with *equity* (his or her own money), *debt* (money borrowed from others), or a combination of the two. Equity is normally taken from savings that have been accumulated over the years. Friends and relatives may also loan money to the proprietor. However, bank loans are the principal source of debt financing for sole proprietors.

Because the sole proprietor *is* the business, particularly in the first few years, personal credit history and financial status determine if and to what extent a bank will lend the business money. (A sole proprietor cannot raise money by selling shares or interests in his or her business, because that would make the shareholders part owners and, consequently, change the business classification. By definition, the proprietor is the *sole* owner of a sole proprietorship.)

Proportionately fewer contractors than design professionals are sole proprietors. This is relatively easy to explain. The work that contractors and subcontractors perform requires a relatively large amount of capital. Most are not able to finance a contracting business by themselves. Professional design services, on the other hand, are less capital-intensive.

It is easier for an individual to provide the capital needed to establish this type of business. See Chapter 9 for more information on financing.

Management

The owner of a sole proprietorship has complete control over, and responsibility for, management decisions. This is true regardless of whether the sole proprietorship is a one-person operation or an operation with a large staff and an articulated management structure. The proprietor may make managerial decisions personally or may delegate that function to one or more employees. In either event, the owner is ultimately accountable for the management of the business.

This freedom appeals to those with an entrepreneurial spirit, who do not want to consult with others, and do not want to have others dictate how they should run their businesses. On the other hand, few people possess all of the talents necessary for successful business management or professional practice. As in most situations, two or more heads are usually better than one.

Liability

Normally the sole proprietor is personally *liable* (legally responsible) for the business' debts. The owner is also liable for the wrongful acts or omissions of any employees or other representatives of his or her business, if their wrongdoing was committed within the scope of their employment duties. For example, suppose that the foreman for a small painting company, while driving to the job site, fails to stop for a red light. He hits another vehicle, damaging it and injuring the other driver. If the sole proprietor of the painting company (the foreman's employer) is not fully insured for the acts of the foreman, his personal assets (such as his house, car, and other property) are subject to claims by the injured party for medical bills, repair costs, and the like.

Obviously, insurance is very important to the successful operation of a sole proprietorship and the personal financial well-being of the proprietor. The owner should investigate General Liability Insurance and surety bonds for the business, as well as Professional Liability ("errors and omissions") Insurance, as needed. (See Chapter 16 for a detailed discussion of the different types of insurance and surety bonds.)

Taxation

The owner of a sole proprietorship must withhold money from his or her employees' wages for federal income tax and social security tax (*Federal Insurance Contribution Act* [FICA]). These funds, and unemployment insurance taxes at both the state and federal level (*Federal Unemployment Tax Act* [FUTA]), are paid by the owner to the IRS and the individual state's department of taxation, on behalf of the employees.

Salaries and wages paid to employees and other business expenses are reported, along with business income, on the sole proprietor's personal tax return(s). The sole proprietor is taxed personally on the net income (receipts minus deductions, depreciation, and credits) of the sole proprietorship.

Continuity

A sole proprietorship ends no later than the death of the owner. In general, a proprietor may assign his or her current contracts and sell the "good will" of the business to another person or entity. The new owner may also be a sole proprietor, but he or she is recognized as distinct from the original proprietor. Thus, upon the assignment or sale of the sole proprietorship, the original business ends and a new one begins.

If the sole proprietor was a licensed design professional who maintained contracts with various clients, those contracts may not normally be assigned or sold, and will terminate upon the proprietor's death. This is because professional relationships with clients are considered personal and cannot be sold or transferred by will. Thus, only the physical assets of the business and the right to accounts receivable can be sold or transferred. The practice itself ends when the proprietor dies.

General Partnership

The *Uniform Partnership Act* is a set of laws governing *partnerships*. This act, which has been adopted largely intact in many states, defines a *general partnership* as "an association of two or more people to carry on a business for profit as co-owners." Typically, the participants in this type of business structure, called *general partners*, share all aspects of the business including assets, profits and losses, liability, and management responsibility. For example, from the title "Smith and Jones, Architects," one could assume that this firm was a partnership and that both Smith and Jones were general partners.

Establishing a General Partnership

A general partnership is established through an oral or written agreement between the partners. Although the law does not require partnership agreements to be in writing, it is prudent to do so to define and record the rights and responsibilities of each partner. The kinds of information that partnership agreements typically contain is shown in Figure 1.1.

If the partnership will be conducting business outside the state in which its principal office is located, it will be required to comply with professional and business licensing or registration laws in the applicable state. This frequently means that both the partnership and the individual partners must comply with the pertinent licensing laws. In addition, state law may require that a partnership's name be registered under the fictitious or trade name.

Financing and Capitalization

Like a sole proprietorship, a partnership is financed with some combination of equity and debt. To obtain equity financing, partners may contribute capital in the form of cash, property, or services. It is often easier for a partnership to raise debt capital than for a sole proprietorship. This is because in a partnership, there are at least two individuals standing behind the credit of the business. Furthermore, some lenders are more willing to lend money to partnerships, since the combination of the partners' talents provides some assurance of success in the business.

Management

Unless their agreement states otherwise, each partner in a general partnership has equal input into the management of the business. As a practical matter, however, in all but the smallest partnerships, one partner is usually designated the *managing partner*, responsible for a wide variety of day-to-day decisions on behalf of the partnership. On major issues, it is common to consult all partners before a decision is made.

Fiduciary Relationship

A *fiduciary relationship* is one of the hallmarks of a general partnership. In a partnership, each partner must put the interests of the partnership ahead of his or her own personal interests. Partners are expected to act in good faith and to disclose any information that might affect the partnership. In particular, a partner may not compete on a personal basis with the partnership, nor take personal advantage of a business opportunity without his or her partners' approval. (For additional information on fiduciary relationships, see Chapter 3, "Basic Agency Law.")

Liability

The general partnership is liable for the wrongful acts of any of its partners, as well as their failure to comply with contractual commitments. Similarly, each partner is fully responsible for his or her own actions to the extent of his or her personal fortune. Furthermore, each partner is also fully responsible for the actions of each of the other partners. A third party may sue one or all of the partners. If they are successful, any one partner may end up paying the full amount of the damages incurred by the third party. Although each partner is individually liable for the entire amount of the partnership's debts and obligations, the partners may, by agreement, *indemnify* (reimburse) each other fully or partially, for payments paid to a third party as a result of another partner's wrongdoing.

Although state laws differ, a general partner is not usually responsible for liabilities of the partnership which existed prior to his or her

---

Information Contained in Partnership Agreements

Partnership agreements usually contain the following information:

- The names and addresses of the general partners
- The address of the partnership's principal place of business
- The business objective of the partnership
- The duration of the partnership
- The amount of cash and personal property that each partner will contribute to the partnership
- The policy of the partnership with respect to borrowing and lending money
- The management responsibilities of each partner
- Procedures for admitting new partners and for withdrawal of existing partners
- Guidelines for compensating partners
- Procedures for paying a partner for his interest in the partnership upon his death or withdrawal from the partnership
- Procedures for resolving disputes between the partnership and individual partners
- Provisions for distributing the business' assets should the business dissolve

Figure 1.1

involvement in the partnership, unless he or she agrees to assume them. By the same token, a retiring partner usually remains liable for partnership debts incurred while he or she was still a partner.

Again, insurance is a very important part of controlling the risk of financial disaster. It is vital for every partnership to investigate General Liability Insurance, surety bonds, and Professional Liability ("errors and omissions") Insurance, as needed. (See Chapter 16.)

Proportionately more design professionals than contractors choose to conduct their businesses as partnerships. Not only does the design professional have a tradition of using partnership organizations but, until recent years, design firms have not needed the liability protection that has made corporate business structures more popular for contractors.

Taxation

If the partnership has employees, it must withhold money from their wages for federal income tax and social security tax (FICA). These funds, and unemployment insurance taxes at both the state and federal level (FUTA), are paid by the partnership to the IRS and the individual state's department of taxation, on behalf of the employees.

The general partnership is required to file an annual tax return with the IRS, reporting its income, capital gains, losses, credits, and deductions. However, this return is for informational purposes only; the partnership itself is not actually taxed on any income. Instead, each partner's share of net income or net loss is reported directly on his or her personal tax return. This is to the partners' benefit in that income is only taxed once (at the personal level, not the partnership level), and partnership losses may be used to offset personal income from other sources.

Continuity

The *Uniform Partnership Act* provides that if a partner withdraws, retires, declares bankruptcy, or dies, the partnership dissolves, unless there is a written agreement that provides otherwise. Dissolution may be desirable in a partnership with just a few partners, since the loss of a key partner may essentially mean the "death" of the partnership's business. In partnerships with a larger number of partners, however, dissolution would probably not be desirable. Most partnership agreements, therefore, contain procedures for continuing the partnership in the event that one or more of the partners leaves. Of course, there must always be at least two partners remaining to continue the partnership.

Other Types of Partnerships

There are other types of business entities that are essentially partnerships. They may have some characteristics that differ slightly from those of a general partnership, but they are legal partnerships nonetheless. These two entities are:

- Joint ventures
- Limited partnerships

Joint Venture

A *joint venture*, like a general partnership, is an association of two or more individuals or entities to conduct business for profit. Joint ventures are a very flexible form of business structure, and they have a wide acceptance in the design and construction communities.

A joint venture is formed simply by agreement of the parties involved. Like a general partnership, a written joint venture (partnership) agreement is normally prepared and signed by the parties. Unlike a general partnership, however, the partners of a joint venture anticipate that the business will be of limited scope and duration. Usually, a joint venture is formed to accomplish a single project or goal, and dissolves when the specific project is completed.

In the construction industry, it is common for architectural firms to form joint ventures to participate on projects that the individual firms would not have been able to undertake alone. This might occur on a very large project, or where, for example, a client in New York wants a Los Angeles-based firm to provide the conceptual design, and a local firm to prepare the construction documents and administer the construction contract. Contractors sometimes form joint ventures to construct unusual or large-scale projects. Further, design professionals and contractors sometimes form joint ventures to deliver design/build services.

Limited Partnership

A *limited partnership* has characteristics of both a general partnership and a corporation. It is composed of both general partners and *limited partners*. The general partners in a limited partnership manage the business and are personally responsible for partnership debts and obligations just like a general partnership. The limited partners, however, are somewhat like shareholders in a corporation (the discussion on corporations immediately follows) in that they contribute capital to the business and share the business' profits. However, in a limited partnership, the limited partners have no decision-making power and control over the day-to-day operations of the business. As such, their liability is limited to the amount of capital they have contributed to the business.

Corporation

A *corporation* is an entity created by statute which exists separate from its owners. Once created, a corporation is responsible for its own debts and obligations. The owners of the corporation are not personally liable for the corporation's debts. Further, each corporation has a centralized management structure and may be established to exist indefinitely. These characteristics are two of the most important reasons why corporations are a widely used form of business structure. Most contractors choose to organize their businesses as corporations largely for these reasons.

Establishing a Corporation

State statutes determine the procedure for establishing a corporation. Although state laws differ, they generally require the *incorporators* (the individuals forming the corporation), to file a standard form called the *articles of incorporation*, or *charter*, with the secretary of state, to be kept as public record. The corporation's name and address, its business objective, the names and addresses of its managers, and information about the type and amount of shares of stock it will issue usually appear in the articles of incorporation.

There are other documents that do not have to be filed with the secretary of state, but which must be prepared and kept with the corporate records. The incorporators must write a set of *bylaws*, which are the internal rules and regulations of the corporation, and must file their meeting minutes with the corporate records.

Once the corporation is created, it must observe several formalities to maintain its corporate status. The owners and managers of the corporation must conduct annual meetings and keep minutes of these meetings with the corporate records. The corporation must adhere to the bylaws, and certain actions must be voted on and approved by the owners and managers. If these formalities are not observed, the business may be viewed by the IRS and the law not as an entity separate from its owners but, instead, as a general partnership. This classification may have negative financial or legal consequences.

A corporation that plans to conduct business outside the state in which it was established must obtain permission from the "foreign states" and register as a "foreign corporation" in each of those states. It must also either establish and maintain an office in each state or appoint a person or entity (frequently the secretary of state) to represent the corporation in that state.

Some states, counties, and municipalities require corporations to obtain professional, trade, and/or business licenses. This generally poses no problem for contractors, but it may be a substantial problem for design professionals. Many states do not allow corporations to offer professional services or may impose substantial restrictions on such services. (See the discussion on professional corporations later in this chapter, and the licensing material in Chapter 11.)

All corporations are required to obtain federal and state employer identification numbers for tax purposes. Businesses that sell personal property items, such as building products and equipment, must also obtain a vendor's identification number from the various states in which they do business. They must also collect and remit applicable state (and perhaps local) sales tax on the items they sell.

Financing and Capitalization

When a corporation is established, provisions are made for dividing the value of the corporation into fractional parts. The total value of the equity in a corporation is called its *stock* or *capitalization*. Each fractional part is known as a *share*. Each owner of the corporation owns a specific number of shares of stock in the corporation. These shares are represented by formal certificates. The owners of these shares are commonly known as *shareholders* or *stockholders*.

Shareholders are required to contribute something of value in return for shares of stock in the corporation. Contributions may be in the form of cash, property (either tangible property, like desks and chairs, or intangible property, like copyrights or patents), services performed for the corporation, or expenses incurred on behalf of the corporation. A shareholder may also donate gifts to the corporation, but may not receive shares or stock in return for gift contributions.

Corporations may also borrow money to finance their business operations. They incur debt and are responsible for its repayment just like a sole proprietor or a partnership.

Management

As the *owners* of the corporation, shareholders must approve all fundamental changes in the corporation, such as the sale of corporate assets, amendments to the articles of organization, issuance of additional stock, and dissolution of the corporation. However, unlike owners of proprietorships or partnerships, the owners of a corporation do not necessarily have managerial control over the daily operations. Rather, they leave this responsibility to two groups:

- Board of directors
- Officers

Board of Directors: The shareholders elect a board of directors. This group of individuals is ultimately responsible for the management of the corporation. State law determines the minimum number of directors (often three) that are required to be on a board. However, a corporation need not have more directors than there are shareholders. With few exceptions, the directors of a corporation do not have to be employees or shareholders of the corporation. Directors are often selected based on their general business experience, their reputation, their skill, as well as the unique point of view that they can bring to the corporation.

Officers: The board of directors elects officers who manage the daily affairs of the business, such as signing contracts in the name of the corporation. Most states require corporations to include in their slate of officers a *president*, a *treasurer*, and a *secretary* or *clerk*. At its discretion, the board of directors may appoint other officers, such as a *vice president(s)*, an *assistant treasurer* and an *assistant secretary*. The officers are employees of the corporation, but are not necessarily shareholders.

Liability

Since a corporation is considered an entity separate from its shareholders, creditors or third parties who sue the corporation cannot make any claims against the shareholders' personal assets. This is because the corporation, and not the shareholders, have incurred the debt or committed the wrongful acts. The exception to this rule is when the claims arise from the shareholder's *intentional* actions or negligence. This is important to architects and other design professionals who personally perform professional services on behalf of the corporations for which they work. These individuals may be personally liable for their actions and might not be protected under the corporate structure. In all other instances, the most a shareholder could lose in a claim against the corporation is his or her own capital contribution to the corporation (the value of the shareholder's stock). This is because if the corporation went out of business and all of its property was taken by creditors, the stock would be worthless, and the shareholder would have lost his or her investment.

A corporation does have some advantage over other types of business structures when it comes to liability. Its status as a separate entity insulates its assets from the wrongful acts and omissions of its shareholders. For example, if a shareholder has an accident while driving to the store, the corporation will not be liable for the injuries or damage that result. The corporation does not incur liability as a result of the shareholder's acts.

As with sole proprietorships and partnerships, insurance is a very important aspect of the successful operation of the corporation. It is imperative for every corporation to investigate General Liability Insurance, surety bonds, and Professional Liability ("errors and omissions") Insurance, as needed. (Refer to Chapter 16 for a discussion of specific types of insurance and bonds.)

Taxation

Because a corporation is considered an entity separate from its owners, it is required to file an annual tax return and pay taxes on its net income. Such income may be retained by the corporation for legitimate future needs, but normally all or part of it is paid to shareholders. The total net income of the corporation is divided by the number of shares outstanding. The amount to be paid per share is called a *dividend*. Shareholders must report their dividends on their personal income tax returns and pay taxes on them. This results in *double taxation*, because the corporation's income is actually taxed twice: once when the corporation pays taxes on its net income, and again when the shareholders pay taxes on their dividends.

To avoid double taxation, some corporations have elected to be taxed like partnerships, in accordance with the provisions of Subchapter S of the Internal Revenue Code. Such corporations are referred to as "S" *corporations*. An S corporation, like a partnership, does not pay any tax on its income. Instead, the S corporation's net income is considered to be the income of its shareholders, who will report it on their personal tax returns. By electing S corporation status, a business can enjoy the usual advantages of corporate status, such as perpetual existence and limited liability for its owners, while avoiding taxation at the corporate level. In all other respects, an S corporation is exactly like a regular corporation.

A corporation must meet all of the requirements of Subchapter S of the Internal Revenue Code to be eligible for S-corporation status. Specifically, the corporation may not have more than 35 shareholders. These shareholders must be either individuals, estates, or certain types of trusts. Shareholders may not be nonresident aliens.

A corporation can elect S-corporation status temporarily. For instance, a corporation may elect this status one year, and choose to terminate it the next year (subject to specific restrictions) so that it is again taxed as a regular corporation.

Another option available to small business corporations is to create shares of stock called *1244 stock* (referring to Section 1244 of the Internal Revenue Code). In general, if a corporation meets the criteria defining a small business, and declares its intention to create shares of 1244 stock, shareholders will be able to deduct *capital* losses which they may incur if the business fails, from *ordinary* income on their personal tax returns. In the absence of such a provision, the capital gains necessary to offset such losses might not exist.

Continuity

Since corporations are legal entities separate from their owners, it is possible for them to exist indefinitely. Shares of stock may be bought and sold, and shareholders may come and go, without affecting the existence of the corporation.

Corporations can be terminated in one of several ways. First, corporations can cease to exist if a timetable has been set up stating the beginning and end of the corporation's existence. Second, corporations may be terminated by failure to file applicable reports with the state. Third, shareholders may voluntarily decide to terminate the corporation. This is done by filing appropriate documents with the state.

Other Types of Corporations

There are a variety of ways to structure a corporation. Two of the most common types of corporations are:

- Closed corporation
- Professional corporation

Closed Corporation

A *closed corporation* is a small corporation with a limited number of shareholders. Typically, the shareholders of a closed corporation place restrictions on the sale of shares to prevent outsiders from purchasing an interest in the business. The stock is held privately and is not traded on any stock exchange. A small, family-owned construction company might well be set up as a closed corporation to maintain full ownership and management control within the family.

In general, the shareholders are also officers and directors of the closed corporation. This means that unlike a typical large corporation, the shareholders of a closed corporation are directly involved in the management of the business. A closed corporation is established under the same legal requirements applicable to corporations in general.

Professional Corporation

A *professional corporation* is a corporation established by a group of individuals to render a particular professional service. Professional services, such as architecture, engineering, surveying, and landscape architecture, are services that are required by law to be performed by licensed individuals.

State laws govern the incorporation procedures for professional corporations, and place restrictions on how and by whom services may be delivered. Although state laws may differ, typically, the incorporators as well as the stockholders, certain officers (usually the president and vice president), and some or all of the directors, must be licensed in the appropriate discipline(s).

Some states require professional corporations to devise a written plan appearing in either an agreement, the articles of organization, or the bylaws, which enables the corporation to buy back a shareholder's shares in the event that a shareholder leaves the firm, becomes disabled, or dies. This requirement prevents individuals who are not professionals from acquiring stock and gaining control of the corporation.

In many states, professional services may not be offered or performed by corporations unless they are professional corporations established in accordance with the laws of the applicable state. This is a substantial consideration for design professionals who practice, or would like to practice, in more than one state.

Conclusion

In general, no one business structure is inherently better than another. The most important task of a business owner and his or her professional advisors is to evaluate and articulate the potential needs of the business, and to select and properly maintain the business structure that is most appropriate for that organization. See Figure 1.2 for a summary of the characteristics of the three major types of business structures.

Comparing the Three Types of Business Structures

	Sole Proprietorship	General Partnership	Corporation
Establishing the Business	Federal employer identification number required if there are employees. Vendor's identification number required if the business sells personal property. Professional, trade, and business licenses required, depending on state law. Registration of fictitious or trade name required.	Established through agreement between the partners. Both partnership and individual partners must comply with applicable licensing laws. Federal employer identification number required if there are employees. Vendor's identification number required if the business sells personal property. Registration of fictitious or trade name required.	Incorporators file articles of incorporation or charter with the state. Must write bylaws, file minutes and records, conduct annual meetings. Must register as a "foreign" corporation to conduct business in other states. Must obtain professional, trade, and business licenses in some states. Must obtain federal and state employer identification numbers. Must obtain a vendor's identification number if the corporation sells personal property. Must collect and remit sales tax on items that are sold.
Financing and Capitalization	Financed through equity and debt.	Financed through equity and debt.	Owners of corporation purchase shares of stock. (Owners are called shareholders.) Can also incur debt.
Management	Owner (sole proprietor) has complete control over management decisions.	General partners share equal control over management of partnership. There may be a managing partner.	Shareholders must approve all fundamental changes in the corporation. Elect board of directors who have ultimate responsibility for management decisions. Board of directors elects officers to manage daily affairs of the business.

Figure 1.2

	Sole Proprietorship	General Partnership	Corporation
Liability	Owner is personally liable for the business' debts, wrongful acts, and omissions of its employees and representatives.	Partnership is liable for the acts and omissions of any of the partners. Each partner is responsible for their own actions, as well as the actions of the other partners. Partners may indemnify each other. Partners are not responsible for the liability of the partnership which existed before their involvement. Retiring partners are responsible for debts and wrongful acts incurred while a partner.	Corporation is responsible for the debts and wrongful acts of the business. Corporation is protected from the acts of the individual shareholders.
Taxation	Owner withholds employees' wages for federal income tax and social security tax at both the state and federal level. Owner is taxed personally on the net income of the business.	Partnership withholds federal income tax and social security tax from employees' wages. Partnership must file annual tax return for information purposes, but partnership itself is not taxed. Each partner is taxed on his or her share of income or loss – on their personal tax return. Income is taxed just once – at the personal level.	Corporation must file annual tax return, pay taxes on net income. Net income is divided and paid to shareholders in the form of dividends. Shareholders then report dividends on personal tax returns. This results in double taxation.
Continuity	The business ends with the death of the owner.	If a partner leaves, for whatever reason, the partnership dissolves unless there is an agreement stating otherwise.	Corporation may exist indefinitely; shareholders come and go without affecting the existence of the corporation.

Figure 1.2 (cont.)

CHAPTER 2 BASIC CONTRACT LAW

CHAPTER 2 BASIC CONTRACT LAW

Introduction

A *contract* is a promise that the law will enforce. A contract can be a simple oral agreement in which one person promises to work for another in return for payment. Or, a contract can be a complicated document incorporating a variety of supplementary materials and running hundreds of pages in length.

Construction projects cannot be undertaken without contracts. Most construction-related contracts are based on promises to perform services or supply goods and equipment in exchange for payment. Additional issues may be addressed as well, including a time frame for performing the services, the terms and conditions under which the services will be performed, and procedures for settling disputes.

Construction projects usually involve several parties and require several different contracts. On a traditionally structured project, for example, there are separate contracts between the owner and the design professional, the owner and the general contractor, and the general contractor and the subcontractors. These contracts establish the relationships between the parties and define their rights and responsibilities. They should be coordinated so that they are consistent with one another. Inconsistencies may cause problems later in the project. For example, confusion may result if the contractor's agreement instructs him to send payment applications to the architect for review, but the architect's agreement does not mention any such review.

A contract that treats all parties fairly, that covers the basic issues known at the time, and that establishes a method for dealing with unexpected events, helps minimize the delays and disputes that are a part of most construction projects. This chapter will examine the fundamentals of contract law of which all practicing construction professionals should be aware.

Contract Requirements

A legally enforceable contract results from the following elements:

- Mutual assent
- Consideration
- Legal capacity to contract
- A legally permissible objective

These are described below.

Mutual Assent

A contract is not legally enforceable unless two or more parties have agreed to be bound to its terms. This is called *mutual assent.* Parties manifest their mutual assent to contract when one party makes an *offer* and the other party *accepts* that offer. An example of an offer in the construction industry is a *bid.* In the bid, the contractor essentially states, "I offer to build the project described in the contract documents, according to their terms, for 'x' dollars." The owner accepts the offer by awarding the contract to the selected bidder.

Sometimes it is difficult to determine whether an overture is, in fact, an offer, or merely part of the preliminary negotiations. In some cases, what seems to be an offer is really an *invitation to make* an offer. For example, an invitation for bids is not an offer. The owner is not offering to make a contract with all contractors who submit bids. Instead, the invitation is a *request* for offers.

An offer has been made if the party who received it (in legal terms, the *offeree*) has reason to believe from the surrounding circumstances, and from the words and actions of the party who made the offer (called the *offeror*), that the offeror intended to invite the offeree's assent to a contract. An offer creates in the offeree the capability to accept it and make a contract. Once the offeree accepts the offer, a contract has been formed.

A contract cannot be formed unless the offer is accepted *exactly as offered.* Sometimes the response to an offer is neither to accept it as offered nor to reject it completely, but to propose modifications to the original offer. An offer that is modified by the offeree and proposed to the offeror is called a *counteroffer.*

Assume, for example, that an architect offers to design a new home for a potential homeowner. The terms of the architect's offer are that he will perform basic services up to the time when a building permit is issued for a fixed fee of $10,000. In response to that offer, the homeowner says that he will hire the architect if the architect performs services through the actual construction period, and for a fee of $9,500. The homeowner's response is the counteroffer because it contains terms different from those in the architect's original offer. The counteroffer terminates the architect's original offer. That is, the homeowner cannot change his mind and accept the original offer unless the architect agrees to allow it. The counteroffer is, effectively, a new offer that the architect is free to accept or reject.

A series of offers and counteroffers are often part of the preliminary contract negotiations. This process narrows the disputed terms of a potential contract and may eventually bring the parties to agreement and the formation of a legally binding (enforceable) contract.

Unless there are terms in the offer stating otherwise, an offer may be withdrawn at any time prior to acceptance. Consequently, it is common for owners to require that bids be open for acceptance for 30 to 60 days after submission to give the owner time to evaluate them. By the same token, if the architect in the example above decides that he does not want to design the homeowner's house, for whatever reason, he may withdraw his offer at any time prior to receipt of the homeowner's response.

Consideration

A contract is not legally enforceable without *consideration*. Consideration is the price, motive, or benefit that induces the parties to form a contract. For consideration to exist, both parties to the contract must intend to enter into the contract and must be willing to take on the obligations associated with it. Each party furnishes consideration to support his or her part of the deal. In a typical construction contract, for instance, consideration is a promise to pay money in exchange for a promise to perform work. Consideration involves a bargained-for exchange of something of *legal value*.

Performing a physical act that one does not otherwise have to perform, forbearing action that one is otherwise permitted to take, and making a promise all have legal value and constitute good consideration. In other words, the person furnishing the consideration must act, or promise to act, in a circumstance in which he or she does not otherwise have to act. Alternatively, a party may forbear, or promise to forbear, an act that he or she has the legal right to perform. Assume, for example, that a painter promises to paint the house of someone to whom he is not related and to whom he owes no obligations. If that person promises to pay him $3,000 (which is not otherwise owed), the promises have legal value. The parties bargained and undertook to exchange promises to do something that they otherwise had no obligation to do.

There is no consideration if either of the parties intends to give a gift, however. For example, if a college student promises to paint his grandmother's house over the summer out of love and affection for his grandmother, there is no legal consideration involved. The grandmother cannot enforce the promise to paint the house. In this case, only the student is giving something of legal value, while the grandmother has undertaken no obligation in exchange.

Courts generally do not examine the *commercial value* of the consideration. For example, a person could agree to sell his new, four-bedroom house for $10 if he chose to. The $10 is sufficient legal consideration to bind the contract because it represents a bargained-for exchange, although it is probably a commercially inappropriate price for the house.

In some states, there is a presumption that valid consideration exists if a contract is made *under seal*. A contract under seal is created by words like "signed and sealed this ____ day of ____" near the space provided for signatures on a contract form. Contracts under seal do not designate a type of contract so much as a procedure for determining whether or not consideration has been provided. Under old English law, the monarch could make binding contracts simply with the impression of his seal of office placed in wax on the contract. The seal was a substitute for legal consideration. Today in most states, however, there must be legal consideration even for contracts under seal.

Legal Capacity to Contract

A contract is not enforceable if one or more of the parties who entered into it did not have the *legal capacity* to do so. *Incapacity* arises if a party, at the time the contract was made, was either:

- Under age (less than 18 or 21 years old, depending on state law)
- Intoxicated so as to be incapable of understanding the nature of his or her action
- Mentally ill
- Under legal guardianship

In the construction industry, incapacity may occur if a corporation attempts to enter into a contract when it is not empowered by law (if it is not within the proper powers of the corporation as authorized by the state) to do so. For example, in many states, general business corporations are not entitled to practice architecture or engineering. A corporation located in one of these states would therefore not be legally capable of contracting to perform architectural services for a business or a member of the general public.

Legally Permissible Objective

Finally, the law will not enforce a contract whose objective violates a *statute* (a law enacted by the local, state, or federal government) or the *common law* (law derived from the law of England and developed through court decisions). A contract that requires one or more of the parties to perform an illegal act, therefore, is not enforceable. An example of an illegal contract is one which requires an architect to disregard fire codes or other building safety regulations.

Types of Contracts

There are two basic types of contracts with which construction professionals should be familiar:

- Express contracts
- Implied contracts

These contracts are examined in detail below.

Express Contracts

When the parties express their mutual assent through words, either written or spoken, they create an *express contract*. Most contracts for design and construction services are express contracts. Generally, contracts do not have to be in writing to be legally enforceable. However, if the parties disagree over the terms of an oral contract, it is much more difficult for them to prove that their interpretation or recollection of the terms is correct. If a dispute develops and the parties seek to enforce the contract through *arbitration* (presenting the dispute to a private party who will make a decision with which the parties agree to abide), or *litigation* (in a court of law), it is much easier to prove the existence and terms of an agreement that has been recorded in writing.

Contracts Required to be in Writing

In some states, certain contracts are *required* by law to be in writing. These laws are commonly referred to as *statutes of frauds*. The original statute of frauds was enacted in England in 1677 and was designed to prevent fraud and perjury by persons involved in specific types of transactions. The various laws enforced today have the same general goal. Of course, regardless of the requirements of the statute of frauds, it makes good sense to reduce all contracts to writing.

A contract does not have to be a single formal written document to meet the requirements of the statute of frauds. Any "writing," whether an informal letter, (called a *letter contract*), a memo, or a receipt, as long as it is signed by, or on behalf of, the party to be *charged* (bound by the terms of the contract), is sufficient to bind that party to the express terms of the contract. The writing may also consist of several documents, if they each indicate that they relate to the same transaction. Figure 2.1 lists the types of construction industry contracts that are required by the statute of frauds to be in writing.

Implied Contracts

Implied contracts are created when a promise is *inferred* from the parties' conduct rather than from their express words. For example, assume that a homeowner says to a roofer, "I will pay you $8,000 to put a new shingle roof on my house." The roofer says nothing in response but proceeds to install the new roof. The roofer and the homeowner have an implied contract which is enforceable in court should the homeowner fail to pay the roofer upon completion of the project.

Quasi-Contracts

Sometimes situations arise in which one party has received a benefit but is under no legal obligation to pay for it or to take action in response because the parties did not agree to form a contract. For example, construction contracts normally require that the contractor give the owner written notice of changed conditions before performing any work for which he or she intends to be compensated in an amount beyond the contract sum. If written notice is not provided, but the extra work is performed anyway, the owner receives the benefits of the work but is under no obligation to pay extra for it.

In a situation such as this, the law may impose on the party who received the benefit (the owner) an obligation to pay the party who gave the benefit (the contractor), despite the absence of an actual contract between them. This is called a *quasi-contract.*

Although the parties did not intend to form a contract, the owner would be enriched at the expense of the contractor if he is not forced to pay for the work. Quasi-contracts are not *true* contracts, because there is no mutual assent between the parties. In the interest of fairness, however, courts will often act as though a contract does exist which requires each party to do what is fair under the circumstances.

To enforce a quasi-contract, Party A must show that: 1) he rendered services that benefited Party B; 2) he rendered the services with the expectation of being paid (he did not volunteer); and 3) to allow Party

Types of Contracts Required to Be in Writing

There are several types of contracts relevant to the construction industry which must be in writing. These are:

- Contracts in which a party promises to assume responsibility for the debt of another person
- Contracts in which an *executor* (one who promises to execute another's will) promises to pay the debts of a deceased person's estate out of his (the executor's) own funds
- Contracts for the sale of land, or any interest in land
- Contracts that *cannot* be (as opposed to "are not intended to be") performed within one year

Figure 2.1

B to benefit without compensating Party A would result in the unjust enrichment of Party B at the expense of Party A. If these three conditions are *not* met, then a party will not be forced to pay for the changes or additions. This is illustrated in the case entitled *Coffee Pot Plaza Partnership v. Arrow Air Conditioning and Refrigeration, Inc.* (see Figure 2.2).

Modifying Contracts

A *modification* is a change to a contract which is made after the contract has been *executed* (signed by both parties). A *change order* is an example of a modification to a construction contract whereby the parties agree to a different price, different services, or perhaps a different quantity of materials. The parties are free to modify the agreement, but all the elements necessary for contract formation must be present in order for the contract to be enforceable as modified.

The only element that is likely to be missing from a modified contract is consideration. For example, a contractor could claim that certain work was not required under the original contract, and subsequently refuse to perform such work unless a modification is agreed upon which would provide him with additional money. The owner may agree to pay the contractor to perform the work. If the owner does not pay, however, the contractor may not be able to enforce the modification, due to lack of consideration. If the work was, in fact, required by the original contract, the contractor already had a duty to perform it. By promising it again, he gave nothing of legal value in exchange for the owner's promise to pay. The contractor had a *prior existing duty* to do the work and provided no new consideration for the modification.

Coffee Pot Plaza Partnership v. Arrow Air Conditioning & Refrigeration, Inc., 412 So. 2d 883 (Fla.Dist.Ct.App. 1982)

An owner of a supermarket in Florida divided his property into two stores. The owner occupied one store, and his tenant rented out the other. The tenant made an agreement with a contractor for the installation and repair of refrigeration equipment so that he could operate a meat market. The tenant then breached his agreement with the contractor by failing to pay for services. At the same time, the tenant did not pay rent to the owner.

The owner terminated the tenant's lease because he failed to pay rent. The owner then tried unsuccessfully to find another tenant to lease the meat market. Meanwhile, the contractor sued the owner for payment for the services performed for the tenant. The contractor claimed that the owner received the benefit of the refrigeration equipment installed on his property and therefore had a *quasi-contract* with the contractor. He argued that the owner would be unjustly enriched at the contractor's expense if he did not pay for the services, even though they were performed at the request of the tenant.

The court ruled in favor of the owner, citing that the owner was not unjustly enriched because he did not request the contractor's services and, because the owner was unable to find a new tenant, he did not receive rent nor benefit from the contractor's services.

Figure 2.2

Interpreting Contracts

One of the reasons for reducing a contract to writing is to document what the parties have agreed to should questions or disputes arise. When this happens, it is necessary to look to the contract to interpret its meaning. Therefore, it is important when preparing contracts to make the contract language as explicit as possible. Parties can then avert or resolve most disputes simply by referring back to the terms of their contracts.

There are two basic ways to interpret written contracts:

- Look to the contract's intent
- Apply standard rules of interpretation

These are described in detail in the following sections.

Intent

A fundamental rule of contract law is that parties are generally free to contract on terms to which they can agree. For this reason, when a judge or arbitrator interprets a contract, he or she first attempts to determine the parties' intent. To determine intent, judges and arbitrators look to the wording of the contract, to customs in the industry, to common usage of terms, to the parties' conduct, and to surrounding circumstances. If a judge or arbitrator cannot discern the parties' intent, then he or she will apply other *standard rules of interpretation* to the contract, or fill in any incomplete portions as deemed proper.

Standard Rules of Interpretation

The following rules of contract interpretation have been developed over the years based on decisions made in cases where contracts were interpreted by the courts. These are summarized in Figure 2.3, and discussed in more detail in the following sections.

Parol Evidence Rule

If the parties' agreement consists of several writings and conversations, the court must determine the scope of the contract, that is, which communications between the parties are part of the contract.

First, the court asks whether the parties intended the written contract to be a final expression of the terms of their agreement. In other words, does the writing represent the agreement in full? If there are many writings made on different dates, the court will examine them as a whole.

Most construction contracts contain an *integration clause* which states that the written agreement is complete, and that all oral and other agreements are integrated, or merged, into it. This means that no matter what the parties discussed, their final agreement is represented by the writing in the contract, for better or for worse. If a contract contains an integration clause, most courts will not listen to testimony about oral agreements that purport to *vary or contradict* the terms of the written agreement.

If there is no integration clause in the contract, but the court decides that the writing(s) is, in fact, intended to be a final expression of the parties' agreement, the *parol evidence rule* is applied. Parol evidence is oral or written evidence of statements made by the parties *prior to, or simultaneously with*, the signing of the contract, and which *vary or contradict* the writing in the contract. The parol evidence rule requires that the court exclude (refuse to consider) parol evidence when interpreting the contract.

This rule is important in the construction industry because oral promises and representations are frequently made by various parties but are never included in the written contracts. Sometimes they are harmless sales pitches, but other times they are material statements about the quality of a product, warranties, or some other aspect intended to induce the other party to sign the contract. If such statements or promises are not included in the written contracts and are, in fact, contradicted in such contracts, the parol evidence rule will likely preclude the court from

Standard Rules of Interpretation

1. The court shall refuse to consider parol evidence (oral or written evidence of statements made prior to, or simultaneously with, the contract, which vary or contradict the contract).
2. Courts shall not infer terms that are different from those in the parties' contract.
3. All the writings that are part of a contract shall be interpreted as a unit.
4. Words shall be interpreted consistently throughout the contract.
5. Words shall be accorded their plain meaning whenever possible.
6. If a term used in a contract has a particular meaning in the construction industry, that meaning must be applied when interpreting the contract.
7. Technical terms used in the construction industry and incorporated into the contract shall be given their technical or trade meaning.
8. Specific provisions shall prevail over general terms.
9. When there are inconsistencies, handwritten insertions take precedence over typewritten insertions; typewritten insertions take precedence over terms printed in the form.
10. When there are differences among drawings, more recent drawings take precedence over older drawings. More detailed drawings take precedence over less detailed drawings. Large-scale drawings take precedence over small-scale drawings made at the same time. Specifications usually prevail over drawings.
11. Words ("ten thousand dollars") take precedence over figures ("$11,000").
12. Some contracts contain statements that establish the order of precedence among documents. When there is no preestablished order, inconsistencies must be resolved by reading the contract as a whole.
13. When a contract provision has more than one reasonable interpretation, the ambiguity shall be resolved in favor of the party who *did not* draft it.
14. If a number of items are included in a contract, but another item of the same type is omitted, that item shall remain excluded from the contract, even though it seems natural to include it.
15. Obvious clerical errors shall not affect the proper interpretation of a contract.
16. Words that have been erased or crossed out shall be ignored.
17. Titles and headings shall not take precedence over terms in the contract.

Figure 2.3

hearing testimony about them. Parties must take care to include all important promises and representations in their written contracts if they intend the other party to live up to them.

However, the parol evidence rule does *not* preclude a court from hearing testimony about the facts and circumstances surrounding the *making* of the agreement and the meaning of the written provisions. Neither does an integration clause preclude such testimony if it helps to explain, rather than vary or contradict, the terms of the written agreement.

Implicit Terms and Conditions

Courts are generally not allowed to infer terms that differ from those in the parties' express contract. Regardless of what the court *thinks* the parties should have said, it is what they *actually* said that matters.

There are certain terms, however, that are frequently omitted from the express language of most construction contracts. Regardless, courts regularly infer promises on the part of the owner, design professional, and contractor. Owners implicitly promise contractors:

- Cooperation and noninterference
- Access to the construction site
- Fair interpretation of the contract documents
- Accuracy in the information provided
- Adequacy of drawings and specifications

Conversely, contractors implicitly promise owners:

- Quality workmanship and materials
- Timely payment of bills
- Compliance with building codes and applicable laws
- Adherence to the progress schedule

Design professionals implicitly promise to exercise that degree of skill and care which can reasonably be expected from similar professionals practicing in the community.

Contract as a Whole

In a court of law, a contract is to be interpreted as a whole, and all writings that are part of the contract are interpreted as a unit. In other words, one part of a document cannot be interpreted without considering the terms of the entire document. It is inappropriate to take a word, sentence, or paragraph out of context and suggest a meaning for it on that limited basis.

Consistency in Usage

Words that have one meaning in one part of the contract must be given that same meaning when used in another part of that same contract.

Plain Meaning

In interpreting contracts, words are accorded their *plain meaning* whenever possible. If words have an ordinary meaning known throughout the country, a court cannot look beyond the written contract itself to determine the parties' intended meaning for those words. If the court has applied the plain meaning rule and the contract language is still unclear, then the court may look at other evidence. For example, the court may assess surrounding facts and circumstances, preliminary negotiations, the parties' prior statements and conduct, and customary usage in the trade.

Custom and Usage

The custom and usage of terms or phrases in the construction industry may provide a definition or fill in a gap in the contract documents. For example, if a term used in a contract has a particular meaning in the construction industry, then that meaning must be applied when interpreting the contract, even though the term may not mean the same thing to a layperson.

Technical Terms

Technical terms and words used in the construction industry and incorporated into the contract language are normally given their technical or trade meaning.

Specific v. General Terms

Specific provisions prevail over general terms. For example, it is common for construction contracts to declare that the owner will communicate with the contractor via the design professional. That general statement applies unless there is a specific statement to the contrary. If the contract also stated that the owner would communicate directly with the contractor concerning insurance matters, that specific statement would prevail and the owner would not need to communicate with the contractor via the design professional on insurance matters.

Printed Terms

When a preprinted form contract contains contradictory typewritten and handwritten insertions, the handwritten insertions prevail over typewritten insertions. In turn, the typewritten provisions take precedence over terms that are preprinted in the form. The reasoning behind this rule is that handwritten and typewritten changes in a contract indicate that the parties took an active role in drafting the contract and, therefore, those changes should be interpreted as the parties' true intent.

Drawings

Construction drawings are usually considered part of the construction contract documents. Therefore, the rules of contract interpretation apply to interpreting construction drawings as well.

When there are differences among drawings, the more recent drawings take precedence over older drawings. More detailed drawings prevail over less detailed drawings made at the same time. Additionally, larger-scale drawings predominate over smaller-scale drawings made at the same time.

If there are inconsistencies between a project's drawings and specifications, and there is no rule of interpretation or order of precedence within the specifications themselves, courts generally hold that the specifications prevail over the drawings. There does not seem to be a particular reason why. Perhaps judges, who deal with words all the time, do not consider drawings as important as specifications.

Dollar Amounts

When there is a discrepancy between references to dollar amounts, written words ("ten thousand dollars") prevail over numerical figures ("$11,000").

Inconsistencies Between Documents

It can be very difficult to resolve inconsistencies among the various documents when a contract consists of more than one document, as construction contracts usually do. All contract documents produced by the American Institute of Architects (AIA) are deemed to be complementary. That is, there is no preestablished order of precedence among the documents for the purpose of resolving inconsistencies. When these documents are used, inconsistencies must be resolved based on other rules of interpretation, such as reading the contract as a whole. Sometimes, however, the parties will include in their contract an explicit statement concerning the order of precedence. When this statement is present, it may not be possible to read the documents as a whole.

Consider the following language:

> *In the event of conflict or discrepancies in the Contract Documents, the priority level of interpretation shall be as follows:*
>
> *Priority Level*
>
> 1. *Agreement*
> 2. *Addenda (last taking precedence)*
> 3. *Special Conditions*
> 4. *Supplementary Conditions*
> 5. *General Conditions*
> 6. *Drawings and Specifications*

If this language were included in the contract, it would be necessary to decide that a provision in the General Conditions was correct, and a conflicting note or indication in the drawings was incorrect, regardless of how that decision might have been made had the contract been considered as a whole.

Interpretation Against the Drafter

Sometimes a contract provision has more than one reasonable interpretation. Generally, the ambiguity is resolved in favor of the party who *did not* draft it.

For example, imagine a provision which stated that payment would be made "upon payment by the owner, no later than April 15th." If April 15th arrives, but the owner has not paid, is payment due under this contract? The provision is ambiguous and would normally be interpreted against the party who drafted it. If the drafter was the party with the obligation to pay, that party would probably be ordered to do so on April 15th, despite other possible interpretations of the phrase. The rationale behind this rule is that the drafter had the opportunity to make the document clear. If the writing is ambiguous, the drafter should bear the consequences. This rule does not apply, however, if both parties negotiate the terms, or if each party was represented by a lawyer.

"Expression of One Excludes Another"

If a number of items are mentioned specifically in a contract, but another item of the same type is not mentioned, that item will be excluded from the contract, even though it seems natural to include the omitted item. For instance, if a contract states that a contractor will be excused from performance in case of fire, hurricane, or flood, will he be excused if there is a blizzard? A blizzard will not excuse the contractor from performance because it was not mentioned specifically, even though it is also considered a "natural disaster."

Clerical Errors and Omissions

Obvious clerical errors will have no effect on the proper interpretation of a contract.

Erasures and Cross-outs

Words that have been erased or crossed out in the contract will be ignored.

Titles and Headings

Titles and headings may assist interpretation by making it easier to locate provisions in the contract, but they will not be a controlling factor and will not be given the same weight as the terms within the contract.

Breach of Contract

A principal function of a contract is to establish and govern the relationship between two parties in certain situations or transactions. Normally, contracts include provisions for resolving disputes between the parties about their performance. When a party fails in some way to perform in accordance with the terms of the contract, that failure is called a *breach of contract.*

Theoretically, there are a number of potential breach of contract claims that may arise on a construction project. Some of the more prevalent claims by contractors against owners result when:

- The owner does not allow the contractor to begin performance after the contract has been awarded
- The owner wrongfully terminates the contractor's right to perform the balance of the contract
- The owner fails to pay the contractor for work that has been performed
- The owner increases the contractor's cost during performance in unanticipated ways

Potential breaches of contract which an owner may claim against a contractor arise when:

- The contractor fails to begin work
- The contractor fails to complete the project
- The contractor fails to pay subcontractors or suppliers

Regardless of the parties involved, in each case there are three conditions that must exist for liability to be imposed based on a party's breach of contract:

- A valid express or implied contract must exist between the parties
- One party must have failed substantially to perform one or more contractually defined duties or obligations
- The other party must have suffered a loss or injury as a result of the first party's failure to perform its duties

The absence of even one of these conditions is sufficient to preclude liability for breach of contract. If there is a breach of contract, however, a party's right to payment is based on the extent of the breach, that is, whether a party has *partially* or *substantially* performed the obligations in his or her contract. This is discussed below.

Partial Performance

If a party partially fulfills his or her contract but does not fully and completely perform all of the obligations, he or she generally has no right to recover any part of the contract price from the other party.

Substantial Performance

The principle that partial performance equals breach of contract can lead to very harsh results. A contractor, for example, might perform a large percentage of the construction required under his or her contract, but not be entitled to any payment if he did not finish completely. Therefore, over the years, the doctrine of *substantial performance* developed in the courts. It provides that a party who has *substantially* performed most of its contractual obligations is entitled to be paid at least a portion of the contract price. That portion would normally be equal to the full contract price minus the amount it will cost the other party to have the job completed as promised under the contract.

For example, a contractor who completes all of the construction required by the contract documents, except for minor "touch-up" items and submission of required paperwork to close out the job, has probably substantially performed (or substantially completed) the project. That contractor would normally be entitled to be paid the full contract price minus the cost to the owner of the remaining work. The doctrine of substantial performance is particularly important in the construction industry, where contracts contain thousands of very detailed requirements that are virtually impossible for an individual to perform fully and completely.

Anticipatory Breach of Contract

An *anticipatory breach of contract* occurs when one party to a contract indicates to the other party, prior to the expected date of performance, that he does not intend to perform under the contract. Not all state laws recognize this concept because, in fact, there is no failure to perform until the time for performance arrives. Under these circumstances, what should the other party do if it is capable and willing to perform its side of the bargain? For example, should the architect continue to perform services after the owner has stated that bills will not be paid when due? Does the architect have to wait until the due date for payment has passed before suing for breach of contract? Many states have laws that will allow the architect to sue immediately upon learning of the impending breach. The following example explains the options that are available when one party repudiates the contract before the time for performance.

Assume that on April 1, Joe the architect contracts with Mary the homeowner to design an addition to Mary's home. The contract states that design documents are to be completed by June 1. On May 1, Joe tells Mary that he has not and does not intend to start the project. Joe has not actually breached the contract yet because he has until June 1 to perform. However, he has made it clear to Mary that he is not going to provide the services for which he contracted. Does Mary have to wait until June 1 to sue Joe for breach of contract? Since Mary is sure that Joe is not going to perform, can she try to get someone else to do the job?

Mary has three options, depending on state law. She can: 1) sue Joe immediately; 2) assume the contract is cancelled, find someone else to do the job, and then sue Joe for the costs of hiring someone else (If Mary sues Joe, she is entitled to the difference between Joe's contract price and the actual cost of hiring another architect to design the addition.); or 3) await the time for performance, thereby giving Joe a chance to change his mind and perform. If Joe still does not perform, Mary may then sue him for breach of contract. However, the third option is not always the wisest. By letting time pass and doing nothing to facilitate the completion of her project, Mary has made no effort to *mitigate*, or lessen, her potential damages. A court may look unfavorably on this, as will be seen later in this chapter.

Defenses

A *defense* is an explanation or reason why a person or entity should not be held liable (legally responsible) as a result of claims made against them by another party. Often, the best defense to a breach of contract claim is that an enforceable contract never existed. The basis for this defense may be that the agreement was invalid because it did not meet all of the requirements for forming a valid, legally enforceable contract.

In other situations, an event which occurs after the contract was formed may be a valid defense to claims of nonperformance. For example, insolvency or bankruptcy, and death or disability, are valid excuses for a party's failure to comply with the terms of the contract. Other defenses include:

- Failure to comply with the statute of frauds
- Duress
- Mistake
- Unconscionability
- Misrepresentation
- Impossibility
- Impracticability
- Statutes of limitations

Each of these is discussed below.

Failure to Comply with the Statute of Frauds

As described earlier in this chapter, the state statutes of frauds require contracts for certain types of transactions to be in writing and signed by the parties. In most states, failure to comply with the statute of frauds makes a contract *voidable*. This means that an agreement has been made, and may be performed voluntarily by the parties, but cannot be enforced against the will of the party who did not sign it.

Assume, for example, that Jones agrees to buy Smith's office building for $800,000. Jones and Smith, for whatever reason, do not put their agreement into writing (typically a purchase and sale agreement would be executed in writing). Smith then refuses to complete the sale. Thereafter, Jones sues Smith and requests that the court force Smith to complete the sale of his building. Smith's defense is that the oral agreement between them is unenforceable because it contemplates the sale of an interest in land, and is therefore required by the statute of frauds to be in writing. Based on this fact, Smith would probably win. A contract does exist, because the parties mutually agreed to its formation, but it is unenforceable because it was not in writing, as specifically required in this case by the statute of frauds.

Duress

Contracts executed under duress are not enforceable. The law will not enforce a contract that one party was physically or verbally coerced to sign.

Mistake

A contract is not enforceable if both of the parties misunderstood its material (substantial) terms at the time they entered into it. Assume, for example, that two parties contract for the sale of a specific building and discover that in actuality, they had two different buildings in mind. In effect, a contract was never created because the parties had no mutual understanding of the terms. In this case, one party could cite mistake as a defense and could terminate its obligation to perform under the contract.

Unconscionability

When one party takes advantage of another party's inexperience, or does not give a fair explanation of the contract terms, *unconscionability* may be cited as a defense to a breach of contract claim. Courts will not enforce a contract that is fundamentally unfair, or that "shocks the court's conscience." This could occur, perhaps, if an experienced interior designer took unfair advantage of a first-time homeowner by charging a full professional design fee *and* reselling at retail *plus a markup*, furnishings that were purchased at wholesale.

Misrepresentation

Misrepresentation occurs when one party knowingly provides false information that is important to the formation of the contract in circumstances in which the other party relies on such information. For example, to induce the sale of their home, a couple might tell prospective buyers that the basement has never leaked. But, in fact, the basement has leaked regularly, and the prospective buyers come to rely on this false statement in making their decision to buy the home. The buyers then discover the truth – after signing the purchase and sale agreement but before closing the sale – and they choose not to proceed with the deal. If the sellers sue for breach of contract, the buyers could cite misrepresentation as their defense.

Impossibility

With respect to contract performance, *impossibility* refers to *physical* impossibility. Courts are divided over whether or not impossibility of performance is a viable defense to a breach of contract claim. Some courts allow impossibility as a defense if the circumstances which make performance impossible were not foreseeable at the time the parties made the contract. Other courts allow impossibility as a defense if the task was not only very difficult, but impractical as well (see Figure 2.4, *Miller v. City of Broken Arrow*).

Still other courts will not entertain impossibility of performance if the individual being sued entered a situation that he or she knew was inherently risky. For example, assume that an experienced contractor

***Miller v. City of Broken Arrow*, 660 F2d 450 (10th Cir. 1981), cert. denied, 455 US 1020 (1982)**

The city of Broken Arrow hired a contractor and an engineer to work on a sewer project. During the course of construction, the engineer issued a change order directing the contractor to use crushed stone to make the muddy and unstable ground firm. The contractor used the crushed stone, but found that the ground was still not firm. He complained to the engineer, who told him to continue using the stone. The contractor completed most of the project but could not finish because the ground was still unstable. At that point, the engineer advised the city to terminate its agreement with the contractor because the contractor failed to complete the project. The city did so and sued the contractor.

The contractor filed a counterclaim stating that it was impossible for him to stabilize the ground using the crushed stone specified by the engineer. The court held that, in order to recover under a theory of *impossibility of performance*, a plaintiff must show actual difficulty in performance, as well as *impracticability* (generally defined as extreme and unreasonable difficulty, loss, injury, or expense). The court said that a contractor should be able to rely on an engineer's plans and specifications. Since it was not only difficult for the contractor to complete the project according to engineer's plans, but impractical as well, the contractor was entitled to recover from the city the balance of the contract price, in addition to lost profits and delay damages.

Figure 2.4

agrees to build a prototype design—one that has never been built before. The contractor then promises that the project, when constructed, will work as planned. The contractor may be liable for breach of contract if the project does not work, even if such performance turns out to be physically impossible under state-of-the-art technology. Courts that do not allow such a defense reason that the contractor knowingly undertook the risks and should have provided for this outcome in the contract terms and conditions. Therefore, the contractor should not be saved by the court from the consequences of that risk.

Impracticability

If it is not possible to perform some task or fulfill an obligation at a reasonable cost, such performance is said to be *impracticable* (although not necessarily physically impossible). For example, assume that an oil embargo severely restricted crude oil imports and drove the price of petroleum and petroleum-based products to exorbitant levels. A contractor with a contract to supply such products for a fixed price does not have an impossible task since, after all, the products are still being supplied. He may have an impracticable task, however, if the cost of these products is higher than what he is charging the client for them. If so, such impracticability would be a defense to a claim for breach of contract if the contractor failed to provide the products as originally promised.

Statutes of Limitations

Each state has statutes that establish the period of time during which a person can sue for a breach of contract. Generally, the period is between four and six years from the date of the breach (depending on the state). If the statutory period is six years, and a party who breached a contract is sued seven years after the date of the breach, that party may cite the statute of limitations as a defense to the breach of contract claim.

In some states, the applicable statute of limitations for certain kinds of contracts is substantially longer than for others. For example, where the limitations period for a typical contract might be six years, the limitations period for a contract under seal might be twenty years. Each party should assess the statutes of limitations in effect in their state before signing any contract.

Remedies

When one party fails substantially to live up to the promises made in the contract, the other party is entitled to have that breach repaired. Courts or arbitration panels are frequently called upon to force the party who breached the contract to *remedy* the situation. Often, contracts themselves specify the manner in which applicable breaches are to be remedied. Under these circumstances, courts and arbitrators will normally rule in accordance with the contract requirements. If contract provisions do not address breaches, the court or arbitrators generally rely on the applicable local, state, or federal law governing remedies for breach of contract. The two basic forms of remedy for a breach of contract are:

- Money awards
- Equitable remedies

The following examines these two types of remedies.

Money Awards

When one party breaches a contract, the other party is entitled to receive a *money*, or *monetary*, *award* with respect to the *damages* (loss or injury) they have sustained. The purpose of a money award is to compensate the victim (the nonbreaching party), as opposed to punishing the wrongdoer. Money awards that punish the wrongdoer are called *punitive damages* and cannot be collected for a breach of contract.

Types of Money Awards

There are three types of money awards possible in breach of contract claims:

- Direct damages
- Consequential damages
- Restitution

Direct Damages: Direct damages pay the plaintiff (the nonbreaching party) what he or she would have received had there been no breach and both parties performed properly under the contract. This type of damages is also referred to as the plaintiff's *expectation interest*, since it provides what was to have been expected under the contract. There are formulas used to calculate the proper amount of direct damages. Different formulas are used in different situations.

Consequential Damages: Unlike direct damages, consequential damages are granted in compensation for those losses that are caused *indirectly* by the breach of contract. Since these damages are not a clear and direct consequence of the breach of contract, the plaintiff must show that the loss was reasonably foreseeable and was contemplated by the parties when they contracted.

For example, if a contractor is delayed in completing an office building, he might realize that the building owner will incur additional rent at its present location during the delay period. It might not be clear, however, that failure to move into the new facility on time will cause the owner to be passed over for consideration on a large new contract. As a result, the owner will rapidly lose customers and will eventually be driven out of business. Losses related to the business' failure are considered consequential losses.

Restitution: Frequently a court will award a sum of money intended to put the parties back to the financial position they were in before the contract was formed. This type of award, which restores both parties to their *status quo*, is called *restitution*.

Assume, for example, that under this legal theory, the contractor seeks to recover the reasonable value of the goods and services supplied by him for the project. This theory is particularly attractive to a contractor in a fixed price contract that is inadequate to cover expenses and will result in a loss. If restitution is made, the contractor will at least recover the reasonable value of the materials and services supplied.

Why should a contractor who has entered into an unprofitable contract be able to recover fully for the goods and services provided? One reason is to prevent unjust enrichment of the owner, who will benefit from those goods and services. Additionally, since the owner breached the contract, it is only fair that he should lose the advantage given to him by the favorable contract.

To support his claim for restitution, a contractor must establish that the owner's breach was substantial, for example, where the owner failed to make progress payments or insisted on an excessive number of changes.

Restitution is measured objectively. This means that the contractor can recover an amount which equals what the owner would reasonably have had to pay to purchase the materials and services from one in the contractor's position at the time and place the services were rendered. Even if the contractor actually paid a higher amount for such materials and services, he will not be able to recover more than was reasonable to pay for them in the circumstances. Further, the contractor cannot recover profits for work that has not been performed and for expenses incurred in preparing to work.

Mitigating Damages

A basic principle of law is that people who sue for damages due to a breach of contract must show the court that they have made a reasonable effort to *mitigate*, or minimize, their losses. A party who makes no effort to minimize his or her losses is not prohibited from suing, but the court may impose a limit on the amount of damages he or she can recover (in proportion to what the court feels could have been avoided).

For example, a contractor who sues an owner for damages he suffered because of delay, must show that he made an effort to minimize his losses. If he can demonstrate that he transferred some employees to another job during the delay, he has shown that he made an effort to mitigate his damages. The case entitled *Smith v. Erftmier* (Figure 2.5) is an example of one owner's attempt to mitigate his damages.

Liquidated Damages

In many cases, actual damages for breach of contract, if any, will be very difficult to ascertain precisely. To avoid potential disputes and problems in proving actual damages, parties often agree to put a *liquidated damages* clause in their contracts. This clause is intended to preestablish the amount of damages to be paid in the event that one of the parties breaches the contract. For example, a liquidated damage clause may establish the amount that the contractor will pay to the owner if the contractor does not comply with the schedule set forth in the contract, and subsequently delays completion of the project.

***Smith v. Erftmier*, 210 Neb. 486, 315 NW2d 445 (1982)**

A contractor was hired to build a grain storage complex in Nebraska. This structure included storage bins which cracked because of defects in the construction. The owner used the bins despite the defects, but sued the contractor for damages.

The contractor claimed that the owner's use of the bins showed that he accepted them, and that his acceptance precluded him from making any claim for damages. The court held that the owner's use of the bins did not preclude his claim. In fact, such use actually *mitigated* (lessened) his damages, to everyone's advantage: The grain had to be stored, and if the owner had not used the bins, he would have had to rent storage space, thereby increasing his damages.

Figure 2.5

Courts will enforce liquidated damage clauses if the amount of damages is reasonable (at the time the contract was executed) in light of the damages that could be expected to result if a breach did occur.

If a reasonable amount has been established as liquidated damages for delay, for example, it is not necessary to provide for a bonus of any sort for early completion. By the same token, if the damages set for delay are unreasonable and constitute a penalty, they will not be enforceable regardless of whether or not a bonus for early completion is included in the contract.

Equitable Remedies

Equity is broadly defined as any court-ordered remedy other than money awards. Equitable remedies are determined based on what is fair and just, and normally will only be ordered when money awards are inadequate compensation to the injured party. Equitable remedies include:

- Specific performance
- Injunctions

Specific Performance

Specific performance is a type of equitable relief whereby a court orders a party to perform specifically as promised in the contract. For example, picture an agreement for the purchase and sale of a custom residence on a spectacular piece of property. If the seller refuses to complete the sale, the buyer may feel that money damages are inadequate; only the house and land will do. He may ask the court to order the seller to perform, that is, to sell the house. If the seller does not perform as ordered, the court can hold him in contempt of court and he may be arrested and fined or imprisoned.

Typically, parties to construction disputes request money damages rather than specific performance. Moreover, judges are reluctant to grant specific performance in construction cases, even if requested. People generally regard lawsuits as a last resort. By the time disputes among owners, contractors, and design professionals reach the courts, it is obvious to all parties that they cannot continue to work together. Often, the last thing anyone wants is to prolong the relationship. An unpaid contractor or architect wants to be paid, but may not want to remain working on the project. Similarly, an owner damaged by the default of a contractor would most likely want to collect money damages and hire another contractor to complete the project. Therefore, courts generally will not force a contractor or design professional to stay at work on a project by requiring specific performance of the contract.

Injunctions

An injunction is a court order instructing a party to stop, or to refrain from stopping, some activity. For instance, an owner may ask a court to issue an injunction to make a contractor stop working on a project that he has refused to build in accordance with the design documents. If the contractor ignores the injunction to stop work on the project, he can be arrested and fined or imprisoned.

Third-Party Beneficiaries

When two people form a contract for the benefit of a person who is not a party to the contract, that person is referred to as a *third-party beneficiary* of the contract. Generally, the third party has the right to demand that the parties to the contract perform their obligations under it.

Assume, for example, that John owes Bill $6,000. Subsequently, John contracts to paint Paul's house, in return for which Paul agrees to pay John's debt to Bill. If John paints Paul's house, but Paul does not pay Bill, Bill may have the right to sue Paul for the $6,000. John and Paul intended that Bill be paid and that the contract was primarily for Bill's benefit. Therefore, Bill is an intended third-party beneficiary of the contract and can sue to enforce it. Of course, Bill may also sue John for the amount of the original debt.

In the construction industry, third-party beneficiary contracts typically exist when the owner engages separate contractors for different work on the same project, although cases arise between contractors and architects as well, as Figure 2.6 illustrates. Often, the separate contracts specifically state the duties owed by each contractor to the others. Such duties might include the duty not to interfere with or destroy the others' work. Generally, if one contractor does interfere with another's work, the injured contractor can sue the interfering contractor, alleging breach of contract. This is possible, even though there is no contract between them, because each was to benefit from the other's contract with the owner.

Assignment, Delegation, and Novation

An *assignment* is a transfer of rights. Under contract law, if one party wants to transfer his interests in a contract to another party, he may do so by assigning the contract to them. An assignment does not, however, discharge the original party's obligations under the contract. Although the purpose of an assignment is for a third party to take over and perform as a substitute for one of the original parties, if the third party fails to perform, the original party who assigned his rights can be held liable.

Prichard Brothers, Inc. v. *Grady* Co., 407 NW2d 423 (Minn.App. 1987)

An owner contracted with an architect for design services. The owner then hired a contractor to build the project based on the architect's drawings and specifications. After the building was completed, the contractor sued the architect for professional negligence, claiming that the architect had caused delays in construction which resulted in increased costs and lost earnings.

The court held that the contractor could not sue the architect for negligence when the claim actually arose from an alleged breach of a contract. Furthermore, the contractor's claim was based on the architect's failure to perform duties under the contract between the architect and owner. Since the contractor had no contract with the architect, he could not sue the architect for breach of contract.

The court stated that the contractor could, however, sue under a different theory: namely, that the contractor was a *third-party beneficiary* of the owner-architect agreement. In other words, the contractor was to benefit from the architect's proper performance and, therefore, was a beneficiary of, and had the right to enforce, the obligations of the architect to the owner. The contractor alleged that if the architect had performed its services properly, the contractor would not have suffered damages for delay and lost earnings.

Figure 2.6

Assignments frequently arise in connection with an owner's financing for a construction project. The lender may want some security in the event that the owner defaults on his or her obligations to repay the loan. The lender can minimize its risk if it can get the building completed, and it can do that best if the architect and contractor remain on the job. To ensure that they do, the lender can require the owner, with the agreement of the other parties, to assign its contracts with the architect and the contractor, to the lender if, and only if, the owner defaults. The lender is entitled to enforce the owner's rights under the contracts, but the owner remains liable for his obligations.

When people refer to an assignment they typically refer inaccurately to both the transfer of *rights* and the transfer of *duties*. To be accurate, the transfer of duties under a contract is properly referred to as *delegation*. When a new party promises to perform the original party's duties under the contract, that is delegation. Generally, the original party remains liable for the proper performance of such duties.

If a party to a contract wants to be completely relieved of its obligations under a contract, they would have to ask the other party to agree to release them and accept a new party as a substitute. That is called a *novation*. It is sufficient to discharge the original party in favor of the substitute. If the substitute performs improperly, the original party is no longer at risk of being held liable.

In a sense, assignment and delegation are similar to third-party beneficiary contracts because they involve performance for the benefit of a third party. The difference is that with assignment or delegation, such performance is not contemplated until after the original contract has been executed. With third-party beneficiary contracts, the original intent of the contract is to perform for the benefit of the third party.

Uniform Commercial Code

The Uniform Commercial Code (UCC) is a model law developed to govern commercial transactions. Every state except Louisiana has enacted the UCC, virtually intact, as its own law.

Article 2 of the UCC governs contracts for the sale of goods, and is of most concern to contractors. It is intended to apply only where the parties have failed to include certain specific provisions in their contract. Accordingly, the terms of Article 2 can be varied by the terms of the parties' contract, out of respect for the basic principle of contract law that parties have the freedom to contract as they choose.

In the construction industry, Article 2 of the UCC applies principally to a manufacturer's or supplier's sale of goods to contractors or subcontractors. For instance, Article 2 can be applied to a contractor's purchase of lumber or bathroom fixtures from a supplier. A basic rule under Article 2 is that an agreement to buy goods costing $500 or more must be in writing to be enforceable. Article 2 also covers the parties' obligations, rights, and remedies, including specifics on delivery terms and *warranties* (warranties will be discussed shortly). It also establishes the procedures to follow when goods supplied are not as required under the contract.

The UCC does not apply to all construction contracts. For example, general contractors promise their clients not only building materials and equipment, but also the labor and services necessary to erect a structure from those materials and equipment. The courts have generally ruled that Article 2 of the UCC does not apply to contracts which predominately require services, even though they also contemplate the sale of goods. Therefore, Article 2 does not apply to a great majority of construction contracts, such as contracts between owners and architects, owners and contractors, and contractors and subcontractors.

Warranties

The UCC includes a section on warranties to define the type and quality of goods that a seller must deliver to a buyer (or a manufacturer to a contractor). A warranty is a legally enforceable assurance of quality or performance of a product or work, or the duration of satisfactory performance. Warranties are a type of contract, and are enforced in the same way. The warranties outlined in the UCC include:

- Express warranty
- Implied warranty of merchantability
- Warranty of fitness for a particular purpose

Express Warranty

Warranties are descriptions of goods and include promises or factual statements (not opinions) relating to them which are intended to induce a buyer to enter a contract. *Express warranties* are often made in a written contract, although it is not necessary that they be in writing. For example, a seller's verbal statements and conduct with respect to the goods for sale are also considered warranties. They may consist of samples or models, plans or blueprints, reference to an official standard, or reference to the quality of goods sent to the buyer in previous transactions.

Implied Warranty of Merchantability

The UCC also provides that specific warranties will be assumed (implied) to be part of sales of goods in certain situations. In other words, the seller does not have to express such warranties; they may apply in any case.

Under the UCC, an *implied warranty of merchantability* is given by all merchants who sell goods. It requires that the goods be *merchantable*. That means they must:

- Not be objectionable to those in the trade (for goods of the nature described in the contract)
- Be fit for the ordinary purpose for which such goods are used
- Be uniform in kind, quality, and quantity (subject to variations allowed by the agreement)
- Be adequately contained, packaged, and labeled, as the agreement may require
- Conform to the promises and statements of fact made on the container or label
- Be of fair average quality for the type of goods (with respect to goods of which any unit or part may substitute for any other in meeting an order)

A seller may disclaim, or avoid being bound by, an implied warranty of merchantability. To be effective, the disclaimer must be either oral or written. If written, it must be conspicuous, and must include the word "merchantable."

Warranty of Fitness for a Particular Purpose

A *warranty of fitness for a particular purpose* is imposed on sellers of goods by the UCC whenever the seller, at the time of contract, has reason to believe that the goods will be used for a particular purpose and that the buyer is relying on the seller's skill or judgment to select or furnish suitable goods.

This type of warranty often comes into play when a contractor consults a manufacturers' representative about selection of a product for a project. The contractor would likely describe the intended usage and ask which product is best suited for that situation. Presumably, the representative is an expert with his products and will recommend the correct one. The contractor probably relies heavily on this representative's advice when deciding whether or not to buy the product. It is therefore very likely that an implied warranty of fitness for a particular purpose would apply.

A disclaimer of a warranty of fitness for a particular purpose *cannot* be oral; it must be in writing, and the language must be conspicuous. No specific language is required, but the UCC suggests the following statement:

> *There are no warranties which extend beyond the description on the face hereof.*

Conclusion

The construction industry is based on a network of contractual arrangements between individuals skilled in the various aspects of design and construction. There is no universal method of governing the relationships between parties on a construction project. They come together voluntarily to accomplish a goal, and they must document the rules by which they agree to conduct their affairs. Therefore, it is important for all construction professionals to understand how contracts are made, how they are broken, and the possible consequences for doing so.

CHAPTER 3
BASIC AGENCY LAW

CHAPTER 3 BASIC AGENCY LAW

Introduction

Most businesses today could not function without engaging one or more people to deal with third parties on their behalf. The business entity, whether a sole proprietorship, a partnership, or a corporation, is known as the *principal*. The individual or entity who acts on behalf of, and under the control and direction of, the principal when dealing with third parties is known as an *agent*. The relationship between the principal and agent is referred to as an *agency relationship*.

Agency relationships are prevalent in the construction industry. Some examples include: materials and equipment purchasers as agents of the general contractor; the president of an office building development corporation as an agent of the corporation; a partner in a large architectural firm as an agent of the other partner(s) in the firm; and even the architectural partnership itself as an agent of its client during the construction phase of the client's project. Without one or more agents to conduct the affairs of the business, firms in the construction industry would be substantially more limited in the scope, quality, and effectiveness of their operations than is presently the case. This chapter examines the fundamentals of agency law and its importance in the construction industry.

Participants in an Agency Relationship

The Principal

A principal, whether a sole proprietorship, a partnership, or corporation, hires agents to conduct business on its behalf. In this way, the principal can effectively be in many places at once. The principal also benefits from the knowledge and skills that the agent brings, which would not otherwise be available to the principal. The principal controls and directs the agent, and must reasonably compensate the agent for services rendered. The principal must also reimburse the agent for out-of-pocket expenses incurred on the principal's behalf. Similarly, if the agent is sued by a third party for acts that the principal authorized, the principal must *indemnify* (reimburse) the agent for expenses incurred in defending the lawsuit, including any money that is paid to the party who sued. The principal reaps the benefits of the agent's actions and is likewise expected to bear the responsibility.

The Agent

The agent is a *fiduciary* to the principal. This means that the agent has implicitly promised to act in the best interests of the principal, regardless of whether or not such action is in the agent's best interest. For example, imagine that a staff architect for a large firm must fly from Boston to Los Angeles to visit a project site. A ticket on Airline A costs $425. A ticket on Airline B costs $480 but gives the architect credit on his personal frequent flyer bonus plan. As a fiduciary for the firm, the staff architect must fly on Airline A because purchasing the lesser priced ticket would save the firm money, and is therefore in the firm's best interest.

An agent must act in good faith and be obedient to, and use reasonable care toward, the principal. An agent must disclose to the principal all information that is relevant to the agency relationship. For example, if an employee's job includes researching the characteristics and prices of windows and purchasing a quantity of windows for her company, and her father happens to own a window manufacturing company, the employee should inform her boss of that fact. This disclosure may affect how the company evaluates the window selections made by the employee.

An agent represents the principal and has the legal power to bind the principal to business arrangements with third parties (people or entities other than the agent and principal). Those third parties, in turn, expect that the agent is acting on behalf of the principal, and conduct their affairs in reliance on the actions and commitments of the agent. For example, when the president of a plumbing contracting corporation agrees to contract with a general contractor on behalf of the corporation, he is acting as an agent of the corporation and, as such, his signature is enough to bind the corporation (the principal) to the contract with the general contractor, as long as the plumbing corporation's name is properly disclosed. In turn, the general contractor can rely on the agent's action and subsequently proceed knowing that a contract exists.

Creating Agency Relationships

An agency relationship is basically a contractual relationship, and is created and enforced just like any other contractual relationship. (See Chapter 2, "Basic Contract Law," for more information on contracts.) Usually, there are no special or formal requirements. For instance, a written agreement is not necessary to establish an agency relationship; the parties' oral agreement or conduct is normally sufficient.

The contractual relationship between the principal and agent normally defines the nature and scope of the actions which the agent is authorized to undertake on behalf of the principal. Once the agency relationship is established, the agent is required to act on behalf of the principal in accordance with the *authority* granted to him or her.

Authority

An agent who acts within the scope of his or her authority may bind the principal to contracts or other commitments. Conversely, a principal is *not* liable for contracts or commitments made on his or her behalf *unless* the agent acted within the scope of his or her authority.

There are two basic types of authority:

- Actual authority
- Apparent authority

Actual Authority

Actual authority is created directly by the contractual agreement of the principal and agent, and exists independent of the knowledge, assumptions, or actions of third parties. Actual authority may either be *express* or *implied.*

A principal confers *express authority* by defining for the agent the nature and scope of the duties that the agent is to perform. Sometimes this takes the form of a written job description or employment contract.

The agent also has *implied authority* to undertake those tasks and perform those actions that are not specifically authorized, but which are reasonably necessary if the agent is to accomplish tasks that were expressly authorized. Whenever possible, the agent should obtain authorization from the principal before performing acts that are not expressly authorized. Often, however, a principal will give its agent express authority to act according to his or her own judgment, especially in emergencies.

To understand actual authority, consider a construction project manager who is instructed by his general contractor employer to visit the site of a planned project on which the firm wants to submit a bid. The project is in a major city more than 1,000 miles from the contractor's home office. The project manager has express authority to go to the other city, and has implied authority to purchase an airline ticket at the contractor's expense to get there. Even though the contractor/employer never directly authorized the project manager to buy an airline ticket, such authority was implied, since an airline ticket is reasonably necessary to accomplish the express objective.

Apparent Authority

Apparent authority exists when a principal's written or spoken words or conduct lead a third party to reasonably believe that the principal consents to the acts performed on his or her behalf by the agent. Unlike actual authority, which the principal confers directly on the agent without any participation by third parties, apparent authority exists as a result of a third party's *perception* of the actions of the principal and agent (regardless of what they have actually agreed on).

Assume, for example, that an architect tells a client that he wants to enter into a contract to provide architectural services. The architect states that he has a designer on his staff and that this designer will be the project manager under the contract. Shortly thereafter, the designer meets with the client and presents him with a contract which they both sign. The designer's signature is written beneath the name of the architect's business. The designer's title, "Vice President," appears next to his signature.

In this situation, the client could reasonably believe that the designer had authority from the architect to sign the contract. Further, the client could reasonably believe that the designer signed the contract in his capacity as a representative of the architect. Even if the architect did not give the designer express authority to sign the contract, the architect is bound to it, because the architect's statements to the client concerning the designer's involvement in the project indicated to the client that the architect intended the designer to act as his agent, and that the architect intended to enter into the contract.

In the case entitled *Guarantee Electric Co. v. Big Rivers Electric Corp.* (Figure 3.1), a subcontractor and an engineer wound up in court as a result of confusion over apparent and actual authority. If it is unclear to a third party whether or not a principal intends to be represented by an agent, the third party should check with the principal prior to dealing with that purported agent. In contract disputes, an ounce of prevention is worth a pound of cure.

Ratification of an Agent's Actions

In the normal course of events, the principal and agent agree on the scope of the agent's authority, compensation, and other matters *before* the agent undertakes any action on behalf of the principal. As a practical matter, however, this is not always possible. Sometimes it is necessary for the principal to grant the agent authority to perform some action *after* the agent has already done so. This after-the-fact-authorization by the principal is known as *ratification.*

A valid ratification has three parts. First, the principal must know, at the time of ratification, all of the important facts with respect to the acts to be ratified. Second, the agent must have intended to represent the principal at the time the acts were performed. Third, the principal must have been in existence at the time of the agent's act. This means that the principal, if an individual, must have been living at the time the agent performed the acts. If the principal is a business entity, such as a corporation, the entity must have been in existence at the time the agent purportedly acted on its behalf. For example, a corporation that was not officially formed until June 1, 1988, may not ratify a contract that was signed on its behalf by an agent on April 1, 1988.

A principal may indicate ratification by silence, by words, or by conduct. If an agent signs a contract on behalf of his or her principal, but has no authority to do so, it is presumed that the principal ratifies, or approves

***Guarantee Electric Co. v. Big Rivers Electric Corp.*, 669 F.Supp. 1371 (WD Ky. 1987)**

An owner entered into a contract with an engineer for the supervision and construction management of a coal-fire electrical generating unit. When a portion of the work on the project was suspended, a subcontractor on the project sued the engineer, claiming that the engineer was negligent because he failed to tell the subcontractor, before commencement of the work, that construction had been suspended.

The subcontractor claimed that an *apparent* agency relationship existed between him and the engineer. He made this claim based on a letter from the engineer to bidders which stated that the engineer was the owner's representative. The subcontractor argued that, as the owner's agent, the engineer had a duty to tell him about the work suspension.

The court said that an apparent agency relationship must be based on the words and conduct of a principal (the owner) to a third party (the subcontractor), and not on an agent's (the engineer's) representations. The letter from the engineer to the bidders, therefore, did not establish an agency relationship. If the *owner* had indicated to the subcontractor that the engineer was acting on his behalf, then the subcontractor would have a valid claim.

In this case, the court declared that the engineer was not an agent of the owner but, rather, an independent contractor, based on the fact that the owner did not directly control how the engineer performed its services, even though they had to be approved by the owner.

Figure 3.1

of, the agent's action unless the principal disaffirms the contract immediately upon learning about it. If a principal chooses to ratify a contract that he or she did not give the agent authority to sign, the principal must ratify the entire agreement. The principal cannot ratify some provisions and eliminate others.

Assume that a saleswoman for a carpet manufacturer executed a contract on behalf of the manufacturer to supply $500,000 worth of carpet. She is only authorized to sign contracts up to $200,000. When she returns to the office it is apparent that she also granted the purchaser favorable payment terms and promised an early delivery date. The manufacturer may either ratify the contract as is, or refuse to honor it. The manufacturer may not decide that it wants to change the delivery date or modify the payment provisions before ratifyingthe contract.

Further, a principal may only authorize acts or agreements that could have been authorized at the time the act was done or the agreement was made. For example, a corporation may not ratify illegal contracts purportedly entered into on its behalf by an officer of the corporation.

Disputes Over Authority

Some of the most fertile ground for disputes during the construction phase of a project is over the actual and apparent authority of the various representatives involved. This includes the architect, who is the owner's agent during the construction phase, other agent(s) of the owner (sometimes known as a *clerk-of-the-works*), and the project superintendent, who is the agent of the contractor.

The standard form contracts commonly used today are quite explicit about the authority of these representatives. For example, AIA Document B141, 1987 edition, which is the typical agreement between owner and architect, defines the duties and scope of authority of the architect during the construction phase. The pertinent provisions are duplicated in the general conditions of the agreement between owner and contractor (AIA Document A201, 1987 edition), so the contractor has notice of the architect's actual authority. Such authority is fairly broad, but has definite limits which the general conditions state may not be extended without the written consent of the owner, architect, and contractor.

The limits should be clear to all concerned. However, in the high-pressure environment of a construction site, the niceties sometimes get lost. Consider an architect who frequently instructs the contractor's superintendent to perform additional work, to delete work, or otherwise to change the work required by the contract. In every case relatively little money or time is involved. In no case is any change ever rejected by the owner or payment refused.

Then the architect instructs the contractor to make a change which involves a different design appearance and about $10,000 in extra cost. The contractor does the work and asks for payment. The owner, however, dislikes the design and does not have $10,000. He then refuses to pay on the basis that the architect had no authority to order such changes.

Although the architect had no *actual* authority under the contract, he did have *apparent* authority. Because the owner never objected to the changes made by the architect and always payed for them, the contractor was led to believe that the architect had the authority to order such changes.

This scenario is potentially hazardous for all parties involved. The owner may have inadvertently authorized, and become liable to pay, the contractor for work that he did not want. The architect may have exceeded his authority and become personally liable to pay for the work that he instructed the contractor to perform, as well as for the cost of remedying the resulting situation. The contractor may not be entitled to payment on the basis that he had actual notice in the contract of the limits of the architect's authority, combined with the statement that the contract could not be extended without the owner's *written* approval. Clearly, none of these results is desirable and could have been avoided if the participants were aware of the implications of their actions.

Employment Relationships

There are two types of employment relationships, which may or may not be agency relationships. It is important to discuss these because the type of employment relationship that is created between two parties is important in defining their liabilities. They are:

- Master-Servant
- Employer-Independent Contractor

Master-Servant Relationship

The terms *master* and *servant* describe the parties to an employment relationship between employer and employee. The master controls not only the servant's overall conduct and performance, but the details of the servant's work. In other words, the master normally provides the workplace facilities, determines what jobs will be done, and dictates how and when those jobs will be performed. Individuals who are on the payroll of a company as employees normally have a master-servant relationship with their employer. In essentially all cases, the master-servant relationship is an agency relationship wherein the servant, as an employee (agent), has been granted certain authority to act on behalf of the employer (principal).

Liability of the Master and Servant

It is important for both the master and servant to know what potential liability they face as a result of their relationship. Is the servant always responsible for his or her own actions? Or is the master liable for everything that the servant does? Can the master shift liability to the servant?

Servants are generally liable for their own actions. If a backhoe operator, out of carelessness, hits someone with his machine, he could expect to be held accountable to the injured person. Similarly, if a registered architect negligently prepares construction documents and places his professional seal on them for his firm, he will be responsible if a wall subsequently collapses as a result of improper design.

Servants are not usually liable for the acts of their master. Consider the situation in which the buyer in a large painting contracting company signs a purchase order on the company's form for a shipment of paint. Subsequently, the company fails to pay for the paint. The buyer is not personally liable because it was the company (master) who breached the agreement. Design professionals in similar circumstances could expect similar results.

However, there have been several cases where servants were held responsible for the actions of their masters. Such cases are relatively rare, however, and usually involve very egregious behavior by the company, which the president or other top official in charge of the company's affairs took no action to stop.

In most cases, the important question is not whether the servant is liable, but whether the the master is also liable for the servant's actions. If the master is liable and has sufficient funds or insurance to pay the injured parties, it will probably not matter if the servant is liable too.

In general, a master is liable for the misconduct of his or her servant if such misconduct was committed within the *scope* of the servant's employment. In other words, as long as the employee does not misuse the authority granted to him or her, either by exceeding such authority or by acting improperly, his or her actions will make the employer liable too. In practice, however, the scope of employment can often be difficult to define. An act done *solely* for the servant's own purposes is *not* within the scope of the servant's employment. Similarly, if the servant substantially deviates from the authorized method, the master is not accountable for the servant's misconduct. Most situations are not that simple, however. If the servant's deviation from the normal and expected method of performance is minor, the master will be held liable for the servant's actions.

The following questions must be answered to determine whether a servant's conduct was within the scope of employment, or whether the conduct was a substantial deviation from such scope:

- Is the act one that is usually done by the servant?
- What was the purpose of the act?
- When and where was the act performed?
- Did the master expect that the act would be performed?
- How far did the servant depart from the usual method of performing an authorized act?

For more information on general liability exposure, see Chapter 4, "Basic Tort Law."

Characteristics of an Employer-Independent Contractor Relationship

The following characteristics indicate the existence of an employer-independent contractor relationship.

- The independent contractor is hired for a short time to do a specific job
- The independent contractor is paid a flat fee for a specific job
- The independent contractor is more skilled in the task than the employer
- The employer does not supervise the details of the independent contractor's work

Figure 3.2

Employer-Independent Contractor Relationship

The relationship between an employer and an independent contractor is also an employment relationship, but it is usually not an agency relationship. Unlike the master-servant relationship, the employer in this type of relationship does *not* have control over the details of an independent contractor's work.

For example, if a firm or individual hires a consultant who is not an employee of that firm, the consultant is normally considered an independent contractor. Similarly, in the construction industry, the relationship between an owner and architect during the *preparation of the design and construction documents* is also an employer-independent contractor relationship. Figure 3.2 lists the factors that often indicate that an employer-independent contractor relationship exists.

Liability of the Employer and Independent Contractor

As long as the employer uses due care and is reasonably careful in selecting the independent contractor, he or she is usually not liable for the subsequent wrongful acts of that contractor. In an owner-architect relationship, the architect is normally an independent contractor whose actions do not subject the owner to liability. For instance, if a visitor to an architect's office is assaulted by one of the architect's assistants, that visitor normally has no claim against the owner who is the architect's client. This is different from the master-servant relationship, where a master is liable if the servant commits a wrongful act within the scope of his employment.

Of course, there are exceptions to this rule. If the independent contractor is hired to do work that is inherently dangerous, and he fails to take precautions to prevent injury to third parties, the employer will be liable for harm that results. In the construction industry context, blasting and demolition are examples of inherently dangerous work.

Often, the key to assessing the potential liability of an employer is to determine whether there is a master-servant relationship or an employer-independent contractor relationship. The major difference between the two is that an employer does not have the right to control the means and methods used by an independent contractor to do a job; a master does. It is irrelevant whether or not the master actually controls the details of performance. If that right to control exists, then the relationship is considered a master-servant relationship. The case *Tsourmas v. Dineff* is an example in which a contractor is found liable for the errors of his subcontractor, after attempting unsuccessfully to use the employer-independent contractor relationship as his defense. (See Figure 3.3.)

It is possible, in certain circumstances, for an independent consultant to be an agent of the employer. For example, the architect is the agent of the owner when dealing with the contractor during the *construction phase* of a project. When the relationship between employer and independent contractor is an agency relationship, the liability that exists in a master-servant relationship is applicable. For more information on liability, see Chapter 4.

Termination of the Agency Relationship

If an agent is given authority for a specific period of time, the agency will automatically terminate when that time period is up. If the parties have not specified when the agency relationship will end, however, it may be terminated by either the principal or agent, upon reasonable notice to the other.

When authority is granted for the performance of a specific act, the authority terminates when that act is completed. A substantial change in circumstances (such as the principal's bankruptcy), or a breach of the agent's fiduciary duty to the principal, will also terminate the agency relationship.

If actual authority has, in fact, terminated, but the principal's words or conduct continue to convey to third parties the impression that the agent has authority, the acts of the agent will continue to bind the principal. In that case, there is no actual authority, but there is apparent authority. Apparent authority continues until third parties are put on notice that the agent does not, in fact, have authority to bind the principal.

Conclusion

Agency relationships exist throughout the business world. At one time or another, everyone in the design and construction business has been, is, or will be an agent or a principal. It is important, therefore, that design and construction professionals understand the basics of agency law so that they do not unwittingly create liability for themselves or their business associates.

Tsourmas v. Dineff, 161 Ill.App.3d 897, 515 NE2d 743 (1987)

A contractor was hired to remodel a building. He performed all of the carpentry work, but hired subcontractors to do other work, including the installation of a gas heater. When the remodeling project was complete, the owners moved into the building. The owners were later injured from breathing carbon monoxide that was found leaking from the gas heater. They sued the contractor for installing the gas heater improperly.

The contractor argued that a subcontractor installed the gas heater, and that he should not be held liable for the subcontractor's errors. The contractor alleged that the subcontractor was an independent contractor and, therefore, he did not have direct control over the subcontractor's performance.

The court held that, although the contractor's labor on the project was limited to carpentry, he acted in the capacity of a general contractor in that he made decisions on the project, hired the subcontractors without the owner's approval, and instructed them how to do their work. As a general contractor, he had a duty to protect anyone who lawfully entered the property. He also had the authority to control his subcontractors. That authority makes a general contractor vicariously liable for the acts of subcontractors. The court ruled that a general contractor may not use the "independent contractor" defense when one of his subcontractors causes injuries to a third party.

Figure 3.3

CHAPTER 4
BASIC TORT LAW

Chapter 4 Basic Tort Law

Introduction

One of the functions of law in society is to preserve the peace and to protect the citizens, their property, and their economic interests. If a person or business entity interferes with another's interests, they may be in violation of criminal law, civil law, or both. Interferences with the interests of the general public are *criminal* violations and are prosecuted by the state. Interferences with personal interests are *civil* violations, and the injured individual or business entity may sue the wrongdoer for redress of their injuries. These civil wrongs are known as *torts*.

There are two basic types of torts:

- Negligence
- Intentional torts

This chapter will examine these two legal principles and their importance in the construction industry.

Negligence

Generally, a person is *negligent* when he or she fails to exercise the degree of care that could reasonably be expected of a prudent and careful person under similar circumstances. Individuals with special training or skills are judged against the standard of other individuals with similar training or skills. For instance, an architect or engineer is expected to perform professional services as well as could reasonably be expected of other architects or engineers practicing in that same community. Negligence is the most common tort claim against design professionals.

Elements of Negligence

There are four elements of a basic negligence case which must be proven before an aggrieved party can be compensated for their alleged injuries. They are:

- Existence of a duty
- Substandard performance in response to that duty
- Proof that the substandard performance was the actual and legal (*proximate*) cause of the alleged injury
- Damages must have been incurred

Each of these elements is discussed below.

Existence of a Duty

A *duty* is an obligation to do something for another person. Duties are created in a variety of ways. Legal statutes can create duties. For instance, under state laws governing the licensing of architects, an architect has a duty to keep confidential a client's proprietary information which the architect may acquire in the course of designing a project. As fiduciaries, architects also have a duty to put their clients' interests above their own.

Duties are also created by contracts. In fact, this is a very important function of contracts. Without a contractually created statement of duties, there would be no general understanding of the rights and responsibilities of each party from project to project. One owner might want the architect to perform cost estimating services, while another would not. One owner may ask the architect to hire a traffic consultant, while another would not. One project may require services related to the removal of asbestos or hazardous materials, while another does not. In each case, the contract would make this clear. In *City of Columbus v. Clark-Dietz & Associates-Engineers, Inc.*, an engineer's contract with an owner provided for supervision of construction. However, the contract language limited the scope of the engineer's duty. (See Figure 4.1.) Contracts that do not clearly define the scope of the parties' obligations to each other create fertile ground for disputes.

By the same token, a lawsuit claiming negligence in the absence of a legally recognizable duty would be dismissed. For example, if an architect is hired by the owner, but was not contractually obligated to review shop drawings during the construction phase of a project, the architect cannot be held liable for failing to point out errors on the contractor's shop drawings.

***City of Columbus v. Clark-Dietz & Associates-Engineers, Inc.*, 550 F.Supp. 610 (N.D. Miss. 1982)**

An engineer and a contractor were hired to design and construct a waste water treatment plant. The project involved building a levee to surround the construction site. The levee failed and water flooded the site, causing damage and delaying the project. The owner sued the contractor and engineer. The court determined that the engineer was responsible for the levee's failure because of negligent design.

This case is especially noteworthy for its ruling concerning the engineer's supervisory responsibilities. There were areas of the levee which had not failed completely, but which were constructed improperly and had to be corrected. The court concluded that these deficiencies resulted from faulty construction work and not from improper design. The contractor, however, argued that the engineer should be fully responsible because he had a duty to supervise the project.

Based on its review of the engineer's contract with the owner, the court determined that the engineer's duty was to "observe the general progress of the work and not to make continuous and exhaustive inspections." Under the contract, the engineer was not required to inspect every step of the contractor's work. The court held that the engineer owed no duty to supervision to the contractor other than to exercise reasonable care when it provided instructions and conducted tests at the job site.

Figure 4.1

More than 30 years ago, it was generally accepted that the obligations and duties of parties to a contract (those in *privity of contract*) were owed *only* to each other, and not to third persons outside the contract. Since then, court decisions have changed this rule. Today most parties to a contract *do* have duties to third parties who might foreseeably be affected by the performance of that contract but are not themselves parties to it. This means that construction professionals may be liable for damages to others with whom they have no contractual relationship. In *Council of Co-Owners Atlantis Condominium, Inc. v. Whiting Turner Contracting Co.*, the owners of a condominium building were entitled to damages from an architect and contractor, even though they had no contract with the construction professionals. (See Figure 4.2.)

Substandard Performance of the Duty

If a duty exists, the law requires a certain level of performance in response to that duty. With respect to the professional negligence of an engineer, for example, the aggrieved party must show that the engineer did not perform as well as could reasonably have been expected of other engineers practicing in the same community. This level of performance is known as the *professional standard of care*. The common law (that body of law brought to this country from England in the 17th century and subsequently modified by court decisions) would not ask more of a professional, whether he or she were an engineer, lawyer, doctor, or accountant. The law does not expect perfection, merely careful and

---

***Council of Co-Owners Atlantis Condominium, Inc. v. Whiting-Turner Contracting Co.*, 517 A.2d 336 (Md. 1986)**

A developer hired an architect and a contractor to design and construct a condominium building. After the building was completed, the owners of the condominium units discovered that the electrical work did not comply with the building code and was not installed in accordance with the plans and specifications. In particular, the utility shafts were not lined with fire resistant wallboards.

The owners of the units sued the contractor and architect, alleging that the deficiencies in the electrical system could cause a fire. They wanted to be paid the costs involved in correcting the system.

The contractor and architect argued that they owed no duty of care to the owners because they did not have a contract with any of the owners; they had contracted with the developer. They also contended that the owners had suffered only economic loss (the cost of correction) and not personal injury or property damage (after all, there was only a *threat* of such damages).

The court ruled in favor of the owners, citing an exception to the rule of *privity of contract* (where an actual contractual relationship is a prerequisite to a lawsuit), in cases involving inherently dangerous construction. In other words, it was not necessary for the owners to have a contractual relationship with the contractor and architect in order to sue them because it was possible that they could be injured by the defects in the future.

Contractors and architects must take into consideration all the people who may be at risk of serious personal injury because of latent and hazardous conditions in a building. In this case, the court allowed the owners to sue for economic loss, reasoning that it was better to recover damages and fix the defect, than to wait until someone was injured before awarding them the costs of correction.

Figure 4.2

skillful performance of duties consistent with what can be reasonably expected of like professionals practicing in similar circumstances.

The courts will, however, enforce a promise to provide a higher level of performance if that is what the parties have agreed to in their contract. A design professional could promise his or her client perfection, or otherwise guarantee a specific result (such as, "This basement will never leak."). The law would not have required this, but the courts will enforce such promises. Thus, design professionals (and others) should be careful that they do not unwittingly agree to a higher standard of performance than is otherwise required by law.

It is sometimes difficult to determine whether or not specific actions were undertaken with the appropriate level of skill and care. In general, laypeople (nonprofessionals) are considered incapable of determining whether or not a professional's performance was adequate or substandard. In the event of a dispute in court, each side must provide *expert testimony* on this issue. Expert testimony is a declaration made by a qualified, experienced professional with regard to the standard of practice in the community and in the circumstances at hand. The expert must express his or her opinion as to whether or not such a standard was met in a particular case. If it is found that the professional performed his or her duties with the appropriate level of professional skill and care, he or she will not be found negligent.

Not all professional negligence cases require expert testimony, however, as is shown in the case *M.J. Womack, Inc. v. State House of Representatives* (Figure 4.3). Whether or not expert testimony is necessary to establish an architect's liability is a factual determination that hinges on whether or not jurors would have a common sense understanding of the requirements of the architect's performance.

M.J. *Womack, Inc. v. State House of Representatives*, 509 So.2d 62 (La. 1987)

A contractor was hired to renovate a state capitol building. The contractor's agreement provided for a bonus if the work was completed earlier than required. During construction, the contractor discovered a structural brace that did not appear on the architect's plans. Because the architect decided the brace should remain where it was, he had to redesign around it, causing the project to be delayed.

The contractor did not complete the project on time because of the design-related delay. The contractor sued the architect for negligence and alleged that the architect's delay caused him to lose the early-completion bonus.

A trial court decided in favor of the architect because the contractor did not present *expert testimony* on the issues in the case. However, a higher court reversed the trial court's ruling.

Generally, in cases dealing with professional standards of care, expert testimony is needed because the law assumes that laypeople are incapable of knowing whether or not a professional's (the architect's) performance was adequate or substandard. In this case, however, the court held that an expert was not necessary because laypeople could understand that tall buildings sometimes need structural braces, and that an architect should be aware of any braces that existed and know their location before designing a renovation.

Figure 4.3

Causation

Causation is the connection between specific acts and specific injuries, and must be proven in cases of negligence. There are two types of causation:

- Actual causation
- Legal causation (sometimes called *proximate* causation)

To determine if there is *actual* causation, simply ask: "But for the action, would the injury have occurred?" But for the fact that the structural calculations were incorrect, would the beam have failed? If the answer is no, there is actual causation between the erroneous calculations and the failed beam. Then it is necessary to determine whether or not the harm that occurred was a reasonably foreseeable consequence of the wrongful act(s) (that is, whether or not the miscalculations were the *legal* or *proximate* cause of the beam's failure).

How closely related is the alleged wrongful act to the harm? What other events contributed to the alleged injury? Sometimes it will become evident that the injury was really caused by an outside, intervening event, rather than by the allegedly wrongful acts. At some point in a particular case, the courts may decide that a given action was just too far removed from the harm to be considered its legal cause.

Assume, for example, that an engineer designed a reinforced concrete parking garage. After it was constructed, he was informed that a truck had entered the garage one day and, as it was driven to an upper level, the floor collapsed. The interior columns punched through the concrete floor slab. The engineer checked his calculations. He was horrified to discover that he had made an error and had not properly specified the size of the lateral reinforcing steel. Investigation of the rubble, however, indicated that the lateral steel had not been overstressed. In fact, the reason the columns had punched through the slab was that the contractor had failed to install properly specified diagonal steel reinforcing around the column heads.

Did the engineer have a duty to design the steel reinforcing properly? Yes. Did his performance meet the professional standard of care? No, because the lateral steel was wrongly sized. Was the substandard performance the cause of the damages? No. The failure was caused by the contractor's omission of diagonal reinforcing. Therefore, there is no causation and the engineer is not negligent.

Damages

Finally, there must be some injury (generally referred to as *damages*) that the law can remedy. If a person's property is damaged because someone drove a truck negligently and ran into it, the law can order the negligent party to pay money sufficient to repair the damages. On the other hand, if two friends in a business deal begin to treat a third in an unreasonable and uncivil manner, causing hurt feelings and a deterioration of the friendship, a court cannot remedy the hurt feelings by ordering the parties to be friends again. The broken friendship is not something to which a price tag can be affixed. In this case, because the element of damages is missing, there is no valid claim of negligence.

Defenses to Claims of Negligence

Two defenses to negligence claims which are particularly relevant to the design and construction industry are:

- Statute of limitations/statute of repose
- Immunity

Statute of Limitations/Statute of Repose

Each state has its own *statute of limitations*, which specifies the amount of time available to an injured party to make a legal claim against another party. If the time has expired, the injured party's claim may be defeated, no matter how valid the claim might otherwise have been. These statutes are designed to eliminate "stale" claims about which memories have faded, documents have been lost or destroyed, and witnesses and other involved parties have moved away or died. A statute of limitations period typically begins when the injured party knows, or reasonably should have known, of his or her injury.

There have been a number of *statutes of repose* passed in various states specifically to limit the time within which to bring suits related to the design and construction of improvements to real property. After a project is completed and opened for use, construction professionals move on to other projects and the owners take control of their buildings. Soon it becomes very difficult to determine whether some problem is related to a defect in construction or is the result of poor maintenance. After all, there is no law or agreement that automatically allows a design professional or contractor to enter a structure they have designed or built, after it is occupied, to evaluate how it is being maintained and operated. Even if they knew that the owner was improperly maintaining the structure, they generally have no power to correct the matter. It is, therefore, unfair to hold them liable for defects beyond a reasonable time during which the owner may discover that the defect(s) exist.

Statutes of limitations and statutes of repose work together as follows. Assume that the state statute of limitations declares that an injured party has three years to sue a party for the damages caused by their negligence. Of course, knowing the length of time in which claims may be viable is meaningless without knowing when that time period begins. Therefore, the statute also provides that the limitations period begins on the date when the injured party became aware of, or should have become aware of, the negligence. Thus, claims made more than three years after the negligence is discovered could be defeated under the statute of limitations.

Next, assume that a structural engineer made a negligent error in 1989 in his calculations for a building foundation. The building is substantially completed in 1992. There may not be any physical evidence of the error until six or seven years have passed. If the damage is discovered and the negligence identified in 1999, the owner (the injured party) would have until the year 2002 to sue – 13 years after the error was made!

Statutes of repose establish an outside time limit on a person's exposure to liability. They are based on the premise that, say, six years after a structure is substantially completed, all claims related to it may be defeated as untimely, regardless of whether or not the "injured" party knew of the negligence at that time. In the situation described in the previous paragraph, if there was a six-year statute of repose, the claim for the engineer's negligence would be untimely after 1998 according to that statute, even though it would have been allowed under the statute of limitations until the year 2002 because the owner did not know about it until 1999.

Immunity

Immunity is another defense commonly used in the construction industry for claims of negligence. For example, architects and engineers operating under the American Institute of Architects (AIA) or the Engineers Joint

Contract Documents Committee (EJCDC) standard form contracts are granted immunity from the results of decisions made in good faith when they assume the role of arbitrator or judge in disputes between the owner and contractor.

Remedies for Injuries Due to Negligence

As stated earlier, for a claim to be valid, there must be damages or an injury that the law can remedy. The monetary award, or remedy, that a court may order a guilty party to pay for injuries suffered by another is also commonly known as *damages*.

Damages

There are basically three types of damages:

- Compensatory damages
- Consequential damages
- Punitive damages

Compensatory damages are intended to compensate for a loss. In the event of bodily injury, compensatory damages would typically cover medical expenses. In addition, the injured party could seek damages for pain and suffering.

It is also common for the owner to recover compensatory damages from a design professional for the cost of repairing a building that failed because it was negligently designed. Such damages compensate the owner for economic harm that directly resulted from the design professional's negligence.

Consequential damages are awarded for costs that result *indirectly* from an individual's or entity's negligence. For instance, if a project is not completed at the agreed-upon time because of a design professional's negligence, the owner would not be able to rent the property as soon as anticipated and would suffer loss of rental income. This loss is not a direct result of the delay, but is one of the consequences of that delay.

In addition to compensatory and consequential damages, an injured party may be able to recover *punitive damages*. Such damages are intended to punish past wrongful behavior and deter similar behavior in the future. Punitive damages are seldom awarded in construction cases.

Factors that May Offset Damages

An individual who is injured by another's negligence is not always entitled to the full amount of the claimed damages. It is possible that the injured party was also negligent and was responsible to some degree for his or her own damages. If so, such responsibility may partially or completely offset the liability of the other negligent party. This is referred to as *contributory negligence*, and occurs when the injured party is also "at fault" in the situation, by contributing to his or her own injury.

Each state has its own body of law regarding contributory negligence. For example, in some states, if the aggrieved party is even 1% at fault, he or she cannot recover anything whatsoever from the other party for his or her damages. In other states, an aggrieved party may be entitled to have the other party pay the full amount of the damages *unless* the aggrieved party was more than 50% at fault.

Comparative negligence is similar to contributory negligence. In states with comparative negligence statutes, the injured party's own negligence does not automatically preclude an award of damages from the other party. The concept allows for an allocation of damages according to the relative (comparative) fault of the parties. For instance, an injured party who is 20% at fault may recover from the other party only 80% of the damages he or she suffered.

Some states allow a pure form of comparative negligence whereby an injured party can recover that percentage of damages caused by the other party, regardless of the injured party's own degree of fault. Thus, even if the injured party was 95% at fault, he or she could still recover 5% of its damages from the other party. Other states have adopted a limited comparative approach whereby the injured party is able to recover damages, but only up to the point at which his or her own negligence is equal to or greater than the accused party's. Beyond that, no recovery is possible. Thus, an injured party who was found to be 51% responsible for his or her own injuries would not be able to recover anything under this state law.

Intentional Torts

Intentional torts are wrongful acts performed on purpose by a party, which result in harm to another. For example, trespassing on another's land or knowingly harming another's property are intentional torts. With regard to trespassing, the intentional act is walking across the property. It is not necessary that the person know that he has no right to be there or that he "intended" to trespass.

To win a lawsuit for an intentional tort, the aggrieved party must show not only that there was wrongful behavior, but that he or she suffered damages, either from physical or economic harm. (In the trespassing example, an injury of nominal scope is presumed to occur as a result of the intrusion. This is an unusual situation arising out of traditional attitudes toward the sanctity of real estate.)

There are several types of intentional torts. Some of those that are pertinent to construction industry situations include:

- Intentional misrepresentation
- Defamation
- Interference with contracts
- Strict liability

These are discussed below.

Intentional Misrepresentation

Misrepresentation is a legal term for what is commonly known as lying. A person is guilty of misrepresentation when:

- He states a material fact (not an opinion)
- that he knows is false
- to induce another party to rely on such a false statement, and
- that other party actually does rely on the statement, to his or her detriment.

For example, an architect who intentionally underestimates the cost of a project to induce the client to continue with the project could be found liable for misrepresentation. Further, if an engineer tells a client that a specific contractor is incompetent to build the project so that another contractor, with whom the engineer has business ties, will get the job, that is also misrepresentation.

The misrepresented fact must be *material*. In other words, it must relate to a serious, substantial aspect of the matter at hand. Also, the listener must reasonably rely on it. In the construction industry, misrepresentation claims are made most often against surveyors and geotechnical engineers who allegedly misstate information about the boundaries and subsurface conditions of property. The owner might then claim that he or she never would have embarked on the project had the truth been known.

In an intentional misrepresentation, the speaker knows he or she is not telling the truth. This is different from situations in which a person misrepresents something that he honestly believes is true. Such a person may be found guilty of *negligent misrepresentation* if he did not use reasonable care in discovering the truth about statements of "fact" on which he intended his listeners to rely.

Defamation

A person has been defamed when the written or spoken language of another tends to hold him up to scorn and ridicule in the eyes of a substantial number of respectable people in his community. The speech or writing must have been presented as a statement of fact, and not just the opinion of the speaker or writer. Defamatory language that is written or printed is called *libel.* Defamatory language that is spoken is called *slander*.

The defamatory statement must be published, that is, it must be communicated orally or in writing to a third party. If an architect sends a nasty letter about a contractor to that contractor, there is no defamation because no information has been communicated to a third party. The letter would be considered published only if a third person reads it.

Additionally, the defamatory statement must be understood to refer to the particular contractor. A letter complaining about contractors or building defects in general, without mentioning specific people or projects, is probably not specific enough. Fortunately, in most states, truth is a defense to libel and slander: If the statement is true, there is no basis to sue for defamation.

Interference with Contractual and Business Relationships

A person who intentionally interferes with a contractual relationship between two other people in order to procure business for him or herself (or for other self-serving purposes) may be found liable for the tort of *interference with contract.*

There must be an existing contract between two parties with which the third party intentionally interferes. For example, an engineer who would like to provide services on a project for which another engineer is already engaged may go to the client and try to have the first engineer fired, or otherwise removed, so that he can have the commission. If the first engineer is fired, he could seek damages equivalent to the amount he would have gained from his contract had the second engineer not interfered.

A related tort is *interference with advantageous business relationships*. This is based on the same concept as interference with contractual relations, except that a current contract does not need to exist between the parties. This tort may arise, for example, where a contractor has, over the years, built all of the new construction for a small industrial company. Between projects there are no contracts. If another contractor goes to the owner at one of those times and attempts to destroy the business relationship between the owner and the first contractor so that he may obtain all future business, the second contractor has probably interfered improperly with the business relationship, entitling the original contractor to damages.

In some cases, a party may be protected from claims of defamation or interference by *qualified privilege*. A qualified privilege protects statements made by one party to another with whom he or she shares a common interest, and who he or she reasonably believes has a right to know the matters discussed. This is illustrated in the case entitled *Alfred A. Altimont, Inc. v. Chatelain, Samperton and Nolan* (see Figure 4.4). However, if the party who made the statements acted with malice, the information is not protected by the qualified privilege, and the statement may be slanderous or libelous.

Strict Liability

Ordinarily, a person is not liable unless he or she is somehow "at fault" by failing to act as well as could reasonably be expected, based on a standard of care. In recent years, however, a body of law has emerged which holds people liable even in circumstances in which they were not negligent or otherwise "at fault." Such liability without fault is known as *strict liability*.

If a person engages in abnormally dangerous or hazardous activities that threaten others and put them at undue risk, he or she will be liable for injuries suffered by those who were endangered, whether or not all possible care was taken while performing the dangerous activity. For instance, if a contractor is blasting for highway construction, and an abutting property owner is injured, the contractor will be liable, even though he was conducting the blasting operations with all due care. The rationale is that if a person, acting for his or her own purposes, wishes to put society at abnormal or excessive risk, he or she will be liable for the damages that result.

Principles of strict liability also apply to manufacturers of products. Generally, manufacturers and suppliers, including wholesalers and

***Alfred A. Altimont, Inc. v. Chatelain, Samperton & Nolan*, 374 A2d 284 (DC 1977)**

A general contractor was hired to work on a church. When the work was behind schedule, the architect on the project wrote a letter to the contractor's surety reporting the slow progress being made. The delay continued in any event, and so the architect wrote a second letter to the surety stating that the contractor was "extremely negligent in his duties."

The contractor sued the architect for libel and interference with business relations, alleging that the architect's statement to the surety caused him to lose business because "of a loss of bonding capacity."

The court ruled in favor of the architect. The letters from the architect to the surety were protected by qualified privilege. As the church's agent during construction, the architect had a duty to protect the church's interests, which he did by contacting the surety regarding the contractor's slow progress. The architect's actions were not malicious because he was not plotting to harm the contractor. Finally, the architect's evidence showed that it was the usual practice of an architect to keep a surety informed of a contractor's progress.

Figure 4.4

retailers, who are engaged in the business of selling products, are liable for physical harm caused to a user or consumer of the product if the product is defective and unreasonably dangerous. The manufacturer/supplier may also be responsible for physical harm to a user/consumer's property caused by a defective product that they sold. A user/consumer can recover damages from the seller as long as the product's condition has not substantially changed from the condition it was in when it was sold.

The manufacturer/supplier is liable to the user/consumer regardless of fault. The rationale is that manufacturers are better able to detect and prevent defects in their products and are better able to bear financial losses caused by injuries from their defective products than the innocent user/consumer.

Strict liability is not unlimited, however. A manufacturer/supplier of a defective product is liable only for physical injury to people and their property. Over the years, decisions in various cases have precluded people who suffer only economic loss, such as lost profits or delay damages, from recovering any money from the manufacturer/supplier of the product that caused the loss.

Clearly, the law of strict liability most often applies with regard to manufacturers/suppliers of materials and equipment used in buildings and other construction projects. So far, strict liability has not been extensively applied to design professionals. There have been some instances where designers and builders of mass-produced homes or other buildings (for example, projects completed on a design/build basis) have been held liable without fault. These projects have more of the characteristics of a manufactured product than the normal one-of-a-kind building. So far, however, that has been the exception and not the rule.

Multiple Defendants

An owner whose building leaks or has failed in some way may not know whether the problems are the result of an incorrect design or improper construction. The owner may want to sue the architect, engineer, and contractor in one lawsuit to determine whether one, two, or all three are liable for the resulting damages. There is no shortage of potential defendants in lawsuits relating to problems with the design and construction of a building. A typical project may involve the owner, lender, contractor, construction manager, surety, subcontractors, suppliers, architect, engineers, interior designers, and cost estimators, to name just a few.

One of the aspects of tort law which distinguishes it from contract law is this potential for *multiple defendants* to a lawsuit. The more defendants there are, the more likely it is that one will be held liable. This is generally good from the injured party's point of view. Unfortunately, the more defendants there are, the more complicated and expensive it is to conduct a trial to determine who, if anyone, really is responsible. If the damages suffered by the owner, for example, total $100,000, that amount is constant whether he sues one person or five. In other words, the owner cannot ask the court to order each of five defendants to pay him $100,000, when the total amount of his damages equals $100,000.

For that reason, the injured party (in this example, the owner) should assess realistically the possibility of obtaining a judgment against each potentially liable party. The owner should also assess the financial resources of each potentially liable party. If a party has only minimum responsibility and is not able to pay a very large judgment, it may not be worth the effort and expense to add them to the lawsuit.

Vicarious Liability

An individual or business entity who is held legally responsible to someone, not because they did or failed to do something, but because of what another person or entity did or failed to do, is *vicariously liable*. This normally occurs in employment and agency relationships where the actions of the agent, who is primarily liable, make the principal secondarily, or vicariously, liable. (Refer to Chapter 3, "Basic Agency Law," for more information on how the actions of employees, agents, and independent contractors impose liability on employers and other principals.)

Indemnity

In situations where an employee causes damage to persons or property for which the employer must pay, the employer may have a right to *indemnity* from the employee. This means that the employer may have the right to reimbursement by the employee for losses incurred as a result of the employee's wrongdoing. Indemnification is meant to prevent a party who is only secondarily liable from paying for the errors of the party who is primarily liable. Although employers may have a right to indemnity, it is not frequently exercised. Normally, employers pay for the actions of their employees and agents without seeking recompense.

Joint and Several Liability

When two individuals or business entities either 1) act in concert (by agreement) for the purpose of injuring another, or 2) act independently, but their actions cause a single, indivisible injury to another, they are called *joint tortfeasors (wrongdoers)* and are *jointly and severally liable* for the damages they have caused. This means that the injured party is entitled to be paid the full amount of his or her damages from *either* of the parties or from *both*, in any ratio, up to the total amount due.

It is uncommon in construction for two parties to specifically agree to cause injury to another. On the other hand, it is very common for two parties to act independently to create one indivisible injury. For example, an architect who negligently designs the details of a roofing system, and a contractor who builds that system in an unworkmanlike manner (not in accordance with standards in the trade) are jointly responsible when the roof leaks, although it may not be possible to determine exactly which act of negligence caused the roof to leak. In this situation, the owner has one indivisible injury and two joint tortfeasors.

The architect may only be 1% responsible for the problem, while the contractor is 99% responsible. However, the contractor may have gone out of business between the time the roofing work was completed and the date the lawsuit was concluded. If so, the architect may have to pay the owner 100% of the cost of repairing the roof. It is usually not difficult to come up with more than one person with *some* liability on any given construction project. Most often the individual with the "deep pocket" (the greatest financial resources) is the one who pays the bills, regardless of their share of actual liability.

This concept has caused considerable dismay among design professionals, and has been the subject of "tort reform" legislation in many states. Most proposals for reform have proposed that joint tortfeasors be required to pay only for their share of liability.

Contribution

Generally, if the architect in the preceding example paid the owner for 100% of the damages, he would *not* be entitled to demand in court that the contractor pay him for the 99% that was the contractor's responsibility. The idea is that one wrongdoer should not be entitled to aid from the courts to obtain the *contributions* of other wrongdoers.

Today, statutes have been passed in some states to allow contribution among joint tortfeasors in certain situations (normally where negligence alone, and not willful action, is involved).

Conclusion

It is extremely important for everyone in the construction industry, particularly design professionals, to have a basic understanding of tort law. Society expects its citizens to be respectful of others' interests and to conduct their affairs with a reasonable degree of skill and care. Those individuals and businesses who make a reasonable effort to foresee the consequences of their actions and to conduct themselves prudently, however, should not unduly fear their exposure to tort liability.

CHAPTER 5

BASIC REAL PROPERTY LAW

CHAPTER 5 BASIC REAL PROPERTY LAW

Introduction

The design and construction industry in the United States spends a substantial amount of time and effort conceiving and refining methods to develop and improve *real property*. Real property, sometimes referred to as *real estate*, is a term used to describe land and whatever is erected on, growing on, or attached or fixed to, land. A simple vacant lot in a planned subdivision, or a complex structure such as the World Financial Center in Manhattan, are both examples of real property. Real property is bought and sold in local as well as international transactions. Further, real property is frequently changed physically to suit the needs of society.

There are other types of property. In oversimplified terms, everything that is not real property is considered *personal property*. Personal property refers to tangible goods, such as furniture, cars, books, and equipment; and intangible goods, such as money, notes, bonds, and stocks. Personal property is neither land nor a legal interest in land.

Fixtures are a cross between real property and personal property. They are tangible goods that are permanently attached to, or imbedded in, real property such that they are considered to be part of that real property. For example, a furnace or lighting fixtures, which are normally considered personal property, become fixtures when they are attached to the real property so that they cannot be removed without substantial damage to that real property. When real property is bought or sold, fixtures are included in the sale (unless the parties agree otherwise), whereas personal property is not included. This chapter addresses the fundamentals of real property law, with which everyone working in the design and construction industry should be familiar.

Estates in Real Property

An *estate* is a legal right, title, or interest in real property. There are many types of estates in real property. They fall into two broad categories:

- Ownership
- Rights to possess and use real property owned by another

Ownership

The owner of real property is entitled to all rights to use and enjoy their property, including the right to dispose of it by sale or gift. Ownership is often equated with control, either for a period of time or forever.

There are four basic types of ownership:

- Absolute ownership
- Estate for years
- Multiple ownership
- Adverse possession

Each of these is described below.

Absolute Ownership

Absolute ownership is the broadest form of estate in real property. The owner(s) possesses indefinitely all right, title, and interest in the property. The owner(s) has the right to sell it or give it away, in whole or in part, while he or she is alive; or to pass it to heirs or to others at the time of death.

Estate for Years

Any ownership of real property which is less than absolute is an *estate for years*. An estate for years will run for a fixed time period, after which it ends and ownership reverts to another. For example, a person who has a life estate in real property is entitled to possess and use that property for a period of time marked by the life of a person, usually the possessor. Some people have given their homes to their children on the condition that they be granted a life estate in them. Thus, the home is the parents' to possess and use while alive, but reverts to the children upon the parents' deaths. Despite its name, an estate for years could be fixed in days, weeks, months, or other measures of time, regardless of how short or long.

Multiple Ownership

Two or more people or entities may own interests in the same real property at the same time. This is known as multiple ownership. There are three basic forms of multiple ownership of real property. They are distinguished from one another by the extent of each owner's legal rights in the real property. Multiple ownership includes:

- Tenancy in common
- Joint tenancy
- Tenancy by the entirety

Before examining each of these, the reader should understand that there are several meanings of the word *tenant*. In the broadest sense of the word, a tenant is one who possesses real property, under any kind of right or title, for any period of time. The word is used in this way when discussing the different forms of multiple ownership. Later in this chapter, a more restrictive, but more common, usage will be adopted in which a tenant is one who temporarily possesses and uses (rents) another's real property.

Tenancy in Common: A tenancy in common is a form of ownership between two or more people with equal rights to occupy and use all of the real property owned in common. The tenants in common may acquire their rights to the property at different times and by different *deeds*. (A deed is a legal document used to transfer ownership of real property.) Although the owners may all fully occupy and use the property, they do not necessarily have equal shares of ownership. For instance, one person may own one-half of the property and the remaining two owners may each own one-quarter. The deed to the property usually describes the size of the shares owned by each tenant in common. If the deed does not do so, the law presumes that the tenants in common have equal shares of ownership.

For example, if X and Y purchase a piece of real property for $100,000, with X paying $25,000 and Y paying $75,000, the deed would state that X owns a one-quarter interest and Y owns a three-quarter interest in the property. If X and Y each pay $50,000 for the property, they do not need to state in the deed their respective shares of ownership. The law presumes that they each own a one-half interest in the property.

Any owner may sell his or her interest in a tenancy in common with or without the consent of the other owners. In the example above, X may sell his $25,000 share to Z, thereby making Z a co-owner with Y, regardless of Y's feelings in the matter.

If one of the tenants in common chooses to sell his or her interest in the property, but cannot find a buyer who wants less than exclusive possession of all or part of the property, the court may be asked to settle the matter. The court will either split the property, giving each co-owner exclusive rights to a portion of it, or the court may order the entire property sold at an auction, with the proceeds divided among the co-owners in proportion to their shares.

One distinguishing feature of a tenancy in common is that when one owner dies, his or her share does not pass to the surviving owner(s), but instead passes to those named in the decedent's will or to his or her legal heirs if there is no will.

Joint Tenancy: A joint tenancy is a form of ownership in which two or more people own *equal shares* of the real property. Like tenants in common, each co-owner has the right to occupy and use all of the property and, therefore, must share with the other owners.

A joint tenancy can be created by deed. In doing so, all of the owners must acquire title (ownership) by the same deed at the same time. The deed must state explicitly the intention of the parties to own the property as joint tenants. Otherwise, the law presumes that they have created a tenancy in common.

The most important distinguishing feature of a joint tenancy is the *right of survivorship*. This means that when one of the owners dies, his or her share of ownership is split among the surviving owners, thereby increasing the surviving owners' proportionate shares.

For example, if X, Y, and Z own a house as joint tenants, they each have an equal right to live in that house. If X dies, his share of the house automatically passes by right of survivorship to Y and Z, each of whose percentage of ownership thereby increases from 33.33% to 50%. When Y dies, Y's interest passes to Z and Z becomes the sole (100%) owner of the property. At that point, Z is the absolute owner rather than a joint tenant.

A joint tenancy can also be formed when a person dies. If that person has a will that describes how he or she wants the property distributed, such property might be transferred to two or more people as joint tenants. If a person dies *intestate* (without a will), their property will be distributed to the heirs under the laws of the state. Those laws typically require that heirs own the property as tenants in common, not as joint tenants.

Generally, a joint tenancy is terminated if one of the owners sells his or her interest to an outside party (one who is not already an owner). For example, if X and Y are joint tenants, and X sells his interest to Z, thereafter Y and Z are not considered joint tenants, but tenants in common. If there are three or more joint tenants, and one of them transfers his or her interest to an outsider, then the right of survivorship between that individual and the others is terminated, but such right remains intact between the remaining original owners.

Tenancy by the Entirety: A tenancy by the entirety is a special form of ownership that can only exist between a husband and wife. When a married couple owns real property as tenants by the entirety, they share an equal interest in the property, each with a right to occupy and use the entire property. Neither spouse, acting alone, may sell its share of the property as long as the other spouse is alive.

To create a tenancy by the entirety the deed must state explicitly that this is the form of ownership that is intended. If a husband and wife own the same real property but the deed does not describe the form of ownership, the law presumes that a tenancy in common is created.

The parties must actually be married to each other to create a tenancy by the entirety. For instance, if a husband and wife who own real property as tenants by the entirety get divorced, that tenancy by the entirety automatically becomes a tenancy in common. Two unmarried people who own the same real property cannot do so as tenants by the entirety because they are not married. In a dispute, most courts would hold that a joint tenancy was created because the parties apparently intended for the right of survivorship to exist, as described below.

Like a joint tenancy, a tenancy by the entirety carries with it the right of survivorship. When one spouse dies, his or her share of the property passes to the surviving spouse. The feature that distinguishes a tenancy by the entirety from other forms of multiple ownership, however, is a certain amount of protection from creditors. Most states have laws that protect property owned in this form from creditors of one of the co-owners. Although the laws differ in each state, generally, a creditor cannot seize one spouse's share of property owned as a tenant by the entirety for the payment of debts if a) the other spouse does not also owe the debt, and b) the real property is the principal residence of the nondebtor spouse. (Regardless, both spouses are liable for debts incurred for items that are necessary to live, such as food and clothing.) Married people often choose to hold their property in this form because of this feature.

A tenancy by the entirety can be terminated only by 1) the death of either spouse (which leaves the survivor as the sole, or absolute, owner), 2) by divorce, 3) by mutual agreement, or 4) by legal proceedings instituted by a creditor of both the husband and wife.

See Figure 5.1 for a summary of the differences between these three types of multiple ownership.

Adverse Possession

Under the doctrine of adverse possession, a person may come to own real property without purchasing it, and without doing anything more than openly and physically possessing it for a long period of time without regard for the rights of the true owner. Although state laws differ, twenty years is the typical length of time that must pass before the person in physical possession of the land can, in fact, become the owner.

An empty house is a prime candidate for an adverse possessor. If a person other than the true owner moves into the house and continuously uses it as his residence, and the true owner does not evict him, at the end of the statutory time period the possessor becomes the true owner of that property. Because of the concept of adverse possession, many office building owners in large cities commonly close open lobbies or plaza areas to the public one day each year to prevent anyone from acquiring an interest in the property through continuous adverse possession.

Rights to Possess and Use Real Property Owned by Another

An individual or entity may have an estate or interest in real property that he or she does not own. This is accomplished through *tenancy* (what is commonly referred to as *renting*), and *easements*. These two principles are discussed below.

Tenancy

According to the more common, but somewhat restrictive, use of the word, a tenancy is a limited right to occupy and use real property owned by another person. It gives the tenant exclusive possession, for a limited period of time, of all or part of some real property that he or she does not own.

The two most common types of tenancies in effect in the United States today are a *tenancy under a lease* and a *tenancy at will.* A tenancy under a lease is created through a written agreement (a *lease*) between the property owner and the tenant. In addition to specifying the length of the tenancy, the lease normally spells out the rights and responsibilities of each party in a variety of important matters. The tenancy ends automatically on the date specified in the lease. The parties must extend the lease or sign a new lease if they wish the tenancy to continue in that form.

In a tenancy at will, there is generally no written agreement between the property owner and the tenant. Typically, the tenancy is not for any specific period of time, and either party may end the tenancy at any time provided proper notice is given to the other party (usually 30 to 60 days). As a practical matter, rent is often paid weekly or monthly. Such a pattern may imply a minimum term for the tenancy. If so, the tenancy can normally only be terminated at the end of an established interval (such as at the end of the week or month).

Three Types of Multiple Ownership

	Share of Ownership	Survivorship
Tenancy in Common	Varies	Shares pass to heirs
Joint Tenancy	Joint tenants own equal shares	Shares are distributed among the surviving owners
Tenancy by the Entirety	Husband and wife each own all	Surviving spouse becomes absolute owner

Figure 5.1

Easements

An *easement* is the limited right of one person to use another person's property in a particular way or for a particular purpose. For example, utility companies frequently obtain or are granted easements in private property so they may run sewers, electrical distribution lines, or other utilities underground or overhead through the property. They also normally obtain a right of access for repairs and maintenance.

In general, the party who desires an easement in another's property must compensate that property owner for the right to make use of their property. Sometimes, however, easements are implied if they are necessary to make property usable. For example, there may be an implied easement through a neighbor's land to another parcel of land otherwise cut off from public streets. The person or entity who has obtained or been granted the easement has the right to use the property for specific purposes, but otherwise has no right to possess it, to own it, or to sell it.

Rights of Property Owners

Property ownership carries with it several rights, as shown in Figure 5.2. The most basic are the exclusive right to possess the real property, the right to use the property freely, and the right to control who may physically enter the property. Other people have no right to enter real property without the owner's permission. Accordingly, the owner, or others in possession of real property, with the owner's permission, may take action to protect the property against *trespassers*, *intruders*, and *intrusions*.

Trespassers are those who enter private real property owned by others without the owner's permission. Since trespassers have no right to travel on private real property, a property owner is generally not obliged to keep his or her property in a safe condition for the benefit of potential trespassers. Although there are exceptions, when a trespasser or other intruder enters another person's real property, he does so at his own risk. If he is injured while trespassing, the property owner will not be liable

Rights of Property Owners

Property owners have the right to:

- Possess and use their property freely
- Determine who may or may not physically enter the property
- Use the ground beneath the surface of their land
- Possess the air space above their land, in compliance with building codes and zoning laws
- Expect that land in its undeveloped state will be physically supported by adjoining land
- Make reasonable changes in the flow of water that is flooding their land

Figure 5.2

for his injuries. On the other hand, property owners generally are responsible for making their property reasonably safe for invited guests, business customers, and others who have legal use of or access to their property.

Objects as well as people may intrude on real property. A fence belonging to an adjacent land owner which partially stands on another's lot is an intrusion. The second landowner has the right to force the fence owner to remove that part of the fence which stands on the second landowner's lot. Similarly, a landowner has the right to remove a neighbor's tree branches or anything else that overhangs his real property.

Subsurface Rights

Real property owners have the exclusive right to use the ground beneath the surface of their land. For example, a landowner may agree to allow another person or business entity to extract minerals from his land in return for compensation, or he may decide to prohibit such mining activity altogether. There have been a number of instances in which previous mining activity had so damaged the subsurface that large "sink holes" developed, occasionally swallowing whole buildings. It is because of this right that utility companies are required to obtain the easements that were mentioned earlier for the services that they run below ground through private property.

Air Rights

Real property owners have the exclusive right to possess the air space above their land, as long as they comply with building codes and zoning laws. (See Chapter 12 for more information on building regulations. Zoning is discussed in greater detail later in this chapter.) In other words, a person may construct a building on his or her property as tall as the zoning laws and building codes permit.

A property owner may sue a neighboring owner who interferes with his air space. For example, a balcony from a neighbor's house that overhangs a landowner's property interferes with the landowner's exclusive right to possess the air space above his land. The landowner may sue to force his neighbor to remove the part of the balcony that overhangs his property. Similarly, a real property owner may be able to take legal action to protect access to light and air if an adjacent property owner builds or uses his property in a way that cuts off the light and air available to the first owner's property.

In many large cities there are cases in which air rights have been sold or traded. Assume, for example, that the zoning laws of a city allow a building as tall as 40 stories on a certain parcel of land. Located on that land, however, is an historic church that is the equivalent of 7 stories tall. The leaders of the church have no desire to use the air space for additional building as allowed by zoning. They may sell the rights to the unused 33 stories, however, to an adjacent property owner (if allowed by the zoning laws) to enable that owner to build higher than otherwise allowable. Both parties profit from such an arrangement.

Finally, a landowner does not have the right to stop *temporary* intrusions into his or her air space. Airplanes, for example, may fly freely over private property, subject to rules restricting noise and other types of pollution.

Land Support

Real property owners have a right to expect that land in its natural state (without buildings, houses, or structures) will be physically supported by adjoining land. If a landowner excavates his land and causes subsequent damage to adjoining land that is in its natural state (for instance, if the soil subsides), that landowner is liable for resulting damages even if he used special care while excavating.

Land with structures on it is also entitled to physical support from adjoining land. However, this right is conditional. The real property owner whose property is damaged must prove that the neighbor acted carelessly (negligently) when excavating his land, and that the weight of the first owner's building was not the cause of the damage.

In cities, for example, where buildings generally occupy all or most of the property, adjacent excavation would remove the lateral support from a building and its foundations, with potentially catastrophic consequences. Developers, therefore, must take steps to provide temporary support of adjacent land while they construct their building(s), or until they can provide alternative, permanent lateral support.

Water Rights

The law allows a landowner to make reasonable changes in the flow of water that is flooding his or her land. If rainfall or snow has caused flooding, a property landowner may drain his or her land, as long as the drainage does not cause adjacent property to flood. In the construction industry, developers must be careful not to alter the topography of the land in such a way as to flood adjacent property.

Buying and Selling Property

The procedures involved in buying and selling real property are basically the same from case to case, although the complexity of the documentation required can vary considerably. State laws, the needs and desires of the parties, and the nature of the real property (such as whether it is raw land or highly developed downtown property), will determine how a particular sale should be handled. For illustrative purposes, and to avoid confusion, the following material will focus on the sale of a residence.

A *title* is evidence of the rights of ownership. A title to property is transferred when a seller gives a deed to a buyer in return for receipt of the purchase price. This transaction is called a *closing* because it signifies that the sale has been closed, or completed.

There are several steps that must be taken before the sale is closed. First, the seller and potential buyer negotiate the terms of the deal, either directly or through brokers, attorneys, or other professionals trained in this procedure. Such negotiations normally culminate in the seller's acceptance of the buyer's offer to purchase the property at a particular price. The offer will frequently request that a more detailed agreement be executed, in accordance with the terms offered and accepted, and within a limited period of time. This agreement is called the *purchase and sale agreement*.

The Purchase and Sale Agreement

The statute of frauds requires that a contract for the sale of real property be in writing. This written contract is called the purchase and sale agreement. It records the seller's promise to deliver title to the property, and the buyer's promise to pay the purchase price, plus a list of other terms and conditions which vary from transaction to transaction. Once the buyer and seller sign the agreement, they are committed to the transaction.

The law requires that the agreement accurately describe the property, identify the buyer and seller (provide names and addresses), and state the purchase price if agreed upon. Usually this agreement will contain a provision making its enforceability subject to the buyer's success in obtaining a loan, prior to the date set for the closing, for all or part of the money necessary to purchase the property. The closing date, place, and time are often specified in the agreement. The agreement also usually includes a list of fixtures, such as the stove, lights, air conditioning units, and other items that are included with the sale of the property.

Often, purchase and sale agreements contain a clause that requires the seller to *convey marketable title*. In other words, the seller must prove that there is nothing wrong with the title which would cause subsequent potential buyers to question whether or not they possess all rights of ownership as promised, and whether or not they could sell the property at a future time without question that they were the rightful owner. A title is unmarketable if there are gaps or other defects in the recorded history of the title. Previous unsatisfied (outstanding) mortgages, liens, and certain easements on the property are encumbrances that would probably make the title unmarketable. Prior to the closing, the lender's attorney or, if the buyer is not obtaining financing, the buyer's attorney, searches the history of the title to the property (conducts a *title search*) to identify any defects and encumbrances. If the search produces any evidence of problems, the parties should work them out before the closing.

The purchase and sale agreement usually requires the buyer to make a deposit, which is later applied to the purchase price. Either a lawyer or a real estate broker normally holds the deposit in *escrow* (in trust) to be released to the seller at the closing.

The parties to a purchase and sale agreement normally take a variety of actions in reliance on the promises in the agreement. They change their legal situations in many ways as a result. For example, the buyer does not have a right to move onto the property until after the closing date, but the seller does not have the freedom willfully to damage the property prior to that date. Additionally, because the seller is in possession of the property, he must pay the real estate taxes up to the closing date.

The reader should understand that a purchase and sale agreement is a contract, and may be enforced in accordance with the general legal principles of contract law. If either of the parties fails substantially to perform as promised in the purchase and sale agreement, the other party may sue to force the defaulting party to go through with the transaction or to pay for resulting damages. (See Chapter 2 for more information on contract law.)

Transferring the Deed

Ownership of real property is not transferred until the seller gives the buyer a deed to the property. The seller(s) must sign the deed in order for it to be valid. The buyer(s) does not have to sign the deed, but his or her name must be on it. The deed must clearly indicate the seller's intent to transfer ownership, and the property must be described in sufficient detail to identify it. Generally, such a description defines the boundaries of the property and refers to the location in the public land title records where information on the property can be found.

The buyer becomes the legal owner of the real property once the deed is transferred. Thereafter, the buyer should take care to have the deed recorded in the public land title records at the registry of deeds (the name of this office varies from state to state) in the county where the land is located. Such a record gives public notice that the new buyer is the rightful owner of the real property. Recording the deed will also protect the buyer from a variety of potential problems and claims. For example, some sellers have attempted to sell the same property to more than one purchaser at a time. A review of the land title records would probably answer most questions about rightful ownership.

Condominiums

A *condominium* is a system by which parts, or units, of real property are separately owned. Such individual units could be apartments, stores, or offices. Each unit owner obtains all the rights incidental to ownership of real property as described in this chapter. Beyond that, each owner shares with the others rights to the common areas in the building, facilities, or land of the condominium.

In a multi-unit building, the common areas would typically consist of hallways, laundry rooms, elevators, stairways, lobbies, and similar spaces. Each owner shares the facilities of the building, such as the pool, tennis courts, washing machines and dryers, and clubhouse. Further, each condominium unit owner shares with the others ownership of the land on which the building(s) and other facilities are erected. This might include the parking lot, shrubs, trees and other landscape features. Although the individual unit owners have absolute ownership of their units, they own the common areas, facilities, and land as tenants in common. Each tenant in common's share of ownership of such elements is usually the fraction resulting from comparing that unit's cost or area to the total cost or area of all of the units.

Establishing and Maintaining a Condominium

State statutes govern the procedure for establishing and maintaining condominium developments. Typically, a condominium is created when the owner or owners of the land execute and record a deed (referred to as the *master deed*) which states their intention to create a condominium. The deed also contains bylaws, rules, regulations, and restrictions which will apply to the condominium and with which the owners of the individual condominium units must comply. The unit owners may not use their units for a purpose which is prohibited in the master deed. For instance, if the master deed states that the units are for residential use only, an owner cannot open a retail business in his or her unit. The bylaws usually specify the rights and responsibilities of the unit owners with respect to repair and maintenance of the condominium, its facilities, common areas, and grounds.

Over the years, condominium developments have been involved in more than their share of disputes with design professionals and construction contractors. Part of the reason is that construction professionals are hired by, and perform services for, a developer, who normally sells the units upon completion. Frequently, the individual unit owners' expectations are different from those of the initial developer. For example, the developer may want to install inexpensive air conditioning units even though such units may be less efficient than the more expensive models. The design professional specifies, and the contractor builds, what the developer wants. However, the unit owner, faced with high utility bills caused by a relatively inefficient air conditioner, may try to sue the designer or contractor for negligence or breach of warranty. Regardless of the reasons, the fact is that condominium design and construction has been a high-risk business in recent years.

Cooperatives

A *cooperative* is another form of multiple ownership that is somewhat similar to a condominium. In a cooperative, however, a corporation is formed to own the land and the structures on the land. Thus, unlike a condominium, the individual units are not owned separately. In a cooperative, the residents of the units own shares of stock in the corporation in proportion to the size or value of the unit they occupy. They also hold a long-term, renewable lease on their own unit. Cooperatives are very common in New York City, but are less common than condominiums in most other parts of the United States.

Public Control Over Real Property Use

The police powers of the government (those necessary to protect the general health, safety, and welfare of its citizens) give states a limited right to place restrictions on the use of land. The state can control real property use in the following ways:

- Zoning laws
- Special use permits
- Eminent domain

Zoning

Zoning is a form of public control over private land use. Each city, town, or other municipal body which intends to enact zoning regulations must first develop a master plan for the future development of the city, town, or municipality. Thereafter, zoning regulations may be imposed to create districts which will help to implement the goals of the master plan. Each district must have ordinances to control how land is to be used in that district. For example, one district might be designated for residential use while others are designated for commercial or industrial use. Regulations may also restrict architectural designs of buildings to make the district more aesthetically pleasing or to preserve a local architectural character.

Zoning Variance

In certain cases a landowner may be excused from complying with a zoning ordinance. To do so, he or she must apply for and obtain a *variance* from the zoning ordinance. A zoning variance allows the landowner to use his or her land differently from what is specified in the zoning ordinance. Generally, to obtain a variance a landowner must demonstrate to the local zoning board of appeals (the name of this group may vary from one community to another) that he or she will suffer some unique hardship or practical difficulty if forced to comply with the ordinance. For example, an ordinance may require that every building be set back 25 feet or more from each property line. It may be that a given piece of property is shaped such that it would be impossible to build anything usable on that property and still meet the setback requirements. That might render the property useless, and could justify a variance allowing the landowner to build, say, 20 feet from the property lines.

A landowner who is denied a variance may request rezoning of his or her parcel of land. It is highly unlikely that an area would be rezoned based merely on one such request, however, because a rezoning usually affects the whole city or town. The local planning board and zoning board would need to consider the interests of all property owners in the community before approving such a request.

Special Use Permits

Alternatively, a landowner may apply to the local zoning board for a *special use permit.* Such a permit may allow the property owner to use his or her property in a way that would not normally be allowed in the district. For example, it might be possible to get such a permit to operate an office in an otherwise residential district. A special use permit is easier to obtain if the ordinance sets forth examples of specific uses for which a permit may be issued.

Existing uses of land or structures which do not conform to local zoning ordinances are labeled *nonconforming uses.* A nonconforming use is tolerated because it existed prior to the enactment of the ordinance that it violates. For example, a town would not normally force a gas station to close down simply because it is located in a district that was zoned for residential use *after* the gas station was built. Although nonconforming uses are allowed to remain as they are, they may not be expanded or enlarged in any way which is inconsistent with the current zoning of the district in which they are located.

Eminent Domain

Finally, states have the power to take private property from citizens for legitimate public uses, regardless of the wishes of the owner of that real property. This is called taking property by *eminent domain.*

There are very specific rules for conducting such a process, beginning with the local governing authority's declaration regarding the purpose for which the real property is required, and ending with compensation for the private owner for the fair value of the property. In general terms, this procedure is referred to as *due process of law.*

If a state places too many restrictions on land use, they may have effectively taken the property from the owner by rendering it useless for development. It is unconstitutional for the state to deprive a property owner of his or her property without due process of law (that is, without allowing him or her to voice grievances and be heard on the matter, and without compensation). In such a case, the owner may take the initiative to sue the city or town to obtain compensation for the loss of his or her property. This is known as *inverse condemnation*, since the compensation process is initiated by the property owner and not the governing authority.

Conclusion

Real estate ownership and development is a major force in the economy of our country. Designers and builders deal with real property issues every day. It is therefore important that they understand the principles of property ownership and the rights of people and business entities to occupy and use real property.

CHAPTER 6

BASIC INTELLECTUAL PROPERTY LAW

CHAPTER 6 BASIC INTELLECTUAL PROPERTY LAW

Introduction

Intellectual property law protects original ideas, concepts, processes, symbols, and other forms of expression. It is intended to help creators and inventors protect their valuable ideas and inventions, and to prevent others from using and profiting from them without permission. *Copyrights*, *trademarks* and *service marks*, *patents*, and *trade secrets* are all considered intellectual property.

Nearly every business produces material that can be copyrighted or patented. Relatively few, however, know how to obtain and preserve a copyright or patent. Similarly, most businesses use trademarks and have trade secrets, but they are not always aware of the steps that should be taken to preserve their rights to own and use their trademarks and to keep their trade secrets confidential. All too often, a business does not realize the importance of preserving exclusive rights in its intellectual property until a competitor is discovered using the same or a similar trademark, idea, procedure, or invention.

Intellectual property law is important in the construction industry because there are so many ideas, processes, procedures, and inventions that can be protected. An architect's drawings and models can be copyrighted to protect the architect's expression of his or her ideas. Structural systems and engineering features can be patented and engineering drawings copyrighted. A contractor can patent a new and original type of insulation or a new construction technique, and can profit from it by licensing others to use it. Tools and other devices are also eligible for patent protection.

This chapter will identify the various types of protection offered to owners of original material by intellectual property laws, and the basic steps that should be taken to obtain and preserve such protection.

Copyright Law

Copyright law protects the right to exclusive ownership of original works produced by writers, artists, composers, and photographers. Copyright law gives the owner of the copyright five exclusive rights. These rights, listed in Figure 6.1, belong to the copyright owner (this person is normally, but not always, the creator) for the life of the creator, plus fifty years after the creator's death. A person or entity who exercises any of these exclusive rights without the copyright owner's permission has infringed on the owner's copyright.

The penalties assessed against individuals who commit copyright infringement include suing for lost profits that the creator would have made had the other party not violated the copyright, suing for attorney's fees, and an injunction issued by the courts to stop the violator from continuing to profit from the copyrighted work.

States do not have their own copyright laws. Rather, federal law is applied uniformly to every copyright matter under the *Copyright Act of 1976*, Title 17 of the United States Code. This law became effective on January 1, 1978, superseding the former federal copyright law which had been in effect to that date. This law was updated in October, 1988, to be compatible with the Berne Convention for the Protection of Literary and Artistic Works. The amendments made at this time typically relate to provisions other than those which have been mentioned in this chapter. It would be good policy to check the amendments, however, if further questions arise regarding copyright law. For more information on the Berne Convention, see Figure 6.2.

What is Protected Under Copyright Law?

Although copyright protection is often associated with artistic works, works produced in a business office can also be protected under copyright law. For instance, it is possible to copyright correspondence, office manuals and brochures, reports, studies, marketing materials, speeches, computer software, advertising displays, and product labeling. In fact, copyright protection is available to any kind of *original* work (work that has not been copied from another work).

As soon as an idea is put into tangible form, the creator is entitled to copyright protection. For instance, as soon as a design professional has transferred his or her ideas to paper, the design is automatically copyrighted. It is important to realize, however, that copyright law

The Five Exclusive Rights of Copyright Owners

Copyright owners have the exclusive right to:

1. Reproduce the copyrighted work
2. Create variations of, or derivative works based on, the copyrighted work
3. Distribute copies of the copyrighted work to the public
4. Perform the copyrighted work for the public (through broadcasts, dramatizations, and motion pictures)
5. Display the copyrighted work to the public

Figure 6.1

protects only the *expression* of an idea, not the idea itself. Copyright law prohibits copying a drawing in tangible form, but it does not prohibit duplicating the underlying idea.

Businesses that know how easy it is to copyright their work are more likely to do so. This is not to say that all copyright matters are simple. Some can be quite complex and may require legal consultation.

Preserving a Copyright

Although an author or creator of original material is entitled to copyright protection as soon as the idea is put into tangible form, there are certain steps that must be taken so that the copyright is not lost. For example, when material produced in an office is issued, sold, leased, transferred, or rented to individuals outside of that office, the material is considered *published*. Published material is protected by the copyright laws only if the copyright has been *preserved* by placing the proper copyright notice on the material.

Copyright Notice

The minimum proper copyright notice consists of the following three elements:

- The symbol ©, the word *Copyright*, or the abbreviation *Copr.*
- The year that the work was first published
- The name of the copyright owner

The United States Copyright Office has issued regulations that specify where the notice should be placed on the material to be valid. The Copyright Act, however, states that the notice is to be affixed "in such manner and location as to give reasonable notice" of a claim of copyright

The Berne Convention

The United States Congress amended the Copyright Act of 1976 in October, 1988. The amendments became effective on March 1, 1989. The purpose of these amendments was to make the U.S. Copyright Act compatible with the Berne Convention for the Protection of Literary and Artistic Works (known simply as the *Berne Convention*). The Berne Convention dates from 1886 and is the world's most widely adopted copyright treaty.

Seventy-six countries, including the United States, are members of the Berne Convention. (China and the Soviet Union are the only major countries that are not members.) One of the principal provisions of the Berne Treaty prohibits members from imposing "formalities" as conditions to "the enjoyment and the exercise" of copyright. One such "formality" in the U.S. Copyright Act was the requirement for registration as a prerequisite for suit. Congress responded by allowing authors from member countries, other than the United States, to sue in the U.S. without first registering their copyright.

Figure 6.2

ownership. Generally, a copyright will be properly preserved if the notice is placed on the document or object anywhere it is clearly visible.

Copyright Registration

A copyright can be registered with the United States Copyright Office. This is done by filing a completed application form, paying a filing fee, and submitting a copy of the work. The number and nature of the copies that must be filed are specified by the statute and Copyright Office regulations. Usually, each work that is registered requires a separate application and fee.

It is not necessary to register the copyright in order to acquire the five exclusive rights of copyright protection. A copyright notice placed on the material is sufficient to do that. Copyright registration offers other protections, however, and registration is a prerequisite to instituting a federal lawsuit for copyright infringement, should that ever be necessary.

Copyright Ownership

The person who creates an original work is the initial copyright owner. Under the present copyright statute, the creator may transfer some or all of his or her rights to others. For instance, an architect may grant a homeowner/client the right to copy plans for making renovations or additions to their house, but not for use in building a new house.

Once the creator transfers the entire copyright in a work, he or she no longer has any rights with respect to that work. An agreement to transfer an entire copyright should generally be put into writing and signed by the *transferor* (the present owner). Such an agreement is subject to all of the usual rules of contract formation (see Chapter 2, "Basic Contract Law") and, unless a gift is intended, would involve legal consideration of some sort. The written agreement may be recorded in the U.S. Copyright Office, but does not have to be recorded to complete the transfer. However, recording protects the *transferee* (the new owner). Usually, a transferee cannot begin a lawsuit for copyright infringement unless the transfer was executed by written instrument and recorded in the U.S. Copyright Office.

The Copyright Act is clear in stating that transferring ownership of a *physical object* does not automatically transfer ownership of the *copyright* for that object. For example, if an engineer's client wishes to own the original documents prepared by the engineer, such documents can be transferred to the client, even though the engineer retains the copyright in them and, therefore, has the right to control the subsequent use that can be made of them.

Works Made for Hire

Who is the owner of works produced by employees of a business? Generally, the employer is considered the creator *and* the owner of the employees' works, as long as the works were produced within the scope of the job. Works produced within the scope of employment are known as *works made for hire*. This rule applies unless the employer and employees have specifically agreed otherwise. If an engineer employed by a firm writes a computer program to perform the calculations necessary to select steel shapes for a bridge structure that the firm had been hired to design, the firm, and not the individual engineer, would normally be considered the creator and owner of the copyright in the computer program.

To avoid misunderstandings, a business owner should include in his or her employment manual, provisions spelling out the ownership rights for works created by employees in the course of their employment at the business.

Derivative Works

A derivative work is a work based on one or more preexisting works. A design professional who adapts old drawings to a new project has created a derivative work. The copyright in a derivative work is distinct from the copyright in the original work, and technically extends only to the new matter contributed by the creator of the derivative work. To avoid infringing on the copyright of the original creator, the creator of the derivative work should obtain permission from the creator of the original work before making use of the primary materials.

Trademark Law

A *trademark* or *service mark* is a word, name, symbol, device, or any combination thereof which manufacturers and service providers use to identify their goods and services and to distinguish them from the goods and services of others. A trademark or service mark works with the product or service being provided to build business. Once the mark is used, it becomes associated in the public's mind with the product or service and develops into an essential marketing tool.

A trademark can be anything: names, sounds, logos, product packaging, background design, or shapes. Consider, for example, the bulldog symbol of Mack Trucks, or the leaping deer which represents John Deere equipment. The "Mr. Build" trademark was designed to have a specific connotation in the construction business. That is, it is meant to conjure up the image of a known entity in the public's mind, much like the "Century 21" trademark implies a certain level of respectability in the real-estate business. All of these businesses make use of trademarks to set themselves apart from their competitors and to market their products or services more effectively.

Trademark law is based on both federal and state statutes. The Lanham Act, enacted in 1947, has been the primary source of federal trademark law. Congress recently amended the Lanham Act with the Trademark Revision Act (to take effect in November of 1989). The major change to the Lanham Act concerns federal trademark registration, which will be addressed later in this chapter.

Establishing Trademark Ownership

Trademark ownership is established when an individual or business entity simply selects and uses a trademark. The trademark is placed on goods or used in conjunction with the product or services that it identifies and represents. The trademark or service mark symbols, ™ and ℠ , should be put on the product or somehow used with the mark. For example, Steelcase, manufacturers of office furniture and equipment, uses "The Office Environment Company"™ in their advertising. These symbols give notice to the public that the owner has claimed rights in the mark and that the mark should not be used by others.

When the trademark is not placed on the product or is not associated with the product or service, the owner loses his or her rights to the trademark. More extensive rights can be obtained by registering the trademark with the federal government.

Selecting a Trademark

Careful consideration should be given when selecting a mark. The mark should complement and work with the product or service. Often, trademarks are invented words that are not in the dictionary, or are a combination of letters that do not in any way explain the product or service with which they are associated. Examples of these are Exxon and Xerox. Sometimes trademarks are words that are in the dictionary but do not convey any information about the product or service they represent, such as "Apple" Computer.

Sometimes trademarks are nicknames or abbreviations of the product name. For example, "Coke" is an abbreviation of "Coca Cola." Trademarks can also be phrases associated with the product which tell something about the product or its quality. For example, Amoco uses the trademark "Permacolor" for a line of commercial carpets that are warranted to be stain and fade resistant for a period of years. A word that is the commonly known name for the product cannot be a trademark. Surnames can be used as trademarks. However, a surname cannot be registered with the federal government until it has acquired a secondary meaning. In other words, the name must identify the product and not the person.

Conducting a Trademark Search

When selecting a trademark, it is advisable to consult an attorney who specializes in trademark law. Proper selection of a trademark can be crucial to the commercial success of the product or service. Obviously, a manufacturer would not want to identify his or her product with a mark that already identifies another product. In fact, a manufacturer who unknowingly selects a trademark that is confusingly similar to another product's trademark could be sued for *trademark infringement*. The result of such a lawsuit could be an order to the manufacturer to stop using the trademark. This could be devastating to a manufacturer who has used the trademark for a number of years. A product without its trademark is essentially a product without an identity.

To avoid this possibility, a search should be done once a trademark has been tentatively selected to determine whether or not the mark is available for use. Trademark search companies can be hired to determine if a mark is already being used. These companies look for similar marks for related goods or services by reviewing marks previously registered, pending applications for registration, state trademark registries, general directories, trade and telephone directories, and other sources. If a complete search is requested, they may use expensive techniques, such as private investigation.

If the product or service is meant to attract a national market, it is important to conduct a thorough national trademark search. On the other hand, if the product or service is to be sold in one state, and the owner sees no possibility in the distant future of expanding the business into other states, a less extensive search may be conducted. If the product will be sold only in Massachusetts, for example, a trademark search of just the New England states may be sufficient.

Trademark Registration

Like copyright registration, trademark registration affords the trademark owner a number of exclusive rights, as listed in Figure 6.3. However, under the old provisions of the Lanham Act, a trademark could not be registered unless it was actually in use. This was considered risky, from a marketing point of view, since companies would not know whether a trademark was acceptable until the product and its trademark were put into commerce (sold in more than one state). A company might spend a great deal of time and money selecting a trademark only to discover that the United States Patent and Trademark Office had denied registration and the mark could not be used.

The *Trademark Revision Act* permits a business to file an application for registration with the United States Patent and Trademark Office and to reserve a trademark before actually using the mark in interstate commerce. The applicant must state its intent to use the trademark in commerce and must identify the product with which the trademark will be used.

If the Patent and Trademark Office determines the mark is eligible for registration, it is published in the *Official Gazette*. Anyone who thinks he or she would be damaged by registration of the mark may, within thirty days, oppose the registration. If the mark clears the opposition procedure, the Patent and Trademark Office issues a *Notice of Allowance*. The applicant then has six months to use the trademark in commerce and, within this time, must submit a *Statement of Use* to the Patent and Trademark Office. At that point, the Patent and Trademark Office may re-examine the mark. If the mark passes re-examination, it then becomes officially registered.

The six-month period during which the applicant must use its trademark may be extended for another six months if the applicant submits an affidavit (statement under oath) in which he reaffirms his intent to use the mark in commerce. This might be necessary if, for example, the

Trademark Registration Rights

There are many rights attached to a trademark once it is registered. Although a trademark attorney should be consulted about the specific advantages of registration, the following rights are generally established:

1. Registration gives official notice to the public that the mark is owned and in use.
2. Registration creates a presumption of ownership and an exclusive right of use.
3. Registration gives the owner the right to go to federal court to protect his or her rights if another party is trying to claim rights to that trademark.
4. Registration gives the owner the right to claim statutory damages in a trademark infringement suit. Statutory damages include triple the amount of actual damages and, in some cases, attorney's fees.
5. Registration gives *nationwide* protection to a trademark.

Figure 6.3

registered product does not come out on time. Following the second extension, an applicant may get a third extension for up to two years.

When the mark is registered with the federal government, the symbol ® is used with the trademark. As long as the trademark is used continuously, and registration is renewed every 20 years, the registration will last indefinitely. The symbol ® should never be used with an unregistered trademark because this deceives the public. If caught, the owner's registration may be denied.

Trade Names

There may be some confusion over the use of trademarks and service marks versus *trade names*. A trade name is the name selected for a business or corporation. When a trade name is first selected and used, it is generally not considered a trademark or service mark. However, with years of continuous use, a trade name may develop into a trademark. The name could start to identify not only the company, but also the name of the product or service. For example, XEROX Corporation is the trade name. XEROX has, however, also become a trademark for the product. Similarly, VELUX-AMERICA, INC. is the name of the company that makes roof windows and skylights and has registered the name VELUX® as its trademark.

Because a trade name can develop into a trademark, it should be selected with as much care as a trademark. Once a corporate name is selected, a search should be done to see if that name is already being used by another corporation. This is done by inquiring at the proper state office (usually the Office of the Secretary of State) to see if that name is already "registered" (meaning, in this context, registered in the state as a corporate name, not federally registered as a trademark). If the name is not registered, it can be reserved for a limited period of time while the corporation is being formed.

Registering a corporate name prevents other parties from registering and using the same name in that state. It does not prevent someone else from adopting the same or a similar name as a trademark, or using it as a trademark or business name in another state, however. It also does not prevent an unincorporated business, such as a partnership or sole proprietorship, from selecting and using that same name. However, the owners of these other types of business entities should search the state corporate records to see if a corporation has already selected the name they want to use. Although partnerships and sole proprietorships are each a different type of business structure from a corporation, they probably should not use a name that is already being used by a corporation.

Similarly, because a search of state registration of trade names cannot assure that the name is not being used as a trademark for another company, a business owner should also conduct a separate trademark search. If the business will be operated in only one state, then a limited search of telephone directories may be sufficient. If the business will be operated in several states or nationwide, then it would be wise to conduct a complete trademark search.

Patent Law

A patent is a grant made by the United States government to an inventor which gives the inventor the exclusive right to make, use, or sell an invention in the United States for a stated period of time – usually 14 to 17 years from the date the patent was issued, depending on the type of patent. A patent protects the ownership of the basic idea underlying an invention. The invention may be a physical object, such as a machine; a composition of matter (chemical compositions or mixtures of ingredients); or an industrial, technical, or construction process.

States do not have their own patent laws. The Federal Patent Act is applied uniformly for every patent matter. This law was enacted in 1836 and revised in 1953. The 1953 revision did not substantially change the law from its original form.

Individuals and firms in the design and construction industry should have some knowledge of the fundamentals of patent law for two principal reasons: first, to obtain patent protection for an invention by the person or firm. A new type of lighting fixture, a system for reusable forms for concrete construction, or a process or procedure for soil stabilization, may be patentable and subsequently very profitable for the inventor. Second, it is important not to infringe on the patents owned by other individuals or business entities. Use of a patented product or process without authorization (usually accompanied by the payment of license fees) can result in legal difficulties for the infringing person or firm.

Obtaining a Patent

A person or firm who wishes to patent an invention should consult an attorney who specializes in patent law, since the process for obtaining a patent can be complicated. For instance, there are rules which restrict the period of time in which an invention may be displayed to the public prior to application for a patent and still be eligible for patent protection.

Before applying for a patent, a search must be conducted to determine whether the invention is truly different from all existing technology. The search can be done in the United States Patent and Trademark Office by a researcher skilled in patent law, who examines patents on file and under application. If the search is successful (meaning that no similar invention is discovered), an application can then be filed.

The application contains detailed information about the invention and constitutes the basis for the actual patent that is issued. In other words, the application describes the features of the invention so that when a patent is issued, it incorporates or is derived from information in the application. An application for a patent must be filed with the U.S. Patent and Trademark Office within one year from the time that the invention is used publicly, is offered for sale, or is sold in the United States. Patent applications should also be filed in foreign countries if the invention is to be sold abroad. Many countries do not allow a product to be patented if it already has been shown to the public in *any* country. Therefore, if an American inventor hopes to obtain patent(s) from foreign countries, the invention should not be shown publicly, and should not be described in published material or sold prior to obtaining the patent.

Trade Secrets

Most businesses have information that they do not want to make known to the general public. Such information may include customer lists, marketing materials, bookkeeping methods, and office systems. The common law protects such information as long as it has some business value and the business itself takes precautions to keep the information secret. This information is called a *trade secret*.

Trade secrets are typically governed by state law. A trade secret may be a formula, pattern, device, or compilation of information which is used in one's business and which gives the business an advantage over competitors who do not know or use the secret. A trade secret is not just information regarding a single or short-lived event, such as the salary of certain employees, but, rather, a process used continuously in a business' operation. Many construction contractors have estimating systems, procedures, and databases that they have developed for use in their business. This information is probably not known in detail outside the company and is believed to give the contractor an edge in preparing estimates and bids for projects. Such information constitutes a trade secret.

Generally, a trade secret must involve a substantial element of secrecy, such that it would be difficult to discover the trade secret without using improper means. Many states, however, have adopted a broader definition to include information that can be kept secret, has some business value, and is at least somewhat novel or unusual.

Maintaining Secrecy

One way to maintain a trade secret is to draft an agreement to be read and signed by every employee which contains a clause stating that confidential information and trade secrets may not be disclosed to people outside of the company without the company's written permission. The agreement should require new employees to identify obligations, if any, owed to previous employers with regard to trade secrets. The agreement may also require employees to assign or transfer to the company inventions developed during, and within the scope of, their employment.

An employment agreement is evidence that trade secrets exist and shows that the employer is trying to maintain secrecy. In the event an employee discloses a trade secret to an outside party, the employee's signature on such an agreement is evidence of conscious (deliberate) misappropriation. These factors all help to build a strong case for court protection in the event a trade secret is disclosed.

Trade secrets can also be maintained securely by other procedures. A company can put restrictions on visitors by requiring them to sign in and out at a security desk, or by escorting them at all times while they are on the premises. The company can tell employees, on a regular basis, exactly what information or categories of information are trade secrets. The company can lock file cabinets that contain important information, and printed material that contains trade secrets can be stamped with a "Confidential" notice.

Protecting trade secrets should be an ongoing process. As such, special consideration should be given to an employee who is terminated. The employer should remind the employee that his or her obligations not to disclose information continue after termination. The employee should return all written material unless the employer gives him or her permission to copy certain items. It frequently happens in architectural and engineering practices that a departing employee wants to take blueprints of projects on which he or she worked to demonstrate his or her experience to prospective employers or clients. As long as such material is screened by the firm so that trade secrets and confidential client information is not disclosed, there is nothing inherently wrong with this practice.

If the employee signed an employment agreement containing a *noncompetition clause*, the employee should be reminded that this clause prohibits him or her from working for a competitor for a period of time after termination. Generally, courts will enforce noncompetition clauses as long as the restrictions are reasonable and not overly broad.

Obtaining Trade Secrets

If a person obtains a trade secret from a business through improper means, the business can sue that person for any damages that result. Improper means include obtaining the trade secret through another person who one knows has a duty to maintain the secret. An example of this is fraud or bribery.

The legal protection offered to holders of trade secrets is limited, however. A business does not have a valid legal claim against a person who obtains its trade secrets by independent development or through proper means. Thus, if a person obtains a trade secret through independent invention, by starting with the known end product and working backwards to find the secret method which developed the product, or by observing the item in public use or public display, that person will not be held liable for infringing on trade secrets. This frequently happens with client lists. For example, a design or construction executive may leave his or her firm after many years of employment. He or she knows all of the clients of the firm. Even if no lists or printed data of any sort are taken from the firm, a list of clients can easily be resurrected from memory and by using the telephone directory. Doing so probably does not involve the theft of a trade secret.

Liability for trade secret infringement is different from liability for a patent or trademark infringement. A person who independently develops a product without actual knowledge of a similar patented product can be liable for patent infringement. Likewise, a person who uses a trademark which he devised independently, can be liable for trademark infringement. This is because information on patents and trademarks is available to the public, and inventors should research this information before marketing their product. A trade secret is more like a copyright, however, where a person who ends up with a product which is similar to another product whose design is copyrighted has not violated the copyright if he used independent means to develop his or her product.

Trade Secrets as Personal Property

Trade secrets are considered to be the personal property of the owner/employer. As personal property, trade secrets can be transferred or licensed. If a business is sold, a value is given to the trade secrets that are being transferred. This figure enters into the total sale price. For example, client lists and business systems of a company have a value just like desks and chairs, and that value must be accounted for upon sale of the company.

Conclusion

Individuals and firms in the design and construction industry focus their attention on physical things, like real estate, steel, and concrete. Because their main goal is to build and to improve real property, they sometimes forget that they are involved in a very creative business, dealing on a daily basis with ideas, concepts, and intellectual processes. The intellectual side of the design and construction industry is, therefore, sometimes given insufficient attention. This is an unfortunate and often costly mistake. The intellectual property of individuals and business entities in the industry is valuable and should be protected like all of the business' other property and assets.

CHAPTER 7

BASIC LABOR LAW

CHAPTER 7 BASIC LABOR LAW

Introduction

The federal government has enacted a number of laws that regulate the relationship between employers and employees. These are known as *labor laws*. This chapter examines two categories of labor laws. The first is concerned specifically with the construction industry. Laws in this category were enacted primarily to protect contractors, subcontractors, and suppliers who work on construction projects funded with public money. Among other things, these laws regulate wages, work hours, and overtime compensation.

The second category contains laws that are broad in scope and apply to private-sector as well as public-sector employers. These laws forbid discrimination against employees, mandate a safe work environment, and regulate collective bargaining agreements, among other issues.

This chapter is not an exhaustive review or analysis of all federal labor laws. Furthermore, this chapter does not address state labor laws, despite the fact that many states have enacted laws that apply to construction projects funded with state money. An analysis of individual state labor laws is beyond the scope of this book. For the purposes of this chapter, it is sufficient for the reader to know that if a state project receives any federal funding, federal labor laws will apply to that project in addition to the state laws. Sometimes the federal laws supplement state laws and sometimes they conflict with them. Therefore, anyone who administers projects funded with public monies should be aware of the federal statutes described in this chapter as well as the state laws applicable to the particular project.

Federal Labor Statutes

There are two types of federal labor statutes:

- Public-sector construction statutes
- Employer statutes

Each type of statute is examined below.

Public-Sector Construction Statutes

Public-sector construction statutes apply to projects that are in some way funded by federal, state, or local government entities. This chapter addresses the four major federal, public-sector construction statutes:

- The Miller Act
- The Davis-Bacon Act
- The Contract Work Hours and Safety Standards Act
- The Copeland Anti-Kickback Act

Everyone involved in federally funded construction should become familiar with the general requirements of these statutes and be aware that more study will be necessary to understand how these statutes apply to individual projects. Each of these statutes is discussed below.

The Miller Act

State mechanics' and materialmens' lien laws are generally available only to contractors and suppliers who work on, or supply materials to, private-sector construction projects since, in the United States, liens cannot be placed on public property. For the general contractor who has a contract with the federal government, this is usually not a problem, because the contractor is assured that money has been allocated to pay for the construction. Subcontractors and suppliers, however, do not have the same assurance, yet are not able to place liens on the public property as security if the general contractor defaults in his or her payments to them.

In 1894 the *Heard Act* was passed to give subcontractors and suppliers working on federal projects protection in the absence of traditional lien rights. Under this act, contractors on federal projects were required to obtain a *penal bond*, which was a combined *performance* and *labor and materials payment bond.* This bond provided subcontractors and suppliers with a source of funds if the contractor failed to make payments to them. There were procedural problems with the Heard Act, however, and in 1934, Congress repealed the act and replaced it with the Miller Act.

The Miller Act applies to all United States government construction contracts valued at more than $25,000. It requires general contractors on federally funded construction projects to obtain a performance bond and a labor and materials payment bond to protect the interests of subcontractors and suppliers. Performance bonds are written by commercial surety companies who stand behind a general contractor's promise to complete the project in accordance with the terms of the contract. Labor and materials payment bonds are the surety's commitment to pay the subcontractors and suppliers if they have not been paid in full by the general contractor. If the subcontractors' and suppliers' claims exceed the *penal sum* (face amount) of the payment bond, the available money is distributed proportionately to all claimants. Those subcontractors and suppliers who have either a direct contract with the general contractor or with a subcontractor to the general contractor are protected by the bonding requirements of the Miller Act. For more information on bonds, see Chapter 16, "Insurance, Bonds, and Lien Laws."

The Davis-Bacon Act

The Davis-Bacon Act was enacted in 1931 and governs all federally-funded contracts valued at more than $2,000 for the construction, alteration, or repair of public structures or public works. The purpose of this act is to protect the level of prevailing wages in the area.

This act requires that "laborers and mechanics" (blue-collar workers and their assistants and trainees) on federally-funded construction projects be paid no less than the local prevailing wages. Today, such wages typically correspond to union wage rates. The rates do not apply to those who perform white-collar work on a project, such as design professionals or administrative or clerical staff. The United States Secretary of Labor determines the applicable wage rates, which are based on the prevailing rates paid to similar types of employees on similar local projects. A schedule of the applicable wage rates must be included in each solicitation for bids.

The Contract Work Hours and Safety Standards Act

The Contract Work Hours and Safety Standards Act regulates hours of work and overtime pay for laborers and mechanics on federally-funded projects. It applies to all contracts on projects that are valued at more than $2,500.

This act requires that laborers' and mechanics' wages be based on a standard eight-hour work day and a standard forty-hour work week. This means that a laborer or mechanic must be paid overtime for working in excess of eight hours in any day or in excess of forty hours in any work week. The act also requires that overtime be paid at a rate not less than one-and-one-half times the basic rate of pay.

The Copeland Anti-Kickback Act

The Copeland Anti-Kickback Act is a criminal statute forbidding *wage kickbacks* from employees involved in federally-funded projects. This act makes it a criminal offense for an employer to force, intimidate, or threaten a person employed in constructing or repairing public buildings or public works to return or decline to accept any part of his or her compensation for work performed. In short, a contractor cannot force his laborers or mechanics to give him a portion of their wages in order to keep their jobs.

The act applies to projects funded in whole or in part by federal loans or grants. For each contract valued at more than $2,000, contractors and subcontractors are required to file wage statements with the government agency in charge of the project. Such statements certify that the full amount due was paid to the employees, and that no amounts were withheld without the employee's authorization. Violation of the Copeland Anti-Kickback Act may result in a fine of up to $5,000, or imprisonment for up to five years, or both.

Employer Statutes

Employer statutes are federal laws that must be followed by certain types of employers. These statutes are *labor-oriented*, which means that they apply to employers regardless of whether or not the projects involve public or private moneys. Employer statutes include:

- The Occupational Safety and Health Act (OSHA)
- The National Labor Relations Act (NLRA)
- The Fair Labor Standards Act (FLSA)
- Title VII of the Civil Rights Act of 1964
- The Age Discrimination in Employment Act
- The Rehabilitation Act of 1973
- Immigration Reform and Control Act of 1986

These labor statutes are generally applicable to employers in the construction industry. In fact, some provisions are aimed specifically at the construction industry. The statutes mandate minimum wage rates, overtime pay, healthy work environments, and nondiscrimination in employment, in addition to a large variety of other matters. Each statute is discussed below.

The Occupational Safety and Health Act (OSHA)

The Occupational Safety and Health Act (OSHA), passed by Congress in 1970, is a federal law requiring employers to maintain a safe working environment that is free from hazards that could cause death or serious physical illness or injury. Employers who have one or more employees and who are "engaged in a business affecting *interstate commerce*" (the buying, selling, or exchange of goods or property between or among states) are required to comply with this act.

OSHA is administered and enforced by the Occupational Safety and Health Administration, (called simply the Administration) within the United States Department of Labor. The Administration conducts investigations to determine whether or not employers are complying with OSHA's standards. The Administration is permitted to inspect work places without *probable cause* (without determining whether or not there is reason to believe that violations will be found), and without advance notice to the employers. The Administration issues citations and proposes penalties for employers who do not comply with OSHA's standards.

Among other requirements, OSHA imposes health and safety standards on employers by requiring them to follow certain procedures with respect to reporting accidents and maintaining records. (There is a specific set of standards that apply to the construction industry, but they are very detailed and, as such, are beyond the scope of this chapter.) In summary, such procedures include:

Accident Reporting: Employers are required to notify the Administration of workplace accidents that cause an employee's death or result in the hospitalization of five or more employees. The employer must report the accident within 48 hours after it occurs. The report must describe the cause of the accident, as well as the number of deaths and other injuries, if any.

Maintaining Records: Employers are also generally required to keep a record of *occupational injuries*, which are injuries that result from work-related accidents. Minor occupational injuries that require only first aid treatment do not have to be recorded. Serious injuries that result in death or loss of consciousness, however, or those injuries that require more than first aid, must be recorded.

Employers must also record all *occupational illnesses*. An occupational illness is a sickness or disease caused by exposure to toxic substances or other environmental hazards in the workplace. This would include an illness that developed from inhaling or ingesting a toxic substance, such as a construction solvent or other chemical agent.

Although an accident may not be serious enough to report to the Administration, employers are required to post annually within their workplace a report of all occupational injuries and illnesses among employees. The report must be made available to the United States Secretary of Labor and the Secretary of Health and Human Services, at their request. This rule applies only to employers of eleven or more employees.

OSHA recently expanded its activities concerning hazardous materials and hazardous waste, under the theory that every American worker has a "right to know" of the possible hazardous substances in use in their working environment. The new *Hazard Communication Standard (HAZCOM)* requires that workers be trained in how to safely deal with hazardous substances and the treatment to apply in case of emergency.

The new regulations require employers to develop written programs that include:

- A list of hazardous chemicals known to be in use in the workplace
- An explanation of how employees will be informed concerning job-site hazards
- A Material Safety Data Sheet (MSDS) for each hazardous substance in use
- A label on each container that specifies the material, the hazards, and the name of the manufacturer
- The development of a training program for employees who might be exposed to hazardous chemicals while on the job

The National Labor Relations Act (NLRA)

The National Labor Relations Act of 1935 (the *Wagner Act*) is commonly referred to as the NLRA. Its primary purpose is to make the *collective bargaining* process fair, especially for employees and their unions. Collective bargaining is the process by which employers and the representatives of employees (such as unions or other labor organizations) negotiate agreements with respect to wages, hours, and conditions of employment.

In the construction industry, there are *union contractors* and *open shop contractors*. Open shop contractors have not signed contracts with trade unions and do not hire employees from the union hall. Collective bargaining agreements are contracts formed between a union (whose members are typically practitioners of a particular trade) and the various associations representing contractors (as employers), and define the wages, benefits, and terms and conditions of employment for union members. For instance, there are bargaining agreements between contractor associations and the carpenters' union, the plumbers' union, and the masons' union. A contractor who hires a union carpenter is bound to the terms of the standard collective bargaining agreement between the carpenters' union and the various contractor associations. The contractor is bound to the agreement whenever he hires an employee who is a union member.

The NLRA has also established rules with regard to the bargaining process itself. For example, employers and unions are required to bargain in good faith, and employers must supply information to the unions upon request. Employers may not bargain directly with employees who are represented by a union. They must be available to meet with unions and must bargain with the union representatives only. The NLRA specifies topics on which the employer *must* bargain, at the request of the employees' representative. Such topics include wages, hours, arbitration procedures, and work rules. The NLRA restricts the dealings that employers may have with employees and their unions. Some of the restrictions include prohibitions against: threatening employees with loss of their jobs if they join a union; questioning employees about their union activities; interference with organizing a union; and refusing to hire a qualified candidate for employment because of his or her union membership or activities.

In 1947, Congress passed an amendment to the NLRA called the *Taft-Hartley Act*. This act imposed restrictions on unions which parallel those imposed on employers by the NLRA. The purpose of the amendment was to balance the relative duties and obligations of unions and employers.

Under the Taft-Hartley Act, unions are prohibited from coercing or threatening employees to join picket lines or to participate in other union activities. A union cannot attempt to persuade an employer to discharge an employee because the employee has opposed union activities. Unions must also represent employees' grievances fairly and must bargain in good faith with the employer.

The *National Labor Relations Board*, commonly referred to as the *NLRB*, was established under the NLRA to enforce its provisions and to prevent employer and labor organizations from engaging in unfair labor practices. Regional offices of the NLRB are located in major cities throughout the country.

The Fair Labor Standards Act (FLSA)

The Fair Labor Standards Act of 1938, also known as the FLSA, is a federal law establishing minimum wage and hour standards for employers and employees engaged in commerce or in the production of goods for commerce. Employers subject to the FLSA are required to maintain employee payroll records for at least two years. These records must contain the following information: the employee's address, sex, occupation, total hours worked each day and each week, time and dates the employee works, the hourly rate of pay, the amount of overtime compensation for weeks during which the employee works overtime, and total wages paid to the employee for each pay period.

Additionally, this act defines exemption requirements. For instance, executives, administrators, professionals, and outside salespersons are exempt from the minimum wage and hour standards. As professionals, architects and engineers are also normally exempt from the requirements of the FLSA.

The FLSA also requires employers to pay overtime to employees who work more than forty hours in one work week. The overtime wages must be at a rate that equals at least one-and-one-half times the employee's regular wage rate. The FLSA does not require employers to provide sick pay, paid vacations or holidays, fringe benefits, pay raises, discharge notices, or reasons for discharge.

The FLSA includes child labor laws that set minimum age standards for certain types of labor, anti-age discrimination rules, and equal pay provisions. The portion of the FLSA which deals with equal pay is known appropriately as the *Equal Pay Act*. This act prohibits employers from discriminating against employees on the basis of sex. For example, for jobs that require equal skill, effort, and responsibility, and which are performed under similar working conditions, an employer may not pay an employee of one sex at a rate less than what he or she pays an employee of the opposite sex. Unequal wages may be paid, however, based on seniority or merit. To comply with the Equal Pay Act, an employer must raise the wages of the employee making less money, rather than reduce the wages of those employees who are paid more money.

Title VII of the Civil Rights Act of 1964

This section of the Civil Rights Act, popularly referred to by its title number, prohibits employers from discriminating against applicants and employees. Under Title VII, discrimination is defined as treating employees differently with respect to compensation, promotions, job conditions, hiring, or discharge, on the basis of race, color, religion, sex, or national origin.

Generally, employers of 15 or more employees engaged in an industry which is part of, or which affects, interstate commerce, are covered by Title VII. The *Equal Employment Opportunity Commission*, referred to as the *EEOC*, is a federal agency which administers Title VII. The EEOC provides employers with notices containing the requirements of Title VII, as well as the procedures for filing a discrimination complaint. Employers are required to post these notices on the premises.

The Age Discrimination in Employment Act

The Age Discrimination in Employment Act makes it illegal for an employer to discriminate on the basis of age. This act protects applicants and employees over the age of forty. Under this act, it is unlawful for an employer to refuse to hire a person based solely on his or her age, if that applicant is age forty or older. Additionally, an employer cannot discharge an employee based on his or her age, nor may an employer use age as a basis for discriminating against employees with respect to compensation or employment conditions.

The Rehabilitation Act of 1973

The Rehabilitation Act of 1973 is intended to protect the rights of handicapped individuals in employment. The act requires employers who have contracts with the federal government valued in excess of $2,500 to hire and promote handicapped people who are qualified for available jobs. The act makes it unlawful for any employer, program, or activity that receives federal money to discriminate against qualified individuals based on some physical or mental disability.

Immigration Reform and Control Act of 1986

This law prohibits discrimination on the basis of national origin. The law also requires employers to verify that all employees hired after November 6, 1986, are either:

- Citizens or nationals of the United States
- Aliens lawfully admitted for permanent residence
- Aliens authorized by the Immigration and Naturalization Service (INS) to work in the United States

Employees must produce proof that they are eligible to work and must sign a form known as an I-9 Form.

Conclusion

Labor law is a sensitive and complicated area of law. This chapter has addressed some of the more important federal laws applicable to employers in the construction industry. This chapter has not discussed state law in any detail, however, because the laws vary from state to state. Construction professionals are urged to investigate the labor laws applicable in the states in which they conduct business. The first rule to apply to labor law is, "look it up." The second rule is, "never violate the first rule." If these rules are followed, employers (and employees) should be able to conform their behavior to the requirements of the law.

PART II

THE LAW AND THE STAGES OF PROJECT DELIVERY

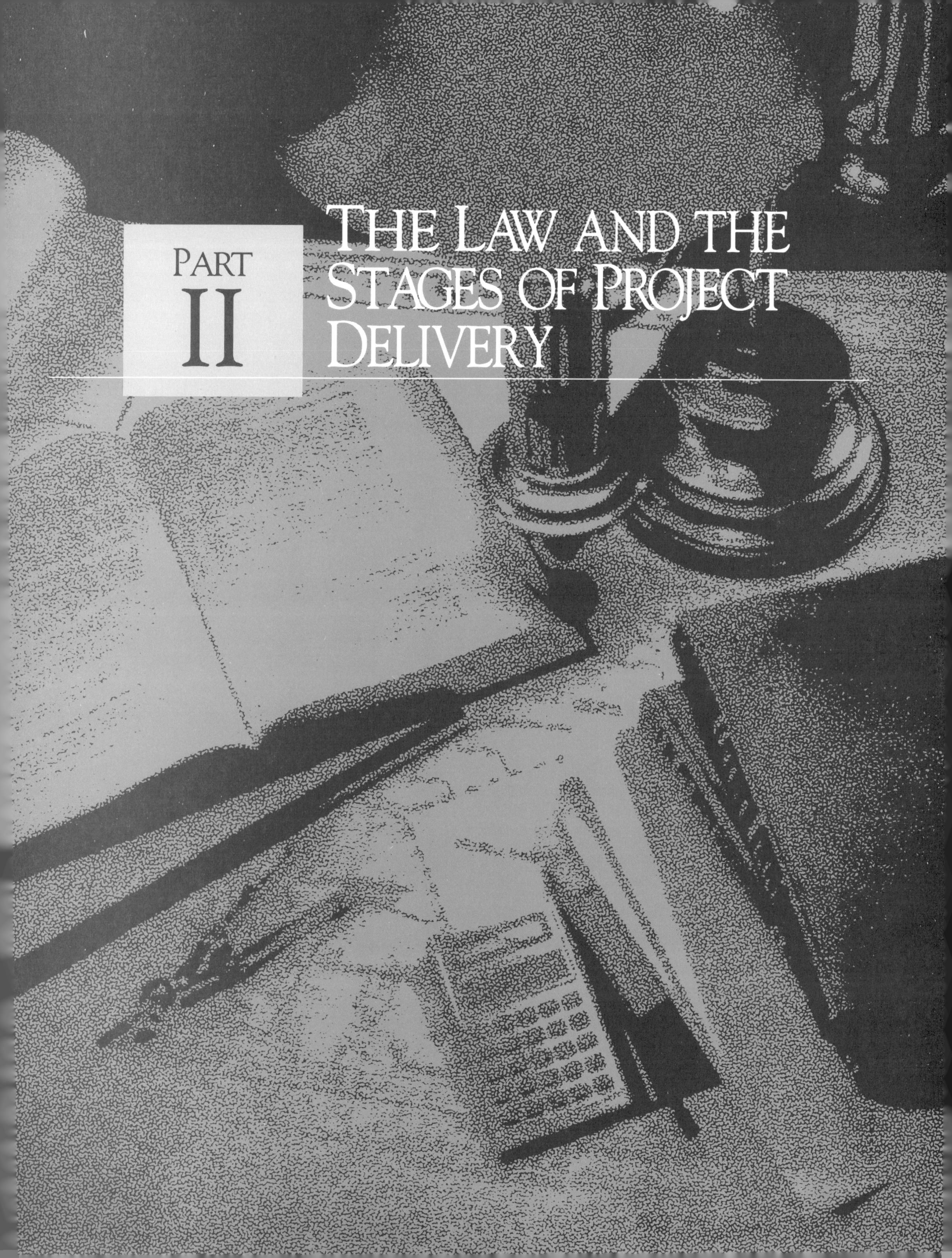

One distinguishing characteristic of design and construction practice is complexity. Every project involves a large number of people – making decisions, producing documents, and modifying the physical environment. As a result, seemingly minor changes can sometimes produce substantially different results. Furthermore, because the number of possible turn of events is nearly infinite, it is frequently very difficult to determine how a particular set of facts fits into the theoretical framework established in Part I. The next eleven chapters explore the situations that commonly arise during a project's development, and the legal principles that govern them.

CHAPTER 8 PROJECT PARTICIPANTS

CHAPTER 8 PROJECT PARTICIPANTS

Introduction

All construction projects require the cooperation of a number of individuals and entities whose common goal is to create, add to, renovate, or otherwise improve, real property. Every participant has their own motivation for engaging in a project and, in turn, contributes their talents and expertise to the special problems of the project. Large projects generally involve many participants, while small projects might require just a few. To understand the construction process and the legal environment in which it operates, it is first necessary to become acquainted with the principal participants on a typical project. These are listed below.

- Owner
- Interim and permanent lenders
- Design professionals
- Construction manager
- General (prime) contractor
- Subcontractors
- Surety
- Manufacturers, materialmen, and suppliers

The chapters that follow discuss the variety of ways in which project participants interact with one another, as well as their rights and responsibilities during the course of a construction project. Each of these participants is introduced below.

The Owner

The individual or business entity who proposes to have a project designed and constructed is generally referred to as the *owner*. This person or entity often owns the property on which the project will be constructed, as well as the actual construction itself. Sometimes, however, the owner is a tenant, meaning that he or she does not own either or both of these elements, but merely possesses the right to use them for a period of time. (See Chapter 5 for more information on real property law.) For our purposes, the term "owner" is intended to designate one of the participants in the development process: the overall client or initiator of the project. The term is not necessarily limited to the actual legal owner of the real property.

The owner has a number of important responsibilities, as shown in Figure 8.1. Owners often delegate some or all of these responsibilities to consultants who are more knowledgeable or proficient in these matters than the owner's staff. There is presently no single, universally accepted way to structure and administer a construction project.

Interim and Permanent Lenders

Financing has been one of the most volatile aspects of construction in the recent past, as interest rates and the value of the U.S. dollar have fluctuated. Partly as a result of such volatility, a wide variety of methods have been developed to finance land acquisition, land development, and construction in the United States. Those who provide financing on a relatively short-term basis are called *interim lenders*. Financing on a long-term basis for a substantial part of the life of the facility is usually provided by *permanent lenders*. The following entities are frequently involved in one type of construction financing or another.

- Commercial Banks
- Savings and Loan Associations (S&Ls)
- Insurance Companies
- Mortgage Companies
- Real Estate Investment Trusts (REITs)

(For additional information on lenders and financing techniques and problems, see Chapter 9, "Financing.")

The Owner's Responsibilities

- Develop information on project needs, goals, and objectives
- Acquire a site and obtain necessary easements
- Secure financing
- Select and hire an architect and/or engineers
- Select and hire a construction manager (if needed)
- Select and hire one or more construction contractors
- Develop a programmatic listing of project elements, spatial requirements and relationships, and other similar matters
- Procure boundary, topographic, and utility surveys of the building site
- Procure soil tests and geotechnical engineering services
- Provide or pay for necessary insurance for property damage and other business- and construction-related risks
- Pay for the required professional and construction services and construction materials and equipment

Figure 8.1

Design Professionals

The term *design professional* is frequently used to refer to a licensed professional who provides design services to owners and others with respect to construction activities and improvements to real property. Included under this title are architects, landscape architects, engineers (of all disciplines), and land surveyors. Although their training, abilities, experience, and the nature of the services they offer may differ, the laws applicable to them are generally the same. For most purposes, the legal rights and obligations that apply to any one design professional will apply to all design professionals.

The Architect

The architect is typically the prime design professional on building projects. As the prime professional, the architect has a direct contract with the owner for the majority of the professional services required on the project.

Architects are licensed by the state based on their demonstrated competence in the art and science of building design. The state has the power to require that only licensed professionals be allowed to design buildings. This authority arises from the state's obligation to protect the health, safety, and welfare of the general public who uses buildings and who may be injured if such buildings are designed incompetently. (See Chapter 11 for more information on licensing.)

On building projects, the architect usually contracts with the owner to provide an array of services, and then consults with other professionals as necessary to fulfill his or her contractual obligations. The architect may hire structural, HVAC (heating, ventilating and air conditioning), plumbing, electrical, and civil engineers as project requirements warrant. It is also reasonably common for an architect to hire consultants in the following disciplines: vertical transportation, lighting, acoustical, fire protection, life safety, and food service.

The architect plays two principal legal roles during the course of a project's development. During the design and bidding phases, the architect is an independent contractor responsible for the means and methods required to perform necessary services. During the construction phase, the architect is the owner's agent, responsible for administration of the contract between the owner and construction contractor. (See Chapter 3, "Basic Agency Law," for additional information on employment relationships.)

The Engineer

Like architects, professional engineers are required to be licensed by the state (sometimes according to the specific engineering discipline, such as civil or electrical, for example). This requirement ensures the health, safety, and welfare of the general public who uses the facilities that they design.

On projects such as manufacturing plants, refineries, sewage treatment plants, highways, bridges, and similar works, engineers are normally the prime design professionals. They contract with the owner and are responsible for an array of services in much the same way an architect is responsible on a building project.

On building projects, engineers generally work under contract to the architect and do not directly contract with the owner. The usual and customary services of structural, HVAC, electrical, and plumbing engineers are considered to be part of the architect's services to the owner. Other engineering disciplines may be added to the project team as necessary. When the architect does hire the engineers, the architect is primarily responsible to the owner for the services performed by the engineers.

Because they rely heavily on the services and advice of engineers, many architects frame their contracts with engineers to commit the engineer to provide the same services (under the same or similar terms and conditions) to the architect that the architect is required to provide to the owner. This reduces the risk that the architect would be required to provide a service to the owner which the engineer had not promised to provide to the architect. Further, it reduces the architect's risk of liability to the owner for an engineering error or omission for which the architect is unable to recover from the engineer. For additional information on the rights and responsibilities of design professionals, see Chapter 10, "Professional Services Agreements."

The Construction Manager

In the 1970s, when inflation was rampant, project owners' costs escalated substantially with each additional month necessary for the construction of their projects. Owners quickly looked for ways to reduce their costs. One way they found was to hire an individual or entity to provide them with administrative assistance. For this task, owners frequently hired general contracting companies who had expanded the nature and extent of the services they offered. Sometimes they were more like business managers with a construction orientation. These individuals or entities were called *construction managers*.

Construction managers convinced owners that they could accelerate the design and construction process by using specific management techniques. For example, they could contract for items with long lead times for delivery well in advance of hiring contractors to begin actual construction on the project. Further, they could manage many separate contractors to overlap design and construction (this process is called *fast-track* project delivery). Acceleration would produce a finished project for the owner in a shorter period of time than the usual design and construction process; and in the construction industry, time is money.

Some construction managers are purely management advisors to the owner. They do not contract directly with any of the various design professionals or construction contractors. More common is the construction manager who performs some of the tasks of a general contractor. This type of construction manager contracts with construction contractors to perform the actual construction services. The primary distinction between a construction manager and a general contractor, however, is that a construction manager performs services during the *pre*construction phase(s) of a project, as well as the actual construction phase. These services are listed in Figure 8.2.

The General (Prime) Contractor

The general, or prime, contractor has a direct contractual relationship with the owner and is responsible to the owner for all construction covered by the contract, whether or not any of it is subcontracted out to others (called *subcontractors*). The general contractor is normally responsible for scheduling and coordinating the work of the subcontractors and for the overall safety and general conditions at the project site. Unlike a construction manager, the general contractor usually has no involvement in the preconstruction phases of a project. (See Chapter 14, "Construction Contracts," for additional information on the responsibilities of the general contractor.)

Most general contractors perform some construction work with their own forces; for example, carpentry and other finish work. They normally subcontract for the majority of the required labor and materials, however, because the subcontractors are more experienced and efficient at the work of their particular trade. It is not uncommon for 80% or more of the total construction work to be performed by subcontractors for the general contractor.

Subcontractors

Subcontractors contract directly with the general contractor to provide labor and materials for the project. They have no direct contractual relationship with either the owner or design professional. Most subcontractors are *trade contractors*, which means they are expert at performing a particular type of construction work, such as plumbing, HVAC, electrical, masonry, and carpentry.

Subcontractors typically assume toward the general contractor all of the duties and obligations that the general contractor assumes toward the owner. Conversely, they are entitled to all of the rights that the general contractor has with respect to the owner. This is important, primarily for the owner, who does not deal with the subcontractors directly, but who

Services of the Construction Manager

Some of the preconstruction services that a construction manager performs include:

- Making preliminary estimates of construction costs
- Evaluating costs of alternate designs and systems
- Advising on the relative feasibility of construction methods, availability of materials, and procurement requirements
- Developing and updating a project schedule
- Advising on the effect of design details on construction feasibility, cost, and schedule
- Making recommendations with respect to safety programs
- Advising on how to divide the construction work among various contractors with respect to coverage, coordination, and union work rules
- Analyzing the types and quantities of labor required, and its availability
- Verifying requirements of equal employment opportunity programs
- Assisting in prequalifying bidders
- Receiving and analyzing bids and making recommendations for the award of the contracts

Figure 8.2

has an interest in seeing that everyone working on the project is working under essentially the same terms and conditions. (Refer to Chapter 15, "Subcontracting," for additional information on this matter.)

Subcontractors are often in the unenviable position of having to finance their own construction work on a short-term basis. They must purchase materials and equipment and perform construction labor and then wait for payment. Often they will not be paid *until* the general contractor is paid by the owner, sometimes not *unless* the owner pays. This is obviously a risky business, since their payment not only depends on the contractor's performance and reliability, but the reliability of the owner as well. Fortunately, the laws in most states permit subcontractors to obtain a *mechanic's lien* on a project to secure payments due. (See Chapter 16 for more information on lien laws.)

The Surety

A surety is a person or entity who promises in writing to make good the debt or default of another. In other words, a *surety* obligates itself to perform services or pay money if the principal debtor does not do so. The evidence of this promise is a *bond*. The principal is the party who is primarily obligated to perform services or pay money to the *creditor*. The creditor is the party to whom the services or money is due. The creditor may look to the surety if the principal defaults.

In the construction industry, the surety is usually a commercial bonding company, the principal is the contractor, and the creditor is the owner. The owner benefits from the bond and pays for it – directly or indirectly – when the contractor includes its costs as part of the fixed price or reimbursable cost on the project.

Manufacturers, Materialmen, and Suppliers

Manufacturers, materialmen, and suppliers sell the necessary construction materials and equipment to the general contractor or subcontractors. Such products range from basic raw materials like gravel or lumber, to finish materials like ceramic tile and paint, to pieces of equipment like refrigeration or water purification units. If a manufacturer, materialman, or supplier has a contract with a general contractor, he or she is considered a subcontractor. If they have a contract with a subcontractor, they are called *sub-subcontractors*. Design professionals specify, or otherwise select for their projects, many products based on the advice and according to the published materials of manufacturers, materialmen, and suppliers. These participants comprise a substantial portion of the construction industry.

Contracts for the sale of products are governed by legal rules that are different from those that apply to contracts for services or labor. As such, manufacturers, materialmen, and suppliers must comply with a different set of laws than those that apply to typical contractors and subcontractors. Refer to Chapter 2, "Basic Contract Law," for more information on this topic.

Conclusion

It is important to understand the relationships that exist among the individuals and entities who participate in construction projects. (These are summarized in Figure 8.3.) Although there are few formal rules governing their interaction, experience has shown that each type of individual or business entity tends to accept basically the same responsibilities and is accorded approximately the same rights from project to project. What they choose to call themselves is important only in so far as they imply that one party has a particular relationship with

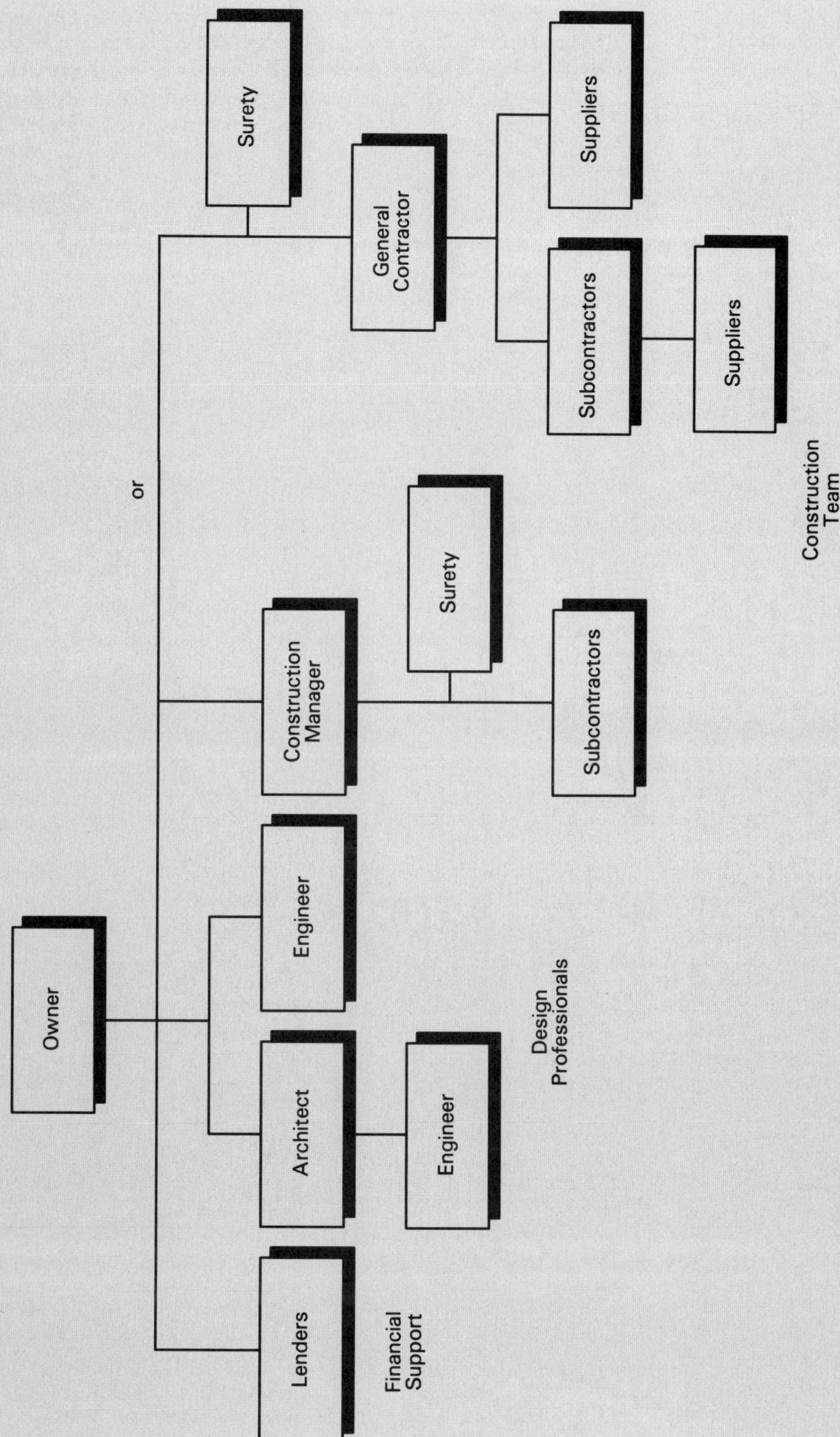

Participants in a Typical Construction Project

Figure 8.3

another party. If the relationship matches that which has been described, for instance, as an owner-contractor relationship, it does not matter that one or both of the parties chooses to be known by some other designation. The chapters that follow focus on the relationships between these participants and their associated rights and responsibilities.

CHAPTER 9 FINANCING

CHAPTER 9 FINANCING

Introduction

Very few owners have sufficient money (*capital*) available to buy real property and to pay for the design and construction services and other expenses associated with improving the property. Owners must either raise money by inviting others to invest in the project as co-owners or, more likely, borrow the money from some other source. This is called *financing*. Even in the unlikely event that an owner had sufficient funds, it is usually advantageous, for both business and tax purposes, for an owner to borrow the money needed for the project.

This chapter examines the ways in which owners finance land acquisition and development, design and construction services, and other costs (such as legal, accounting, and insurance advice) associated with developing real property. Included is a discussion outlining the consequences of an owner's failure to follow through with his or her financial commitments.

There are several types of financing available for construction projects:

- Equity financing
- Debt financing
- Creative financing

These arrangements are shown in Figure 9.1, and are discussed in detail in the following sections.

Equity Financing

In the beginning of a project, the owner must have money available to invest in the venture. Most lenders will not loan money for a construction project if the owner does not have a certain amount of *equity* (ownership interest) in the project, which it stands to lose should the project fail to achieve its financial goals. For example, owners are normally expected to provide approximately 20% to 25% of the total value of the project from their own funds.

An owner who does not have that amount of money available must seek *equity financing*. This means that the owner must invite others to become part owners by investing their money in the project. For example, an individual may form a partnership with another individual, or add a partner(s) to an existing partnership, to develop the project. The new partner contributes funds and increases the owner's equity in the project.

Of course, the new partnership (not the original individual or entity) becomes the owner of the project. (See Chapter 1, "Basic Business Structures," for more information about partnerships.)

Another possibility is for the owner to form a corporation to develop the project. By selling stock, the corporation is able to raise the necessary equity funds. The stockholders have an indirect interest in the project, since they own the corporation and the corporation owns the project. (See Chapter 1 for more information on corporations.)

A sole proprietor cannot obtain equity financing and remain a sole proprietor. Of course, if the sole proprietor has the necessary funds, he or she does not need equity financing. If the sole proprietor incorporates or forms a partnership to raise equity money, he or she, by definition, is no longer a sole proprietor. If the sole proprietor raises money without involving others as part owners, this is called *debt financing*, and is described in the next section.

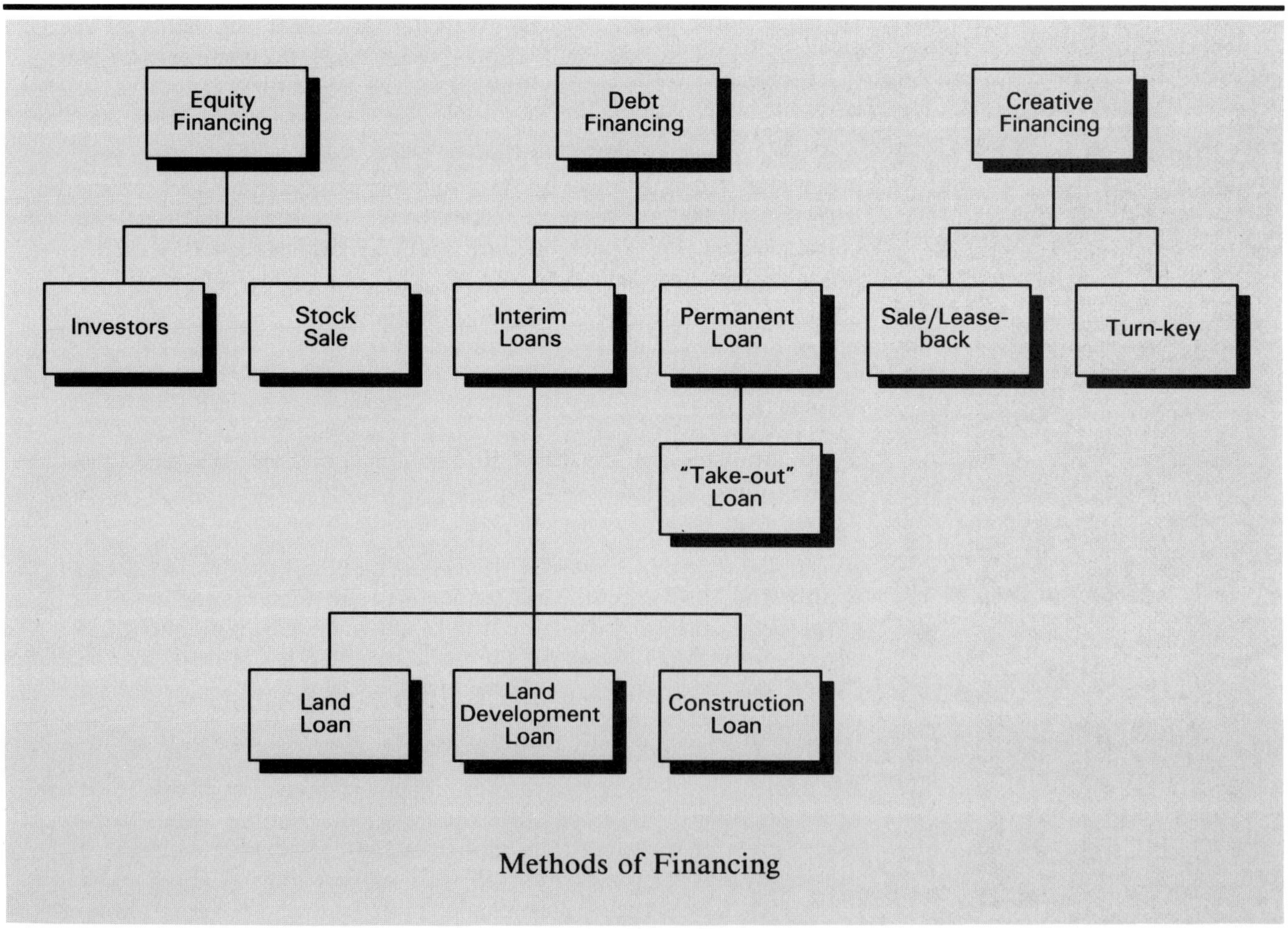

Figure 9.1

Debt Financing

A *debt* is that which is owed to others. If a person or business entity loans an owner money to buy or develop land or to construct a building (with the intention that the owner will repay the money at some point, normally with interest), the owner is paying for the land acquisition, land development, or construction through *debt financing*. The person or business entity who borrows money is referred to as the *borrower* (or *debtor*), because they have incurred a debt. The person or business entity who lends money to a borrower is called the *lender* (or *creditor*), because they have extended credit to the debtor. The owner's equity plus the owner's debt equals the total value of the project.

An owner may finance the amount of the debt through either:

- Interim loans
- Permanent loans

Interim Loans

There are three principal types of interim, or short-term, loans that are used to finance construction projects:

- Land loans
- Land development loans
- Construction loans

These loans are granted for specific activities that are intended to be accomplished in relatively short periods of time. When the activities are completed, the money that was borrowed, plus any interest that has accrued, is repaid to the lender.

Land Loans

A *land loan* is used to finance the purchase of raw or undeveloped land that a developer believes is underutilized and could be improved by building something on it. Most land loans are made by commercial banks. However, savings and loan associations (S&Ls), mortgage companies, and real estate investment trusts (REITs) also make land loans.

The *loan-to-value ratio* for a typical land loan is approximately 50% to 60%. This means that a borrower could expect to be able to borrow approximately 50% to 60% of the *appraised* value of the land (the value of the land if offered for sale to the public). This is a rather low ratio, reflecting the high level of risk involved in turning undeveloped land into usable, money-making property. It is quite possible that the developer will have spent the money borrowed to acquire the land and then find, for example, that there is no demand for the intended project. With no income from the project, the developer may be unable to repay the loan.

A land loan typically must be repaid within one to three years, at rates of approximately *prime plus three*. This means that the interest rate applicable to the land loan is the *prime interest rate* of the lender plus three percentage points. (The prime interest rate is the rate at which the lender is willing to lend money to its most financially stable and credit-worthy customers.) For example, if the prime rate in a specific month was 9.5% per year, then the rate applicable to the land loan would be 12.5% per year.

Sometimes *installment land purchase contracts* are used as financing devices. These contracts allow the purchaser to pay for the land in installments over time. Title to the property usually remains with the seller during the installment period. (Refer to Chapter 5, "Basic Real Property Law," for additional information about ownership of real property.) After some percentage of the payments has been made (this varies depending on local law), the purchaser has specific equitable interests in the title and the property. For example, after having paid 60% of the total amount due, the purchaser will not forfeit the property simply because a payment was late or skipped.

Land Development Loans

Land development loans are used to finance the costs of *improving* raw land to the point where it is ready to be built upon. This may include the legal costs of rezoning; costs for surveying, planning, and engineering; and costs of physical improvements such as grading, landscaping, installation of streets and utilities, and other preliminary physical improvements. Commercial banks and S&Ls are the principal institutions providing land development loans.

As is the case with land loans, it is quite possible that the owner will have spent the money borrowed to develop the land only to find that there is no demand for the proposed project. Without income from the project, the owner may be unable to repay the loan.

Construction Loans

Finally, construction loans are used to finance the costs associated with the construction phase of the project. Construction loans are generally short-term loans intended to remain outstanding only for the period of construction plus a short time thereafter.

Most construction loans are *interest only* loans. Money is drawn from the lender as construction proceeds, and the borrower pays interest on the outstanding balance. Generally no amount of the payment goes to repay the principal amount of money borrowed. The principal will be paid from the money the owner receives when the completed project is sold, or from the money the owner receives from the *take-out* loan, which will be described later in this chapter.

Construction loans are high-risk loans and, consequently, they carry high, fluctuating interest rates. The business and legal risks that make construction loans so expensive are listed below.

- The only security for the loan is the land and buildings under construction (which, of course, are only partially complete).
- There are a large number of risks during construction, which may or may not be foreseeable. (Refer to Chapter 17, "Sources of Dispute," for an extensive discussion of such risks.)
- Third parties have intervening rights (such as mechanics' liens) which may entitle them to be paid first. The lender will then be paid only if there are sufficient funds left over to do so. (See the sections that follow on *foreclosure*.)
- Zoning and building requirements may change, they may be misinterpreted, or they may be overlooked for whatever reason – all of which will jeopardize the occupancy of the project. If the project cannot be occupied, it cannot earn the money that was intended to be used to pay back the loan.
- Permanent financing may fall through even though it was promised. This could happen if the owner fails to meet the conditions established for granting permanent financing (for instance, requiring that the project be completed on time in accordance with the approved documents, and that it will be able to be used for its intended purpose).
- The finished project may be unmarketable due to poor salesmanship, poorly established priorities, or poor timing.

To protect their interests, lenders normally reserve the right to approve all of the owner's contracts with the contractor and the design professionals. Furthermore, they frequently require the owner to assign the design and construction contracts to them, effective in the event that the owner defaults on the construction loan. (Refer to Chapter 2, "Basic Contract Law," for more information about assigning contracts.) The idea is that if the owner has defaulted on his or her loan obligations, the lender has the right to step into the owner's role under the design and construction contracts, and to complete the project with the minimum amount of delay and expense possible.

Permanent Loans

A *permanent loan* is a long-term financial commitment. It is to a construction owner what a home mortgage is to a homeowner. This type of loan provides the owner with most of the money necessary to purchase the property or project, with the understanding that it will be repaid over a number of years (typically 20 to 40).

Traditionally, permanent loans were made by institutional investors like insurance companies or pension funds. In the past, these loans were frequently written at a fixed interest rate. Today, many lenders are unwilling to commit to a rate for a long period of time. Those who do may also want to share in a project's profits or capital appreciation.

The amount of the permanent loan that a lender will normally be willing to make available is a function of a *capitalized income stream* for the project, not a function of the cost to build or operate it. The lender typically asks the question, "What amount of money would have to be invested at a specific interest rate to produce a specific amount of income per year?" For example, if an office building can generate $500,000 per year in net income (the amount remaining after expenses are paid), and 10% is an appropriate *rate of return* for the use of capital (the profit earned), then the "value" of the building is $5,000,000:

$$\$500{,}000 \div 0.10 = \$5{,}000{,}000$$

(Five million dollars invested at 10% interest would generate $500,000 per year in income.)

A lender might be willing to loan 80% of that "value," or $4,000,000. If the owner can design and construct the project for $4,000,000, no equity money is needed. Even if it will cost $6,000,000 to build the project, the lender will typically not loan more than a percentage of the capitalized income stream, leaving the owner to supply the rest through equity financing.

The permanent financing used to repay the construction loan is called the *take-out* loan. The funds are taken out by the owner in a lump sum at the completion of construction and are used to pay off the construction loan. Then the owner repays the permanent lender in installments over a number of years. Ideally, the application is completed and the commitment for the take-out loan is received prior to the owner's application for the construction loan. In fact, one of the key pieces of information that a construction lender typically wants to know is whether or not there is a take-out commitment. If there is, the lender has some assurance that there will be money available to repay the construction loan when the project is complete.

Creative Financing

There are other more unusual or creative ways to finance construction projects. These include:

- Sale/Lease-back arrangements
- Turn-key projects

Sale/Lease-back Arrangements

In this type of arrangement, the owner obtains construction financing for a project that he or she personally intends to use (as a corporate headquarters, for example). Then a deal is made to sell the structure to another person or company, to a pension fund, or to an insurance company, upon completion of construction. The sale proceeds are used to pay off the construction loan. (This is done in lieu of more conventional permanent financing described above.) Finally, the previous owner leases the facility from the new owner on a long-term basis. The lease payments are calculated to be roughly equal to the principal, interest, and other costs of financing the project through more conventional methods.

This device has also been used by the owners of existing facilities. For instance, an owner may prefer, for whatever reason, to pay off the balance owed on his project by selling the facility and leasing it back. The advantages to this are that the owner may be granted a longer time to repay the loan, he may not have to pay as high an interest rate, or he may enjoy a more favorable tax status.

Turn-key Projects

Some owners do not want to be involved in the design and construction process in the usual way. Instead, they turn to a business entity that is willing to assume the legal and financial responsibility for getting the structure built. The owner essentially tells that firm what he or she wants, then commits to purchase the facility for a specific price, on the condition that it is satisfactorily completed according to criteria that the parties establish. All the owner needs to do is wait until the project is completed and then *turn the key* to the front door to start making use of the facility.

Defaults

A borrower who fails to make a payment, or who fails to act in accordance with other provisions in the loan agreement, is said to be in *default*. To ensure that the borrower will repay the debt, lenders normally reserve the right to sell the borrower's property if he or she defaults. Usually, that property is the land and the building(s) under construction.

A *foreclosure* is a proceeding that a lender may initiate when the borrower defaults on his or her loan. The lender sells the borrower's property in accordance with the terms of the *mortgage document* (the document by which the lender obtained the right to sell the property if the borrower defaulted). Lenders, however, usually prefer to take less drastic action. For example, they may prefer to restructure the loan by agreeing to more favorable terms with which the borrower can comply and which will enable him or her to complete the project.

Remedies

There are several remedies available to lenders when borrowers default on their loans:

- Acceleration of debt
- Workouts
- Foreclosure

These are described in the following sections.

Acceleration of the Debt

Acceleration provisions are common in loan agreements. An acceleration clause allows a lender to demand payment of the entire debt if the borrower misses a payment. In other words, the lender may sue for the outstanding amount without having to wait and see if the borrower defaults on the subsequent payments. Acceleration clauses also allow the lender to stop *disbursing* the loan funds (paying them to the borrower), and to commence foreclosure proceedings.

When a loan agreement does not contain an acceleration clause, a lender has two options, neither of which are very attractive. The lender can sue for the installment that was not paid, and then wait to see whether or not the borrower will default on any other installments. Or, the lender can wait until the entire debt becomes due (this could take years), sue for the entire unpaid amount, and foreclose on the borrower's property. Unfortunately, this could also take years.

Loan documents sometimes allow acceleration due to defaults other than nonpayment. For example, if the circumstances on which the loan was based change to the lender's disadvantage, or if the borrower dies or files for bankruptcy, the lender may be able to claim that the borrower was in default. Other defaults which may warrant acceleration of the debt include construction problems, such as 1) an increase in the project's costs, 2) breach of warranties, 3) delays, or 4) failure to build in accordance with the construction documents. Additionally, if the borrower fails to pay taxes or insurance on the property, the loan documents will normally allow the lender to accelerate the debt.

Workouts

If the borrower defaults on his or her loan, the lender's objective usually is to keep the development of the project moving ahead so that it will be completed on time. Foreclosure proceedings rarely, if ever, further this objective. Foreclosure costs money and causes delays. It requires the lender to find another borrower/developer to continue the project. The new developer and his or her own team of design professionals and contractors must then spend time becoming acquainted with the project, delaying matters further.

Most lenders prefer handling a default with a *workout*. Using this method, the lender restructures the loan agreement to allow for completion of the project without the need to foreclose on the security. In a workout, the lender either keeps the original borrower on the project or substitutes a new borrower. It is usually to the lender's advantage to retain the original borrower (as long as the borrower has the ability and commitment to complete the project), since he or she is most familiar with the project.

The most common reason why borrowers default is lack of funds. If the amount of loan funds that remain undisbursed (not yet paid to the borrower) are enough to finish the project, the lender will usually extend the term of the loan (delay the final repayment date) and will continue funding the project. The lender may choose to reduce the interest rate on the loan, either for a limited period of time, or for the duration of the project. The lender can also increase the amount of the loan. In doing so, however, the lender runs the risk of providing more money to what may already be a failing project.

In return for restructuring the loan, a lender will usually require the borrower to meet certain new obligations. For example, the lender may ask the borrower to provide additional security, such as land, equipment, or supplies. The lender may also require the borrower to provide a personal guarantee (a promise to repay the loan with personal assets if the security provided is insufficient).

If the changes made to the loan documents are going to delay or change the project, the construction lender must also seek the permanent lender's approval. Otherwise, the permanent lender may not be required to fulfill its take-out commitment. In addition, the lender should obtain the general contractor's consent before making any changes in the scope of the work or the construction schedule. The lender and borrower should also take into consideration the interests of others who may be affected by a workout, such as lienholders.

If a lender seriously doubts whether a defaulting borrower will be able to repay the loan, the lender may reduce his or her risk by replacing the defaulting borrower with a new borrower. To do this, the lender must assume ownership of the property which was offered as security for the loan, and then transfer the property to a new borrower, subject to new loan documents (including a new loan agreement and mortgage). The defaulting borrower transfers the property by voluntarily signing the deed over to the lender. (Refer to Chapter 5, "Basic Real Property Law," for a discussion of the sale of real property.) The transfer of the deed is convenient for everyone involved because it is done in lieu of costly and time-consuming foreclosure proceedings.

The defaulting borrower may also transfer the project to a new borrower who must be approved by the lender. The new borrower takes over the project, subject to the existing loan documents. This means that he or she is bound to the provisions of such documents.

Foreclosure

When a borrower defaults, the lender may choose to sell the property to pay off the borrower's debt instead of resolving the situation by a workout. This is called *foreclosure*. The lender has a right to force a sale of the property, because the property usually has been given to the lender conditionally, as security for proper payment by the borrower.

Power of Sale Foreclosure: When the borrower defaults, a lender who has been granted a power of sale under the loan agreement may force a sale of the property without going to court. This is referred to as *power of sale foreclosure*. This method is faster than court proceedings. Since the courts are not involved in this type of foreclosure, however, the sale may be challenged by third parties who claim to have some legal interest in the property.

In general terms, to accomplish a power of sale foreclosure, the lender must give public notice of the sale as prescribed by state law, and then arrange to have the property sold by a sheriff or other public official. The sale is not that simple, however, and the mortgage documents and state laws prescribe various steps that the lender must follow exactly if the sale is to be valid.

The money obtained from the sale is used to pay off the borrower's loan. Excess money, if any, is used to pay other creditors of the borrower if they had liens on the property. (Refer to Chapter 16 for additional information about liens.) The borrower may keep any remaining money.

If the sale proceeds are insufficient to pay the borrower's debt, liens established by other creditors are eliminated by the foreclosure proceeding. Creditors do retain the right to sue the borrower for amounts owed, but there is no security for the debt. As a result, the new owner (the purchaser) obtains the property free of liens.

Judicial Foreclosure: A foreclosure can also be accomplished by going to court. However, judicial foreclosure can sometimes take as long as two years, especially if the borrower *contests* (challenges) the sale. The contractors and others working on the project may abandon the project as a result.

A judicial foreclosure is also more expensive than a power of sale foreclosure because of the attorneys' fees and court costs involved. The advantage of going through the court system, however, is that the title to the property is confirmed by a court judgment. The new owner of the property is usually assured that the court stands behind his or her ownership rights and, therefore, the previous owner and his or her creditors have no legal rights to the property. Most lenders prefer the judicial foreclosure to the power of sale foreclosure because the court supervises the distribution of the proceeds from the sale of the property.

Although the steps to accomplish a judicial foreclosure differ from state to state, generally, the lender first identifies those parties who have liens on the property. The lender then files documents with the court which indicate the borrower's default and affirms the lender's acceleration of the debt. Notice of a hearing is served to all parties, including the borrower, lender, and other creditors who have placed liens on the property. At the hearing, the borrower (and anyone else) may contest the sale. If the court orders a sale, the lender gives notice to the public of the sale. After the sale, the court confirms the sale and determines the rights other lienholders have to any proceeds of the sale which exceed the amount owed to the lender. The borrower may keep any money from the sale in excess of the amount owed to the lender and other lienholders. If the price of the property does not completely cover the borrower's debt to the lender, the lender may seek a *deficiency judgment* from the court. A deficiency judgment gives the lender the right to collect the balance of the amount owed by the borrower.

Conclusion

Financing is an essential part of the project delivery process. Few owners have the necessary resources to finance their projects. Those who do are still better served by borrowing money to pay for the land acquisition and development, design and construction services, and any other project-related expenses.

An owner can obtain equity financing, debt financing (long- or short-term loans), or make other arrangements, such as selling and leasing back the property, or receiving the project on a turn-key basis. Regardless of the type of financing, an owner must adhere to the repayment schedule of the loan. An owner who fails to repay his or her loan according to the terms of the agreement is said to be in default. The consequences of default range from an acceleration of the debt to foreclosure.

CHAPTER 10

PROFESSIONAL SERVICES AGREEMENTS

Chapter 10 Professional Services Agreements

Introduction

One of the first tasks facing an owner is selecting a design professional to plan and design the contemplated project. Depending on the nature of the project, the prime design professional may either be an architect or an engineer. Architects are often prime design professionals on building projects, while engineers are usually prime design professionals on projects like bridges and industrial plants. In either case, the design professional generally becomes the owner's advisor, performing an array of services during the course of the project's development.

The exact description of such services and the terms and conditions under which they will be provided varies from project to project, from one owner to another, and from one design professional to the next. To foster a clear understanding of each parties' rights and obligations, it is important to make a written record of the parties' agreement. This agreement is called a *professional services agreement*. (Refer to Chapter 2, "Basic Contract Law," for additional information on contracts in general.)

Types of Professional Services Agreements

There are several forms that professional services agreements between design professionals and owners may take. They include:

- Oral agreements
- Letter agreements
- Traditional standard form agreements
- Custom agreements

Each of these is discussed below.

Oral Agreements

All too often, design professionals work on the basis of a handshake. There is a verbal agreement between the parties, but little or nothing is put into writing. In fact, many people question whether or not oral agreements are enforceable. Generally, they are. The only exceptions are contracts that are required by the statute of frauds to be in writing. (See Chapter 2 for more information on this statute.)

The problem with oral agreements is in proving the existence of, and *terms* of, the agreement. Often, the parties will only have discussed services and compensation, but not any of the terms and conditions that are normally a part of written contracts. Even honest people can forget exactly what they agreed to, especially when intervening events set up competing interests and cloud the memory. Unfortunately, it is often true, as film mogul Samuel Goldwyn observed, that "Oral contracts aren't worth the paper they're printed on."

Because of their limited coverage and the difficulty in defining their terms and conditions, oral agreements are not suitable for projects that require a substantial number of professional services and associated expenses. Experience has shown that design professionals performing services under oral agreements generally have greater difficulty obtaining payment for services rendered, and find that owners expect them to assume more professional responsibility than what they consider reasonable.

Letter Agreements

Letter agreements are contracts that are prepared in the form of a letter. They are widely used by design professionals and others in the construction industry because they are not very intimidating to clients. Clients who are presented with letter agreements tend to feel reassured that the design professional is not trying to take advantage of their relative unfamiliarity with the industry by demanding that they sign a complex and lengthy contract that they may not fully understand. At the same time, letter agreements generally set forth the major elements of the deal: price, scope of work, payment terms and, often, time for performance or completion of the services. Letter agreements are used primarily for projects involving small dollar amounts, and are often a precursor to a more formal agreement.

The most common drawbacks of letter agreements are: 1) they usually do not include language addressing ways to anticipate and resolve potential problems, and 2) they typically are lopsided in that they describe in some detail what is required of the design professional, but do not state what is required of the owner (other than to pay the bills). It is risky to rely on a less than comprehensive agreement, since even small projects can generate substantial disputes. Nevertheless, the limited amount of writing found in letter agreements is frequently better than none at all.

Traditional Standard Form Agreements

Better still are the traditional standard form agreements produced by the various professional societies, such as the American Institute of Architects (AIA) and the Engineers Joint Contract Documents Committee (EJCDC), which is comprised of representatives from the National Society of Professional Engineers (NSPE), the American Consulting Engineers Council (ACEC), the American Society of Civil Engineers (ASCE), and the Construction Specifications Institute (CSI). The AIA, for example, has been producing forms of agreements for the construction industry for approximately 100 years. These forms have stood the test of time. They are comprehensive without being inordinately lengthy. Although many other groups have begun to publish standard form documents (and some appear to be quite useful), none have a history of court interpretation or an acceptance in the market comparable to that of the AIA.

Standard form agreements attempt to balance the competing interests of the owner and design professional. They also attempt to reflect the legal requirements of a majority of states. In any given case, however, an owner or design professional may be able to negotiate a more favorable arrangement for themselves than what is described in the form, or they may find that the law handles a particular topic differently in their state than in the form. This is not a criticism of the forms but, rather, an acknowledgment of the fact that a generic form cannot fit all circumstances without some modification.

Similarly, the standard form agreements are somewhat inexact in describing the services to be performed by the design professional. This is not necessarily bad, since the professional is being hired primarily for his or her professional skill and judgment. However, the language in the standard form agreements may not accurately reflect the needs of the particular project. Standard forms are best used on projects in which the owner expects the design professional to assume a "traditional" role in designing and administering the construction of the project.

Custom Agreements

Although there is nothing inherently wrong with the traditional standard form agreements, fewer and fewer projects are designed and built using "traditional" procedures. On any given project, considering the parties involved, there may be legitimate reasons to vary from the traditional approach. For example, it may be desirable for design and construction to overlap in sequence, or for the owner to take a more active role during the construction phase. In those cases, a custom agreement may be created, or the traditional form agreements may be tailored to fit the circumstances of the project.

When drafting this type of agreement, the parties must be quite precise in defining and describing the tasks that each will do. This requires forethought and precision in documenting the extent of services that will be provided by the design professional as well as the agreed-upon compensation. If subsequent events change the project requirements, the required services must usually be changed in response. Very little is taken for granted when this type of contract is used. As a result, drafting custom agreements can be very time-consuming.

See Figure 10.1 for a comparison of the advantages and disadvantages of these four types of professional services agreements.

Content of Professional Services Agreements

Although the specific terms of every professional services agreement are different, they should generally address the following issues:

- The design professional's responsibilities
- The owner's responsibilities
- Scheduling
- Methods of compensation
- Payment schedules
- Ownership and use of documents
- Provisions for termination
- Dispute resolution
- Miscellaneous provisions

The standard form agreements generally address each of these issues. If a letter or a custom agreement is used, the following material may provide a checklist of issues to consider (See Appendix B for sample copies of AIA Document B141, *Standard Form of Agreement Between Owner and Architect*, 1987 Edition, and EJCDC Document No. 1910-1, *Standard Form of Agreement Between Owner and Engineer For Professional Services*, 1984 Edition.).

Comparing Professional Services Agreements		
Type of Professional Services Agreement	**Advantages**	**Disadvantages**
Oral Agreement	• Informal agreement; legally enforceable • Can be used for projects involving small dollar amounts	• May be difficult to enforce since there is no written record of the agreement or its terms and conditions • May not be used for contracts that are required by the statute of frauds to be in writing • Design professionals working under this agreement may have greater difficulty obtaining payment; may be expected to assume greater responsibility than under other contract forms
Letter Agreement	• Sets forth the major terms of the agreement, but is not intimidating to clients the way preprinted agreements can be • Used for projects involving small dollar amounts	• Usually does not include language to avoid or resolve disputes • Agreement is typically lopsided; defines the responsibilities of the design professional, but usually does not elaborate on the owner's role
Traditional Standard Form Agreement	• Is comprehensive without being inordinately lengthy • Attempts to balance the interests of the owner and design professional • Attempts to reflect the legal requirements of a majority of states	• State requirements may differ from the terms of the agreement • Description of required design services is inexact and may not reflect the needs of the project • Best used on traditionally structured projects
Custom Agreement	• Used for projects that vary from the traditional format • Are tailored to fit the individual needs of the project	• Requires forethought and precision when drafting • Contract language must be amended if there are changes in the project requirements • Time-consuming to draft

Figure 10.1

The Design Professional's Responsibilities

Basic Services

Those services that the design professional and owner commit to provide and pay for are included in the standard form agreements under the heading *Basic Services*. Basic services are agreed upon at the time the contract is signed, although they may vary from project to project.

Other services may or may not be needed during the course of a project's development. These are generally referred to as *additional services* because the parties have not committed to provide and pay for them at the time the agreement is signed. A basic service on one project may be an additional service on another, depending on the desires of the owner and the needs of the project.

The basic services a design professional typically provides under a traditional professional services agreement are divided into five phases:

- Schematic design
- Design development
- Construction documents
- Bidding or negotiation
- Construction contract administration

These are discussed in the following sections.

Schematic Design: During the schematic design phase of a project (also known as the *study and report phase*), the design professional reviews the owner's program for the project to determine whether or not the program, schedule, and budget are compatible. If they are, he or she then reviews with the owner available approaches to the design and construction of the project. Once the design professional and owner agree on their approach to the project, the design professional prepares drawings and other documents (such as charts, calculations, and written descriptions of the project) to illustrate the scale and relationship of project components, and provides the owner with a preliminary estimate of construction cost.

Design Development: During the design development phase (also known as the *preliminary design phase*), the design professional prepares drawings and other documents (such as outline specifications and design criteria) to describe the size and character of all major elements of the project. These elements include architectural, structural, mechanical, and electrical components. At this time the design professional also advises on adjustments, if any, to the preliminary estimate of cost.

There is no standard list of documents that are required of the design professional by the end of the schematic design and design development phases of services. Documents are developed to suit the needs of the individual project, according to the professional judgment of the design professional and his or her consultants.

Construction Documents: Most of the design professional's time on a project is spent during the construction documents phase (also known as the *final design phase*). The design professional prepares working drawings and specifications describing in detail the requirements for construction of the project. He or she then advises the owner on additional adjustments, if any, to the preliminary estimate of construction cost and helps the owner file the appropriate documents with building officials and others who have jurisdiction over the project.

Further, the design professional assists the owner in preparing bidding information and forms, and in drafting the proposed form of the owner-contractor agreement. Design professionals do not take final responsibility for the agreement between the owner and contractor, however. Since the design professional is not a party to the owner-contractor agreement, the owner's lawyer is the proper person to assume responsibility for such a contract. This is because only attorneys are legally entitled to prepare contracts to which they are not parties. Of course, design professionals can contribute to the content of the contract based on their technical expertise and construction industry experience. In the end, however, the legal content of the agreement should be approved by the owner's attorney.

Bidding or Negotiation: During this phase, the prime design professional and his or her consultants either 1) assist the owner in obtaining bids from contractors interested in constructing the project, and in awarding and preparing a construction contract with the low bidder, or 2) assist the owner in negotiating an agreement with a designated contractor. This may entail explaining or interpreting the construction documents, perhaps making adjustments to them and, finally, advising the owner as to whether or not the contractor's price proposals are reasonable.

Construction Contract Administration: The design professional may also administer the construction contract on behalf of the owner. During this phase, the design professional advises and consults with the owner regarding construction progress and changing circumstances. He or she may issue instructions to the contractor with respect to interpretations, newly discovered site conditions, changes, and other matters.

The prime design professional and his or her major consultants are required to visit the project site to observe the progress and quality of the construction work. The goal of these visits is to protect the owner's interests under the construction contract. The design professional may also determine the amounts owed to the contractor, based in part on the site observations and on the contractor's monthly payment requests, and recommend to the owner that such payments be made. Other duties of the design professional during the construction phase are listed in Figure 10.2.

Predicting and Controlling Construction Costs

Few aspects of a design professional's services are as prone to dispute as the prediction and control of construction costs. Design professionals generally do not guarantee construction cost predictions because they typically do not control 1) the cost of labor, materials, and equipment, 2) the contractors' methods for determining bid prices, and 3) general market forces of supply and demand. The standard form agreements reflect these facts.

This is a very difficult concept for many owners to grasp, since the understanding in most commercial settings is that an estimator is bound to his or her estimate. For example, if a car owner takes his vehicle to a garage for repairs, and the mechanic estimates that it will cost $200 to fix the radiator, the car owner does not expect to be told, when he returns to pick up his car, that the repairs will cost $400. Similarly, construction project owners often do not understand why design professionals cannot also be expected to guarantee the accuracy of their construction cost predictions.

One major difference, of course, is that the auto mechanic who gave the estimate also performed the work. The design professional who predicts construction costs does not normally construct the project and has no control over the pricing methods of those who do. All that most design professionals agree to, and all that the law requires of them in the absence of a specific agreement to the contrary, is the exercise of normal professional skill and care when making cost predictions.

If the owner does ask the design professional to agree to a *fixed limit* (or maximum) of construction cost, the design professional should consider the interrelationship between price, quality, and scope to avoid exceeding that limit. Only one of these three variables can logically be determined at any given time. For example, if price is fixed and cannot be exceeded, then, logically, both the scope and the quality of the construction of the project must be controlled by the design professional. Similarly, if the owner insists on a specific scope for a project, then price is a variable at the mercy of the bidding contractors (since building codes and other regulations will only allow the level of quality to drop to a certain level). Further, most professional liability insurance policies exclude from their coverage claims arising out of construction cost estimates. (See Chapter 16, "Insurance, Bonds, and Lien Laws," for more information on Professional Liability Insurance.)

Responsibilities of the Design Professional During the Construction Phase

- Interpreting the contract documents and evaluating both the owner's and contractor's performance under them
- Rejecting construction work that does not comply with the requirements of the contract documents
- Requiring special testing of construction work, if necessary
- Reviewing and commenting on shop drawings, product data, and samples
- Preparing change orders, if required
- Inspecting the project at the time of *substantial* completion of the construction work
- Receiving, for the owner's records, copies of warranties and other documents that the contractor is required by the owner-contractor agreement to produce
- Inspecting the project at *final* completion of the construction work

Figure 10.2

Additional Services

It may not be possible, in the beginning of a project, to determine whether or not a specific service is needed. If it turns out that a service is required, but has not been considered and provided for in the agreement, problems can arise. Negotiations about price and other matters are frequently difficult to conclude once circumstances have given one of the parties substantial leverage.

To avoid such problems, most standard form professional services agreements include a list of "additional" services and a method for calculating compensation to the design professional for providing them. By their nature, some additional services are to be provided as circumstances of the project dictate, subject to the owner's right to decline to have the design professional provide them. Other additional services are provided only if explicitly ordered by the owner in advance of performance. On many projects, when it is decided at the outset that an "additional" service is required, that service is made part of the basic services required under the contract, even though that particular service would not usually be categorized as such.

The Owner's Responsibilities

In addition to the list of services required of the design professional, professional services agreements outline the owner's responsibilities on the project. The primary responsibility of the owner is to pay for the services provided by the design professional. The owner is also expected to participate in the design of the project by making decisions regarding design and various other matters, and by obtaining approval from those regulatory agencies that have jurisdiction over the project. Beyond these broad duties, there are a number of things that an owner is commonly expected to do in conjunction with, or to facilitate, the provision of services by the design professional. For example, the owner is expected to provide information about his or her project requirements. This would normally include a program describing space requirements and relationships as well as other pertinent criteria that the project is expected to meet. This would also include a construction budget, as well as a schedule or deadline for completing construction.

The owner is usually responsible for providing property surveys showing boundary lines, utilities, topography, easements, the size and location of structures and improvements, and other pertinent physical and legal data about the project site. Similarly, the owner is responsible for obtaining data about subsurface conditions which affect, for example, the load-bearing capacity of the soils. Frequently, the prime design professional is willing to assist the owner in this task.

Finally, the owner is responsible for procuring necessary services that are not normally offered by design professionals. These services can range from materials testing services used to verify whether or not the contractor is meeting the requirements of the contract documents, to legal, accounting, and insurance counseling.

Scheduling

Scheduling, or the timing of performance and completion of the design professional's services, is important. There is normally pressure to complete the services by a deadline or a series of deadlines. Unlike construction contracts, however, professional services contracts do not typically include a provision stating that "time is of the essence of the contract." That phrase has the legal effect of making any delay, no matter how brief, a substantial breach of contract.

In a construction contract, the contractor must build the work specified in the contract documents within the established time. It is a relatively objective process to determine whether or not the contractor has accomplished this by the deadline. With professional services agreements, however, professional judgment is being purchased by the owner, and it is not as easy to measure when the services are complete.

It is therefore considered counterproductive to curtail proper professional deliberation just so that the design professional can meet a specific deadline. Instead, it is common to require design professionals to work to meet deadlines with such speed as due professional care and diligence allow.

Methods of Compensation

There are three common methods of compensating design professionals. They are:

- Stipulated (lump) sum
- Percentage of construction cost
- Hourly rates (or multiples of cost)

These are discussed below.

Stipulated Sum

Under the stipulated sum method, the owner and design professional agree that the design professional will provide certain services for a specific amount of compensation. Owners like this method because it tells them in advance what the design services for their project will cost. Of course, if the scope of services changes substantially, so will the stipulated sum. This method is based on agreement about the *value* of the design professional's services. The cost to provide such services is not directly involved.

Percentage of Construction Cost

The percentage of construction cost method assumes that the value of the design professional's services is directly related to the cost of the construction he or she has designed. Thus, the design professional might be compensated at the rate of, say, 9% of the construction cost of the project. This is frequently a fair and workable method of determining compensation.

In some cases, however, this method may penalize one or the other party. For example, an architect may spend a substantial amount of time and effort designing a project that is easy and inexpensive to construct. While such a project may save the owner money, the design professional ends up with a smaller fee than if the project had been inefficient and expensive to build. Conversely, it may take no more time or effort to select and specify an expensive material than an inexpensive item, and yet the architect will be paid more for choosing the expensive one. Although still widely used, this compensation method is not as well regarded as those that are based more on the value of the design professional's services.

Hourly Rates

When the scope of the required services is unknown or difficult to estimate, hourly rates, or multiples of costs, are frequently used as a basis for determining the design professional's compensation. Using this method, rates are established and included in the agreement for individuals who will perform services (or, alternately, the cost of each individual's salary and benefits per hour is multiplied by some factor to cover overhead and profit). Compensation is then derived by multiplying the time spent by the design professional times the hourly rate (or multiple of costs).

On the one hand, this method is fair because the design professional is compensated for all services rendered, and the owner has only paid for what he or she has received. On the other hand, owners do not know in the beginning of the project how much the design services will ultimately cost, and the design professional has no incentive to work efficiently to reduce costs.

In addition to compensation for services, the owner is frequently expected to reimburse the design professional for many out-of-pocket expenses incurred during the course of performing his or her services (such as the cost of printing the plans and specifications). These are known as *reimbursable expenses*. If the design professional had to include such expenses as part of a fixed amount of compensation, he or she would have to guess at costs and then would have every incentive to hold those costs down to maximize profit, perhaps at the expense of the owner's other interests on the project. Since the expenses would not have been incurred except in the interest of the project, it seems unfair for the owner to pay more than legitimate costs, or for the design professional to remain unreimbursed for costs incurred in the owner's interests. The most equitable solution is for the owner to pay these costs. Payments for the reimbursable expenses are normally made on a monthly basis for expenses incurred in the preceding month.

Payment Schedules

Professional services agreements typically establish a payment schedule for services rendered. If nothing is stated in the agreement, general legal principles normally do not require the owner to make any payment until all contracted services are completed. However, owners should understand that few architects or engineers can afford to carry the expenses associated with performing services for a substantial project for very long without cash flow. It is extremely important, therefore, that they provide interim payments to the design professional during the performance of services under the agreement.

Sometimes partial payments are made as phases of services are completed. More commonly, payments are made on a monthly basis in accordance with the percentage of services completed during the month. For example, if 15% of the total amount of required services was performed in the first month, then 15% of the total amount of compensation would be due at that time. Similarly, if 25% of the total amount of services required was performed by the end of the second month, then 25% of the total amount of compensation would be due, minus the 15% previously paid.

Ownership and Use of Documents

Another sensitive issue in drafting professional services agreements is the ownership and use of documents and other materials (such as drawings, specifications or reports, models, computer disks, and videotapes) produced by the prime design professional and his or her consultants. Most owners feel that after they have paid for such items, they should have the unrestricted right to possess and control them. Unless the provisions of the professional services agreement provide otherwise, that is essentially correct.

Most standard form contracts used today (like those produced by the AIA and the EJCDC) contain provisions establishing the documents as *instruments of the design professional's service*, as opposed to *commercial products*. Such contracts provide that the designer owns the physical documents and that, according to current federal copyright law, the prime design professional, as creator, owns the copyright in the documents as well. (See Chapter 6, "Basic Intellectual Property Law," for additional information on copyrights.)

The standard forms of professional services agreements do, however, allow the owner to use the documents for information and reference with regard to the continued use and occupancy of the project after its completion (whether or not the design professional's services are involved). Furthermore, the owner is typically given the right to use the documents to complete the project with another design professional if the original architect or engineer defaults on his or her agreement.

If the owner wants or needs to use the design professional's documents for broader purposes, the design professional may choose to *license the use* of those materials or ideas. For example, consider a design firm hired to create a prototype design for a retail establishment that is intended to be duplicated across the country. The firm may charge a normal fee for developing the initial design and documentation. It may then negotiate a payment for each subsequent use of the documentation required to develop additional units. Design professionals may also license the use of designs for fixtures, as well as patented construction methods or procedures.

By the same token, design professionals must be careful not to infringe on the copyrights of others. For example, an owner may have a falling out with his design professional, for one reason or another, and want to hire another firm as a replacement. The natural tendency is to pick up the documentation prepared by the first firm and develop it as necessary to complete the project. Depending on the terms of the agreement between the owner and the first design professional, and the reason why the owner broke off the relationship with the original firm (whether or not the firm was at fault), the owner may have no right to use those documents. If the owner cannot use the documents, it is highly unlikely that any other design professional would have the right to use them. An architect or engineer hired under these circumstances must be careful not to infringe on the copyrights of the original design firm. It would be wise for the second firm to demand a statement from the owner that the owner has the right to use the documents for the intended purposes, and will indemnify and hold the second design professional harmless from damages that may result from such use.

Provisions for Termination

Standard form professional services agreements normally contain language providing for the termination of the agreement. These termination provisions are often quite simple and merely reflect basic legal principles related to breaches of contract. For example, it is typical to include a provision which states that either party may terminate the agreement if the other fails substantially to perform (if the other party breaches the agreement). (Refer to Chapter 2, "Basic Contract Law," for more information on breach of contract.)

Additionally, design professional services agreements often contain a termination provision not normally seen in general commercial contracts. This provision gives the owner the option to terminate the contract without penalty in the event that the owner permanently abandons the project.

Dispute Resolution

Most professional services agreements specify procedures to follow for resolution of disputes. The traditional standard form agreements have long stated a preference for arbitration as a means of settling disputes. Of course, litigation is always an option and, in recent years, owners and design professionals have increasingly specified alternative dispute resolution procedures, such as mediation. For more information on arbitration, litigation, and other dispute resolution procedures, see Chapter 18.

Miscellaneous Provisions

In addition to the provisions previously described in this chapter, there are normally a variety of miscellaneous provisions included in professional services agreements to address usual legal concerns. These concerns may include:

- The appropriate starting date for the statute of limitations
- Subrogation of property insurance claims
- Integration of prior agreements
- Disclaimer of any intent to benefit third parties

These are all common contract provisions and are discussed in detail in Chapter 2.

Additionally, many design professionals have also made it standard practice to request provisions relating to liability risk management. For example, the high cost and limited coverage of available Professional Liability Insurance has prompted many design professionals to request that their liability to the owner and others be limited to either 1) available insurance proceeds, 2) the fee, or 3) some agreed upon amount appropriate in the circumstances. Agreement on such provisions is a matter of economic necessity for many design professionals who believe that without them, the potential risks associated with providing services far exceed the rewards.

Similarly, services related to hazards like asbestos, PCBs (polychlorinated biphenyls), and pollution are generally considered to be uninsurable. Design professionals have therefore begun requesting that owners indemnify them (hold them harmless) from liability related to such services performed on the owner's behalf. (See Chapter 4, "Basic Tort Law," for more information on professional liability.) This is a relatively new area of the law, with few established precedents. Individual legal counsel is particularly appropriate on these subjects.

Conclusion

It is extremely important for an owner and design professional to be explicit in stating the terms of the agreement for providing professional services. Although professional services agreements can be oral, it is preferable that they be in writing. On relatively small or uncomplicated projects, a letter agreement may be appropriate. The owner and design professional should remember, however, that the traditional standard form agreements published by the design professionals' organizations have a long history of use and interpretation and a reputation for fairness. They are often appropriate for use on projects without substantial change. In other cases, a custom agreement may be required. Such an agreement may be original or merely a modified version of a traditional standard form agreement. Regardless of the type of agreement that is used, the provisions described in this chapter should be addressed in the agreement. The process of developing and negotiating the professional services agreement is an excellent opportunity to anticipate and avoid potential problems.

CHAPTER 11 LICENSING

CHAPTER 11 LICENSING

Introduction

Businesses and professional practices are highly regulated by state and local governments, primarily through licensing requirements. In general, licenses are intended to protect the public from unqualified trade and professional practitioners and to raise revenue through licensing fees.

The licensing process allows the government to examine the competence of the individuals who intend to practice certain trades (such as construction contracting) and professions (such as architecture and engineering). The government can, therefore, protect the interests of the public by licensing only individuals with a demonstrated level of experience and competence.

Furthermore, fees charged for licenses contribute to the revenue of the state or local government. Many businesses are required to obtain and pay fees for one or more licenses. These fees have no relationship to demonstrable skill or experience in a trade or profession. For example, retail businesses are normally required to obtain and pay for a license to sell goods to consumers.

There are three basic types of licensing relevant to the design and construction industry:

- Professional licensing
- Trade licensing
- Business licensing

This chapter examines the licensing requirements for design professionals, contractors, and design and construction firms, and the legal implications of practicing without being properly licensed.

Professional Licensing

Architects, engineers, and other professionals who perform design services, whether for buildings, roads, power plants, or other facilities, affect the physical and psychological well-being of their clients, as well as the public who will enter, use, or come into contact with the structures they design. Traditionally, states have retained the power, through professional licensing laws, to control and regulate activities affecting the health, safety, and welfare of the public. Generally, all those who present themselves to the public as design professionals, and who offer design services to that public, must be licensed by the appropriate state(s). Each state has established a minimum level of competence that must be shown before an individual can obtain a license. By requiring design professionals to be licensed, the state can ensure that those individuals possess at least this minimum level of competence, and that the general public will, theoretically, be protected, since only those individuals will be permitted to design buildings and engineering works.

It is illegal to practice one of the design professions without being properly licensed in each applicable state, and the penalties for doing so can be severe. Depending on state law, unlicensed individuals may be fined, ordered to discontinue practice, or even jailed. Furthermore, courts are not likely to enforce a contract between parties who are not properly licensed. For instance, if an unlicensed engineer provides design services, and the client does not pay the engineer for services rendered, the state courts are not likely to be sympathetic to the engineer's plea when he or she sues to collect unpaid fees.

Regulating Professional Activity

The definition of those practices or activities for which a license is required varies from state to state. State laws regulating the activities of design professionals may be categorized as *title laws*, *practice laws*, or a *combination* of the two. These three categories are discussed below.

Title Laws

Some states have what are known as title laws. These states regulate the *titles* that practitioners may use to describe themselves. For example, it would be illegal in a state with an architectural title law for an individual to call himself an "architect" unless he were properly licensed as such in that state. Similarly, it is probably improper for that same individual to refer to himself as an "architectural designer," since many title laws also restrict the use of "derivative" names (those that are derived from, or which form a part of, the title "architect"). In *title states* (states that have enacted title laws), it is not what people do that is regulated, but how they represent themselves.

Practice Laws

Another approach that states use to regulate professional activity is to define the specific *tasks or behavior* that may only be undertaken or performed by licensed professionals. By defining restricted *practices*, the state can control and regulate behavior and activities without regard to the titles people use. Most states that have practice laws define the restricted practices quite broadly. For example, an architectural licensing law would normally include all services that relate to the design, construction, enlargement, or alteration of a building. The activities or tasks that constitute the "practice" of architecture or engineering might include:

- Consultations
- Investigations and evaluations
- Preliminary studies
- Aesthetic design
- Preparation of plans, specifications, and other construction documents
- Site development studies
- Administration of construction contracts

Clearly, most or all of these services require that practitioners have formal training and experience. States can monitor and regulate that level of experience through the licensing process. In *practice states*, it is not what people call themselves that is regulated, but the services that they actually perform.

Combined Title and Practice Laws

Quite a large number of states have *combined title and practice laws*, either because the laws explicitly regulate both titles and practices, or because the state regulatory board's and attorney general's history of interpretations of the law have made it clear that both titles and practices are restricted. States with combined laws regulate not only how a practitioner can represent him or herself to the public, but also what services he or she may perform.

Obtaining a Professional License

The rules for obtaining a license to practice architecture or engineering are largely the same from state to state, although states may differ in the importance they place on the licensing criteria. To obtain a license, a person must: 1) meet a minimum age requirement, 2) be of acceptable moral character (generally this means that the person has never been convicted of a serious crime), 3) possess a minimum standard of education, and 4) have experience working under the direction of a licensed practitioner in the applicable field. Having met the initial criteria, each candidate for licensure must then pass an extensive examination and pay the required fees.

Professional Practice by Business Entities

For the most part, professional licenses are only granted to *individuals*, not businesses or other entities. This is because only individuals are able to take and pass the required examination. As such, partnerships or corporations are precluded from obtaining professional licenses.

Therefore, even though an individual architect or engineer may be properly licensed, there may be additional requirements that the business must meet in order to practice the profession. It is important, therefore, to confirm that the business entity, as well as the individual design professionals, meet the applicable requirements for professional practice. The requirements for each type of business structure are discussed below.

Sole Proprietorships

If the sole owner of a business (a sole proprietor) is a properly licensed design professional, there are usually no further requirements with respect to professional licensing in the state where the business is located. The individual *is* the business.

Partnerships

If the business is organized as a partnership, compliance with the professional licensing laws is more complicated. Each state has rules concerning the practice of architecture or engineering by partnerships. Some states require that all partners be professionals licensed in the same discipline, regardless of which state has licensed them. In other states, professional disciplines may be mixed in firms as long as the appropriately licensed people are responsible for the work in their disciplines. For example, it may be appropriate to form a partnership between architects and structural engineers as long as the architects are responsible for the architecture and the structural engineers are responsible for the structural engineering.

It is also common for state law to specify the minimum percentage of partners who must be licensed. In some states that minimum is 100%, which means that all partners must be licensed professionals. These states prohibit licensed professionals from forming partnerships with nonprofessionals. Most states require that more than 50% of the partners in a firm be licensed. By doing so, they are assured that the firm is controlled by licensed professionals. Some states merely require that at least one partner be licensed and that this individual be in charge of the professional services. It is very important that professional partnerships comply with the requirements of the state(s) in which they conduct business.

In any case, state law typically requires that the name of the firm not be misleading. For example, it may be permissible to call a firm "Smith and Jones, Architects and Engineers," if Smith is an architect and Jones is an engineer. It would be improper to call that same firm "Smith and Jones, Architects," when Jones is, in fact, an engineer, and not an architect.

Corporations

There are additional licensing issues facing design professionals who intend to provide services through a corporation. A corporation is an entity separate from its stockholders. In some states, therefore, corporations themselves are required to obtain professional licenses or *certificates of authorization*. However, most states do not license corporations *per se*, and some do not even permit design professionals to offer and perform professional services through corporations.

Those states that allow corporations to practice the design professions mandate that a specific minimum percentage of the directors, officers, and/or stockholders of a given corporation be licensed professionals in one state or another. They reason that only properly licensed professionals should be in control of the quality of the professional services provided by the firm. In such a state, it would be unacceptable to place a nonprofessional in a position in which he or she could dictate to, or overrule the decisions of, a licensed professional. No matter what the requirements are with respect to the composition of the board of directors, officers, or stockholders, states generally require that a properly licensed individual be in charge of the actual professional services performed by the firm. For more information on sole proprietorships, partnerships, and corporations, see Chapter 1, "Basic Business Structures."

Interstate Practice

Although many firms have local practices, it is increasingly common for design professionals to work with clients on projects located outside of their local area. Such clients or projects might be in adjacent states or in the general geographic region. For example, a large number of firms in Boston practice in many or all of the New England states. Of course, some firms have national and even international practices.

Individuals and firms must be properly licensed in each of the states in which they practice. This has become easier over the years due to the efforts of bodies like the *National Council of Architectural Registration Boards* (NCARB). Organizations like the NCARB promote uniformity in state requirements so that reciprocal licensing agreements can be reached. Were it not for such groups, it would be necessary for practitioners to duplicate the examination and licensing process in each separate state.

If the firm is a sole proprietorship and the proprietor is appropriately licensed in the state, there normally are no other requirements. However, in more complex business structures, at least one individual practitioner must be licensed in each applicable state, and other requirements must often be met as described above. This should occur before the person or firm even *offers* to perform services in another state, and certainly before services are actually performed.

If the firm is a partnership or corporation, more investigation is necessary. Even though one or more individuals may be licensed in the new state, and even though the firm is properly established and operates in its home state, there is no guarantee that it will automatically comply with the laws in the new state. For example, in an architect's home state, the law may require that only 51% of the partnership interests in the firm be held by registered architects. Suppose that a senior partner in the firm is licensed and holds a 51% interest, but the junior partner is not licensed and holds a 49% interest. In the new state, the law requires that 2/3 of the interests in a partnership be held by licensed professionals. If the junior partner is not willing to relinquish some of his or her ownership interests in the firm to the senior partner, it might be necessary to bring in a new, licensed, partner to take a 1/3 interest. Then the firm could operate properly in the new state.

If a firm practices as a corporation in its home state, management should be aware that not all states allow for the corporate practice of architecture, engineering, surveying, and other design professions. It is possible that corporate practice is not allowed in the state where the new project is located. Or, it might only be allowed if the corporation is a professional corporation (PC), where all stockholders are licensed professionals. (See Chapter 1 for more information on professional corporations.)

Some firms comply with these restrictions with a little creative business structuring. Assuming that the client is willing, it may be possible to establish a new firm of one or more individuals who are properly licensed in the new state. They may be able to contract to perform the services required by the client and then subcontract some (perhaps a substantial part) of the work to the original firm. If this option is considered, the firm should be careful that they do not establish a new business organization in name only, and without *substance*. Actions that are taken primarily to circumvent licensing laws may result in disciplinary action by the state licensing body.

In some cases, it may not be clear if a firm is "practicing" in a state such that it must comply with the licensing laws of that state. In can be a substantial burden for a firm to become licensed in multiple jurisdictions. A proper determination of whether or not a license is needed is important.

Suppose a firm's only office is in Cincinnati, Ohio. It is contacted by a new client who has heard of the firm's reputation for innovative retail design. The client is headquartered in Chicago, Illinois, but has stores all over the country. The client visits the architect's office in Cincinnati and hires the firm to do some design studies for the client's review. If they are approved, they will be developed, under separate contract, into construction documents for a site near Detroit, Michigan. The architect is presently only licensed in Ohio. Is there a problem?

Since the client solicited the firm, and all services are presently planned to be performed in the architect's office in Cincinnati, no one has performed or offered to perform services anywhere other than in Ohio (where the firm is licensed), despite the fact that the client is based in Illinois and the proposed project is in Michigan. If the project moves ahead such that the architect is attending many meetings in Chicago and performing a substantial amount of work at the proposed site in Detroit, the architect is probably considered to be practicing in multiple jurisdictions. Certainly, if the architect will ultimately be asked to put his or her professional stamp and seal on the documents in Michigan, the firm will have to be licensed in Michigan at that time.

Trade Licensing

It has become more common in recent years for states to require construction contractors to obtain licenses. Presently about one-half of all states have imposed such licensing requirements. The purpose of trade licensing laws is essentially the same as for professional licensing laws: to protect the public from incompetent and dishonest contractors. If only those contractors who have no criminal history and who are at least minimally competent can be licensed, the public health, safety, and welfare will be protected.

Some states have extensive requirements that must be met to obtain a contractor's license. As a rule, however, there are no minimum educational requirements. A contractor who wishes to be licensed normally must demonstrate only a specific amount of experience and must pass an examination.

There may be a variety of negative consequences for unlicensed contractors practicing in jurisdictions that require contractors to be licensed. For example, the contractor may be denied access to the state courts with respect to claims arising out of the construction work. If the client fails to pay, the contractor may not be permitted to sue for the unpaid amount. There could also be criminal penalties for unlicensed practice.

Finally, although a state may have no requirement, many local governments have passed ordinances requiring that contractors be licensed. Contractors are well advised to check local ordinances, even in cases where the state does not require a trade license.

Business Licensing

In addition to professional and trade licenses, there are a variety of licenses applicable to businesses. Many of these business licensing requirements are intended to raise revenue through licensing fees, as opposed to signifying some minimum level of competence. Others have some business or legal purpose unrelated to the skills or level of training of the business owners. For example, license data may be required for statistical reasons, or to comply with certain labor laws.

State Business Licensing Statutes

Most states have statutes requiring unincorporated businesses (such as sole proprietorships and partnerships) that operate under a name other than the name(s) of the owners, to file a certificate listing the names and addresses of the owners of the business. If James Black and Amanda White wanted to call their architecture and interior design business "Unlimited Designs," that would probably be allowed, but they would have to file a *fictitious name certificate* in the city hall of the municipality in which their office is located. This certificate is also known as a *d/b/a certificate* because it is the partnership of Black and White "doing business as" (d/b/a) Unlimited Designs.

If a business is incorporated and intends to operate in states other than the one in which it was originally incorporated, it must file papers and in some cases pay fees in those states to register as a "foreign" corporation. This procedure provides the new state with immediate revenue, as well as a record of important data on the corporation. With this record, the state can verify the amount of income earned in that state, it can determine whether or not tax returns are filed, and can identify the proper person(s) to sue should that ever be necessary. If a corporation does not file as required, it will not be recognized as a valid corporation in the state and may be penalized for conducting business improperly in that state.

Additionally, some states require each business to *register* (enroll with the state) as a business entity and to pay an annual fee for the business license. This fee might be a flat rate, a rate based on a percentage of gross sales, or some other method. This type of business license is essentially a revenue-raising device.

Local Business Licensing Statutes

Local authorities, whether county, city, or town, also frequently impose licensing requirements on businesses that operate within their jurisdictions. The types of licenses required on the local level are also revenue-raising devices, and are in addition to those requirements imposed at the state level.

Conclusion

Licensing is a form of regulation which affects all businesses and professional practices. Tradespeople (contractors) and design professionals are normally required to demonstrate, to the satisfaction of the state, that they are honest, competent, and experienced. They must earn the right to offer their services to the public; and the public hires them partly in reliance on the fact that they have a state license as proof of their competence. In addition to the professional and trade licenses, most business entities must obtain other licenses, which are issued principally as a means to raise revenue and monitor business activity.

In all cases, failure to obtain and maintain required licenses can have serious monetary consequences, and may even result in criminal penalties. Therefore, all professionals, tradespeople, and business owners should educate themselves regarding the licensing requirements in each state and locality in which they offer or perform services, or conduct their businesses.

CHAPTER 12 BUILDING REGULATIONS

CHAPTER 12 BUILDING REGULATIONS

Introduction

The building industry in the United States has been regulated throughout most of this country's history. The Tenth Amendment to the U.S. Constitution gives the states the power to protect the health, safety, and welfare of their citizens. One of the ways this power is exercised is through building regulations.

Building regulations affect every stage of project delivery, from site selection and financing, to design, construction, and operation of the project. The owner and design professionals, in conjunction with the owner's legal counsel, must determine which regulations are applicable to the proposed project. The design must meet the established criteria, and the construction contractor(s) must ultimately build according to the approved construction documents.

There are three main categories of building regulations with which design and construction professionals must comply. They are:

- Land use regulations
- Environmental regulations
- Codes

Each of these is discussed in the sections that follow.

Land Use Regulations

Land use regulation is a generic term used in this chapter to refer primarily to *zoning laws*. In addition to zoning laws, however, local regulations may establish historical districts, waterfront districts, or industrial development zones, to name a few. Land use may also be controlled by private agreements found in deeds or subdivision regulations. In the early development stages of any project, owners and construction professionals should investigate the existence of such restrictions.

Zoning Laws

Zoning laws regulate the way land may be used in specific areas within a community. They restrict the type of structures allowed within a specified area, and impose other limitations such as height, lot coverage, and parking. (See Chapter 5, "Basic Real Property Law," for more information on zoning laws.)

There are no regional or national model zoning codes. One reason is that there are no universally applicable principles of science or physics to codify and implement. Zoning is not grounded in physical laws, but is a construct of human minds. Zoning regulations are intended to help communities implement their individual comprehensive master plans for development and growth. For those reasons, zoning laws are very much a function of their local environments and are not subject to uniform codification the way building codes are.

If a proposed project complies with the restrictions of the local zoning laws, design and construction can proceed. Typically, once physical construction begins on a site, the project will not be stopped because of changes in the applicable zoning laws. There is a risk, however, that the laws will be changed in some way before the project reaches that point. Therefore, owners and design professionals should be alert for news of proposed changes to the applicable zoning laws.

If the proposed project does not comply with the applicable zoning restrictions, it may be possible to obtain permission to proceed at variance with those restrictions. It can be very difficult, time-consuming, and costly to obtain such permissions, however. Frequently, the local community will also impose substantial conditions with which the owner must comply in order to obtain permission. In general, developing a project within the bounds of existing law is normally preferable to seeking a variance or other sort of special permission.

Environmental Regulations

There are a number of federal regulations that have been enacted to protect the environment. The five acts examined in this chapter are:

- National Environmental Policy Act
- Clean Air Act
- Clean Water Act
- Resource Conservation and Recovery Act
- Comprehensive Environmental Response, Compensation, and Liability Act

A full discussion of the extent and complexity of these federal environmental regulations is beyond the scope of this chapter. The following materials merely identify and summarize some of the important aspects of environmental laws of interest to the design and construction industries. It is normally wise to consult an attorney who specializes in environmental law for the details of such rules and regulations.

National Environmental Policy Act

The National Environmental Policy Act (NEPA) was enacted in 1969 (and became effective in 1970) to establish policy and procedures with respect to environmental protection. Among other things, this act established the *Environmental Protection Agency* (EPA). The EPA conducts research, sets standards and guidelines, and enforces federal environmental regulations.

State Implementation Plans

States are permitted by federal law to adopt their own statutes and regulatory programs. State statutes and regulations must be at least as stringent as the federal laws, and many are more stringent. These state initiatives are called *State Implementation Plans* (*SIPs*), and must be reviewed and approved by the EPA. The EPA establishes and administers programs in those states that do not have approved SIPs.

Clean Air Act

The Clean Air Act (CAA) was enacted in 1970 to control air pollution. The CAA required the EPA to establish minimum air quality standards at a level that must be met to protect human health. These standards apply to certain relatively common pollutants such as lead, particulate matter, ozone, hydrocarbons, carbon monoxide, nitrogen oxide, and sulfur dioxide. The EPA has also established less stringent standards for the protection of property and vegetation.

The CAA also regulates hazardous pollutants, including asbestos, beryllium, mercury, vinyl chloride, radon-222, radionuclides and benzene. For example, the CAA regulates the removal and disposal of asbestos. It requires that the EPA be notified before substantial amounts of asbestos are removed from an area, and establishes very specific procedures for doing so.

Everyone involved in design and construction must address the problem of asbestos and other hazardous pollutants. Design professionals are presently unable to obtain Professional Liability Insurance to cover potential claims arising out of or relating to their involvement with asbestos and pollution. Contractors cannot begin demolition or construction on a project until preexisting asbestos has been removed properly, and lenders will not finance projects unless they are convinced that hazardous materials like asbestos have been, or will be, properly removed or contained.

The requirements of the CAA are very extensive and are not mentioned in their entirety here. Owners and operators of facilities should review the CAA with their legal counsel to be certain that they comply with its provisions.

Clean Water Act

The Clean Water Act (CWA) of 1977 was enacted to amend the Federal Water Pollution Control Act (FWPCA) of 1948. This law is intended to control water pollution by regulating both the quality and quantity of pollutants discharged into water. Its goal is to make all waters in the country safe for fishing and swimming.

The CWA prohibits any person or business entity from discharging pollutants into any "navigable body" of water, meaning any waterway or body of water other than a sewer. Pollutants are also very broadly defined in this statute, and include "dredged soil, solid waste, incinerator residue, sewage, garbage, sewage sludge, munitions, chemical wastes, biological materials, radioactive materials, heat-wrecked or discarded equipment, rock, sand, cellar dirt, and industrial, municipal, and agricultural waste."

The CWA contains provisions allowing states to implement their own plans for accomplishing the CWA's goals. State programs must be at least as stringent as the CWA and, in fact, most are more stringent.

National Pollutant Discharge Elimination System

The law recognizes that while it may be impossible to prevent discharges altogether, it is possible to control them. No discharge of pollutants is permitted without a *National Pollutant Discharge Elimination System (NPDES)* permit. The NPDES permit sets forth specific restrictions and requires facilities to report regularly on pollutants and any other substances that it discharges into waters.

Corps of Engineers

Anyone who discharges "dredged" or "fill" material into the waters of the United States must also obtain a permit from the U.S. Army Corps of Engineers (COE). Such permits are required for the construction of bulkheads, backfilling for utility lines, and similar types of activities. For discharges that could produce major changes to water systems, extensive study and documentation of the effects of the changes may be required before a permit will be granted.

The requirements of the CWA are very extensive and have not all been mentioned in this chapter. Facilities owners and operators should review the CWA with their legal advisors to be certain that they comply with its provisions.

Resource Conservation and Recovery Act

The Resource Conservation and Recovery Act (RCRA), passed in 1976, regulates the storage, transportation, and disposal of solid waste and hazardous waste. This act focuses on preventing future problems with uncontrolled waste disposal.

Under the RCRA, solid waste is defined as "any garbage, refuse, sludge from a waste treatment plant, water supply plant, or air pollution control facility, and other discarded material, including solid, liquid, semisolid, or contained gaseous material resulting from industrial, commercial, mining, and agricultural operations and from community activities." Hazardous waste is a special type of solid waste that, "because of its quantity, concentration, or physical, chemical or infectious characteristics, may . . . pose a substantial present or potential hazard to human health."

The RCRA also regulates the management and operation of treatment, storage, and disposal facilities, and requires permits to operate such facilities. The provisions of the RCRA are too numerous and complex to discuss in a single chapter. Anyone who owns or operates a facility that generates solid waste is responsible for that waste and should adhere strictly to the provisions of the RCRA.

Comprehensive Environmental Response, Compensation, and Liability Act

While the RCRA regulates the process of solid waste management and disposal, the Comprehensive Environmental Response, Compensation, and Liability Act (*CERCLA* or *Superfund*) focuses on cleaning sites that have been contaminated with hazardous wastes. CERCLA was enacted in 1980. CERCLA and the RCRA are the two major federal statutes regulating the disposal of hazardous wastes. CERCLA is administered by the EPA and requires those responsible for the presence of hazardous materials on a site either to clean the site or pay the costs of having it cleaned by others.

Under CERCLA, many parties can be held liable for the presence of hazardous wastes. Included are:

- The current property owner (even though the hazardous waste was dumped or generated on the site before they purchased the property)
- Previous property owners (if the hazardous waste was released while they owned the property)
- The party who generated the waste (the party who arranged for the disposal or treatment of a hazardous substance at the site)
- The party who dumped, released, or transported the waste on, or to, the site

Lenders are also very much concerned about hazardous waste because they may be responsible for cleanup costs if the property owner cannot repay his or her loans and the lender forecloses and takes over the property.

Shareholders of a corporation may be personally liable for cleanup costs if they controlled or directed the party responsible for generating the waste. Similarly, corporations may be held responsible for hazardous waste liabilities imposed on their subsidiaries (those business entities substantially owned by the corporation).

The EPA does not have to prove negligence or fault on the part of the allegedly responsible party in order to recover cleanup costs under CERCLA. Principles of strict liability (liability without fault) apply under CERCLA. (See Chapter 4, "Basic Tort Law," for more information on strict liability.) Therefore, it is irrelevant that a property owner purchased a site without knowing about the existence of hazardous waste at the site. Under strict liability principles, that owner is still liable.

Because the risk of responsibility for hazardous waste is so great, everyone involved in real estate transactions should be familiar with hazardous waste laws, particularly when commercial property or property in an industrial or urban area is involved. Fortunately, there is a "third party" or "innocent-purchaser" defense available to purchasers of real property. This is described below.

Superfund Amendments and Reauthorization Act

The *innocent purchaser exemption* is included in the Superfund Amendments and Reauthorization Act (SARA), enacted in 1986. A purchaser can be relieved of liability for cleaning up hazardous waste if he or she can prove that they:

- Exercised due care with respect to site inspection and testing before purchasing the property
- Did not know, and had no reason to know, of the presence of hazardous waste at the time of the purchase

Under SARA, a purchaser will be exempt if he has made "all appropriate inquiries into the previous ownership and uses of the property, consistent with good commercial or customary practice."

Clearly, it is wise to make such "appropriate inquiries." At the least, the potential purchaser should research town records to make sure the property is not, and was not, a waste-generating or disposal site. Town records may not provide all the facts, however, so a potential purchaser may have to test the soils and water at the site. This can be a costly process.

All such inspections and tests should focus on the history of the site. Is there a facility presently on the site, and what are its current and previous uses? What types of waste could possibly have been generated or stored on the site? A potential purchaser should make an examination of the site's treatment facilities, storage tanks, and disposal sites. If waste was generated on the site, was it released onto other property or into a lake? If so, what is the environmental status of those properties? A potential purchaser should also consider the location of the site. Are nearby facilities generating waste? If so, has this waste been released onto the site in question?

The answers to these questions are extremely important. CERCLA imposes joint and several liability on those found guilty of violating its provisions. (Refer to Chapter 4 for additional information on joint and several liability.) Thus, even if a party's role with respect to hazardous waste was minor, he or she can be held responsible for the total costs of cleaning the site.

Super-Liens

Many states have enacted their own superfund statutes to assist in implementing the provisions of CERCLA. Some of these statutes have created *super-liens*, meaning that the state can place a lien on an owner's property to recover the state's costs of cleaning the site of hazardous materials. Super-liens take priority over all other encumbrances on a particular piece of property. (See Chapter 16, "Insurance, Bonds, and Lien Laws," for additional information about liens in general.)

Other Acts

There are other federal environmental regulations that have not been discussed in this chapter, but which may have an impact on the progress and direction of a project. These acts are listed in Figure 12.1. The owner and design professional should explore, with the owner's legal counsel, the requirements of the applicable laws.

Codes

A *code* is a collection of regulations that have been systematically arranged to eliminate internal conflicts and inconsistencies, and to clarify and make the law comprehensive on a particular topic.

Types of Codes

There are a large number of codes that are applicable to the design and construction industry. Some of the more common ones are:

- Building codes
- Housing codes
- Fire and life safety codes
- Architectural barrier codes

These are discussed below.

Building Codes

A building code is the legal minimum requirement established or adopted by a governmental body to regulate the design and construction of buildings. Building codes are the most prevalent of the various codes in existence. They address both administrative procedures, such as permits and inspections, and technical requirements for particular

Additional Federal Environmental Regulations

- The Federal Noise Control Act of 1972
- The Endangered Species Act of 1973
- The Surface Mining Control and Reclaimation Act
- The Coastal Zone Management Act of 1972

Figure 12.1

structures, such as the minimum amount of light and ventilation necessary, the building's ability to support structural loads, and sound control techniques.

Most building codes used today are derived from or incorporate one of three *model codes*. These model codes are shown in Figure 12.2. Each code responds to the needs and physical circumstances (geography and climate) of a particular region of the country. None of these model codes is legally enforceable, however. They are merely guidelines that individual states or communities can adopt in whole or in part. Nonetheless, these codes are valuable resources, and are revised periodically to reflect technological advances and new information gained from academic study and practical experience. They frequently cite or incorporate by reference other related codes, such as mechanical, electrical, plumbing, gas, and energy conservation codes.

State Building Codes: Many states have adopted statewide building codes based on one of the model codes below. States rarely adopt a model code in its entirety, however. Variances in local climatic conditions, as well as past experience, demand that modifications be made before it can be adopted for local use.

A state building code establishes minimum design and construction standards for building projects in that particular state. Such codes may specify the particular *method* of accomplishing a task (called a *prescriptive code*), or they may define the desired (and required) *result* or *goal*, and let the designer achieve it in whatever way possible (called a *performance code*). Performance codes frequently require that specific government standards or private testing laboratory standards (such as those of the *American Society for Testing and Materials* (*ASTM*)) be met. Once incorporated, these additional standards become as much a part of the state codes as if they were originally included in the codes, and they are just as binding.

Local Building Codes: Many states allow local communities (counties, cities, and towns) to enact building codes based on the conditions and circumstances within their boundaries. In general, local codes may not

Model Building Codes

- *Basic Building Code*, published by the Building Officials and Code Administrators (BOCA), and used primarily in the eastern and north-central states
- *Uniform Building Code*, issued by the International Conference of Building Officials (ICBO), and used mainly in the western states
- *Southern Standard Building Codes*, established by the Southern Building Code Congress, for use in the southern states

Figure 12.2

be more lenient than the applicable state code (if one exists). They may be more stringent in specific areas, however, if the local authorities so desire.

Housing Codes

Housing codes are principally intended to establish and regulate the minimum conditions necessary for human habitation. For example, these codes contain provisions dealing with natural light, ventilation, sanitation, heat, cooking equipment, and the maximum number of people allowed in a space of a specific size and characteristic. Housing codes are likely to be more stringent for multiple dwelling units that are located in the same structure (such as in apartment buildings) than in single unit buildings. This is because the consequences of a failure in any building system (such as the plumbing) are greater, so the requirements for the performance of the systems are also greater.

Fire and Life Safety Codes

There is presently no official national standard for protecting people from fires in buildings. The closest thing to such a standard is the *Life Safety Code* promulgated by the *National Fire Protection Association* (NFPA). Life safety codes are primarily concerned with protecting human life when fires occur in buildings. Typical building codes, on the other hand, are concerned primarily with preserving the structural integrity of buildings to prevent or minimize property damage in the event of fire, earthquake, or other disaster. Of course, lives will probably be saved if buildings remain structurally sound in disasters, but the focus of building codes is not on saving lives.

Architectural Barrier Codes

Architectural barrier codes, more appropriately known as *architectural access codes*, are intended to establish minimum requirements for unimpeded access by the physically handicapped to buildings and other structures. For example, these codes establish the dimensional requirements for architectural features such as doorways and ramps, and modifications to ordinary fixtures including sinks, toilets, drinking fountains, and public telephones, so that they are convenient for the physically disabled to use.

Compliance with Codes

One of the reasons why design professionals are hired by owners is to provide a design that complies with applicable laws and building regulations. Most owners believe that failure of a design to comply with a code requirement indicates that the designer was negligent. This is not necessarily true. Codes often conflict with one another, and are subject to interpretation. If a design professional has exercised normal care in discovering, interpreting, and designing in accordance with code requirements, he or she will not usually be liable to the owner, even if it is found that the design does not comply with some aspect of a particular code, rule, or regulation.

Conversely, just because a design complies with the requirements of the building code does not mean that a design professional cannot be found negligent in some respect. The codes specify a *minimum* standard that must be met under all circumstances. It is possible that the minimum standard may be inadequate in a specific case. Design professionals must design for the circumstances at hand and should not rely blindly on the codes. It is certainly possible that a design that complies fully with all applicable codes is inadequate and may render the design professional liable if it fails.

Permits

A permit is a document issued by a governing authority granting the right to construct planned improvements within the time periods specified. It is generally necessary to obtain permits prior to building or substantially renovating habitable structures. Permits may be required for demolition, excavation, foundations, and general building construction, as well as for hazardous waste storage or removal, and development of environmentally protected areas, as discussed earlier in this chapter. If proposed changes or improvements are minor, if they do not affect structural supports, or if they do not otherwise affect public health or safety, a permit may not be required.

The permit-granting process allows local authorities to determine whether or not the applicant has complied with pertinent laws, codes, and regulations. In many localities, it is necessary to submit plans and specifications signed and sealed by licensed architects and/or engineers in order to obtain a permit. Building officials will typically review the proposed plans anyway, but will do so with the assurance that the design is at least minimally competent. Permits are granted if applicable requirements are met. Local officials may not arbitrarily refuse to issue a permit, nor may they apply review criteria or standards that are not expressed in the codes and laws.

After the required plans have been submitted and approved, and the construction process begins, there are normally a series of inspections at the construction site to determine if the construction is being executed according to the plans and in compliance with applicable codes. Once the permit is granted, it is normally transferable to another owner or developer if the property is sold to them. The new owner may then build in accordance with the permit, if it has not expired.

Design professionals and owners should not assume that the existence of a valid building permit shields them from liability if it turns out that the building, although built in accordance with the approved plans, does not comply fully with applicable laws, codes, and regulations. Even though a permit has been issued, there is no assurance that the improvements described in it may actually be constructed, nor does it mean that code officials have somehow guaranteed that everything indicated on the plans is fully in accordance with the codes. Building officials have some power to interpret the codes and regulations, but most building departments are understaffed and incapable of allocating the time and effort required to review all pertinent details and to check the underlying design assumptions and calculations on large-scale projects. Therefore, it is not wise to rely on a building department review to discover and indicate errors in judgment or execution of the design. All those who design and build are required to do so in accordance with applicable laws. If they fail to do so, the building permit may be revoked and construction halted.

Conclusion

Building regulations affect nearly every aspect of the construction process in the United States. Land use regulations, environmental regulations, and codes are all ways in which the health, safety, and welfare of the general public are protected.

Owners are ultimately responsible for their projects. They rely heavily on design professionals to advise them about applicable regulations and to create a design that complies with such regulations. The owner and design professional, in turn, expect that the contractor will build the project in accordance with the design and other applicable laws, codes, and regulations. It is important, therefore, for everyone associated with a construction project to be generally familiar with the regulations described in this chapter so they can act appropriately under the law.

Chapter
13

Bidding

CHAPTER 13 BIDDING

Introduction

Most construction contracts are awarded through the process of *competitive bidding*. This process is based on open competition among all interested and responsible bidders, and is intended to produce the lowest price for the specified work.

The rules for bidding on public-sector construction projects vary, depending on the state in which the project is located, and whether the project is funded by federal, state, or local money. The most important characteristic of the bidding process on all public-sector projects is that it is controlled by legislation, leaving little or no room for flexibility in the selection process. Owners of private-sector construction projects, on the other hand, can solicit bids, negotiate with selected contractors, or do nearly anything they want to procure the required services for their projects.

This chapter examines the competitive bidding process in general. Differences between bidding methods for public- and private-sector projects will be discussed as they arise.

The Competitive Bidding Process

There are a number of steps to the competitive bidding process. They are:

- Issuing invitations or advertisements for bids
- Providing instructions to bidders
- Determining the qualifications of bidders
- Submitting the bids
- Evaluating the bids
- Awarding the contract

Each of these steps is discussed below.

Invitations or Advertisements for Bids

Invitations for bids are normally issued for private-sector projects. They are sent to contractors whom the owner and design professionals believe are reputable and are interested in the type of work involved on the project. Enough contractors should be invited to ensure that the bidding process is competitive – usually six or more firms.

More formal *advertisements for bids* are required by law for public-sector projects. Advertisements must normally be placed in at least one general circulation newspaper (such as the *Los Angeles Times* and the *Boston Globe*) and at least one trade paper (such as *Dodge Reports*). The advertisement must appear a certain number of times (often 2 to 4) and at frequent intervals (once a week) before the date set for receipt of bids. This procedure is intended to alert all interested bidders to the fact that the owner is seeking bids.

Figure 13.1 lists the information that is typically provided in the invitation or advertisement for bids.

It is important to understand the scope and effect of an invitation or advertisement for bids. It is *not* an offer but, rather, a *request* for offers. (See Chapter 2, "Basic Contract Law," for more information on offers.)

The power to accept a bid (the offer) does not end until the time stated in the invitation or advertisement (or a reasonable time thereafter if no deadline is given). By including a date, the owner does not have to worry that the bid will expire or be withdrawn before the owner has had adequate time to evaluate and accept or reject it. A request for bids implies that the owner intends to accept the lowest or best bid (offer). An owner who wants to reserve the right to reject all bids should include this term in the invitation or advertisement.

Contents of an Invitation or Advertisement for Bids

An invitation or advertisement for bids usually contains the following information:

- Name and location of the project
- Name and address of the owner (frequently the awarding authority on public-sector projects) and the prime design professional
- Date, time, and location that bids are due
- Nature, scope, and description of the project
- Whether or not bids will be opened and read publicly
- Where, how, and the cost to obtain bidding documents
- Names and locations of places where bidding documents are available for inspection
- Procedure for submitting bids
- Type and amount of bid security, if any (such as a bid bond for 5% of the bid price)
- Whether or not the owner has the right to waive formalities in the bidding process (usually not on public jobs)
- Basis on which bids are judged (usually to the "lowest responsible" or "lowest and best" bidder)

Figure 13.1

Instructions to Bidders

On private-sector projects, instructions to bidders usually accompany the bidding documents to familiarize bidders with the way in which the bidding process will be conducted for the project. These instructions lay out the rules and regulations applicable to that project. Federal or state law establishes instructions for public-sector projects.

Instructions normally provide that the act of bidding implies certain things. For instance, each bidder is assumed to be familiar with the "local conditions" under which the work will be performed. Such an assumption is useful in deterring claims for unanticipated conditions. This would tend to preclude, for example, a contractor from North Carolina claiming that he was entitled to money in addition to the contract price based on the fact that he did not know the general characteristics of the local soil or the extent and quality of the tradespeople in the local labor force for the project that he bid in Florida.

Additionally, instructions to bidders often require that each bidder base his or her bid on the information in the bidding documents *only*. This is intended to prevent a bidder from basing his or her price on materials or equipment other than those called for in the documents, with the hope that cheaper materials or equipment can be substituted later in the course of the project, after the contractor has been awarded the contract.

Finally, it is common for bidding documents to be issued only in complete sets called *bid packages* (see Figure 13.2). Bid packages are issued to "prime" bidders only. "Prime" bidders are usually general contractors bidding on the entire scope of work on the project. By requiring that the prime bidder accept responsibility for the total bid package, the owner is better able to avoid or defeat claims by the prime contractor that a subcontractor made a mistake in his bid to the prime bidder because he saw only a part of the contract documents and missed some of the work appearing on another part of the documents. When faced with such a claim from the subcontractor, the prime bidder may try to pass it on to the owner, to the owner's detriment.

There are standard forms of instructions to bidders available commercially. The current form published by the AIA is Document A701, *Instructions to Bidders, 1987 Edition*. The EJCDC publishes Document 1910-12, *Guide to the Preparation of Instructions to Bidders, 1983 Edition*. Both of these standard forms are useful documents.

Addenda and Substitutions

Addenda are changes to the construction documents or other proposed contract documents made prior to the bid date. The impetus for such changes may come from the owner, the design professional, or the bidders.

A final check is usually made at the design professional's office once documents have been issued for bid. The design professional may spot a problem area that should be corrected. Or, the owner or design professional may simply change their minds about some facet of the project. Further, one or more bidders may call to the attention of the owner and design professional any part of the construction documents that is unclear or ambiguous and needs clarification. This should be done at least 7 days before the bid date to allow the design professional time to reexamine the materials and, if necessary, prepare amendments and issue the changes to the bidders. Similarly, requests from potential

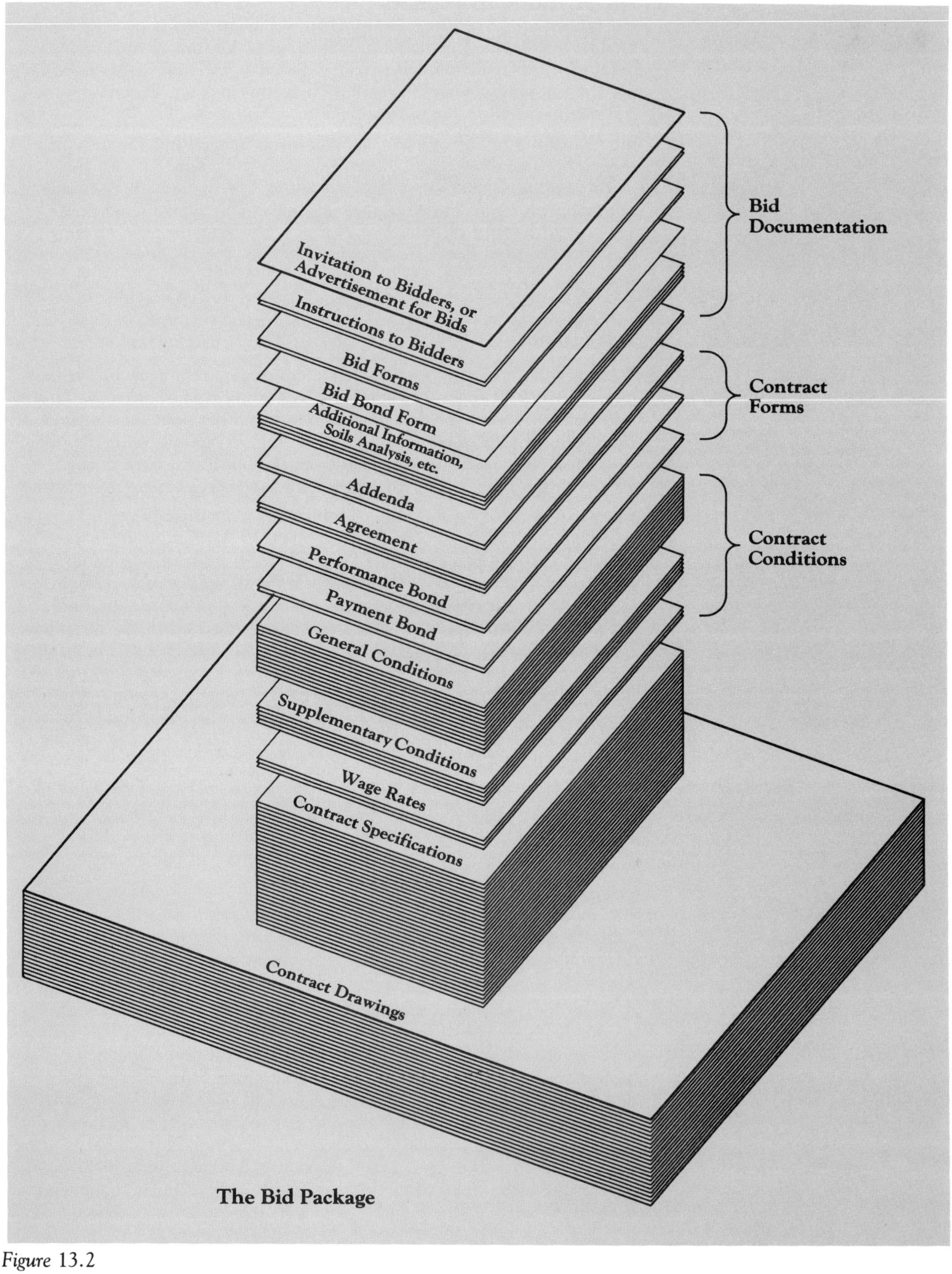
Invitation to Bidders, or
Advertisement for Bids
Instructions to Bidders
Bid Forms
Bid Bond Form
Additional Information,
Soils Analysis, etc.
Addenda
Agreement
Performance Bond
Payment Bond
General Conditions
Supplementary Conditions
Wage Rates
Contract Specifications
Contract Drawings
Bid
Documentation
Contract
Forms
Contract
Conditions
The Bid Package

Figure 13.2

bidders about the *substitution* of one product or material for another should be made approximately 7 to 10 days prior to the bid date.

After checking the documents and reviewing comments about potential ambiguities and proposed substitutions, the design professional should issue *written* addenda clarifying appropriate items and approving substitutions, if any. This should be done no later than 4 days before the bid date to allow bidders time to consider the changes and make appropriate adjustments in their bids, if necessary. It is important that addenda be issued in writing so that all bidders will base their bids on the *same* documentation. Bidders should rely on written addenda only, never on telephone conversations, no matter how formal or with whom.

Qualifications of Bidders

Either before or after the bid date, the owner and design professional may insist that the contractor(s) prove that they are qualified to perform the work necessary to complete the project. The factors normally considered are:

- The nature of past projects undertaken by the contractor
- The capital structure (the assets and financial status) of the contractor's firm
- The nature and condition of the contractor's physical plant (machinery and equipment owned by the contractor)
- Personnel regularly employed by the contractor, especially supervisory personnel
- The quality of past work
- Bonding capacity

Interestingly, a surety company uses the same criteria (with the possible exception of work quality) to decide whether or not to issue bonds for a contractor. Therefore, if a contractor is bondable, chances are he or she is qualified to submit bids.

On public-sector projects, contractors are often required to obtain certificates declaring their eligibility to bid on such projects. Upon application by the contractor, the governmental entity investigates the contractor's qualifications and, if they are adequate, issues a certificate to enable the contractor to submit bids on specific types of public-sector projects. Without this certificate, the contractor is not deemed qualified to bid.

Submitting Bids

Bid Forms

Although there is no inherent reason why a particular bid form must be used, in practice one is usually provided with the bidding documents for each project, and bidders are required to use that particular form. If all bidders are using the same form (regardless of which form that might be) the owner is assured that the bids can be compared on an "apples-to-apples" basis.

The information provided by the contractor on the bid form must either be typed or written in ink. This minimizes the potential for confusion or subsequent manipulation of data. Similarly, the bid price should be stated both in words and in numbers. If there is a discrepancy, the rules of contract interpretation require that words prevail over numbers. (Refer to Chapter 2, "Basic Contract Law," for more information on interpreting contracts.)

Each bid must be signed by an official who is legally authorized to bind the bidder. If the bidder is a corporation, a corporate officer is usually authorized to sign on behalf of the corporation. The corporate seal is then affixed to the bid. In a general partnership, any partner may sign the bid. Of course, a sole proprietor may sign on his or her own behalf.

It is common for the bid form to require that the bidder list all addenda issued, indicating that the bidder received them and based his or her bid on the information contained in them. Unit prices are also often quoted on bid forms. For example, a price might be quoted for a cubic yard of concrete or a lineal foot of pipe or conduit. In fact, there may be two prices for each unit – one an "add to," and the other a "deduct from," the number of units specified in the contract documents.

Unit prices are frequently quoted for excavation and rock removal, since subsurface conditions are so difficult to predict precisely, and the anticipated quantities can vary considerably. Quoting unit prices in the bid establishes a cost reference point for changes that are made during the course of construction.

Finally, disclaimers contained in bidding information are frequently repeated in the bid forms as a record that the bidder has agreed to them. The form also usually provides that the contractor will hold his or her bid open for acceptance for a specified period of time. Without such a provision, the contractor would be entitled to revoke his or her bid any time before the owner accepted it.

In short, the bid is intended to be the contractor's offer to construct the project according to the contract documents and for the bid price. In turn, the owner assumes that acceptance of the bid will bind the contractor to perform his or her promise.

Deadlines

Bids must be submitted before the established deadline. Generally, bids are submitted in sealed envelopes. No telegraphic or verbal bids (by telephone or in person) are acceptable. Bids are opened only after the submission deadline has passed. This protects the confidentiality of each bid and prevents bidders from manipulating their bids in any way based on disclosure of others' bids.

If a bid is submitted late, it may be rejected outright. On private-sector projects, the owner and design professional may make an exception and receive a late bid depending on: 1) how late the bid is, and 2) whether or not the bidder could conceivably achieve any competitive advantage by submitting the bid after the deadline. (In general, these same considerations are applied to all exceptions to the bidding rules that are made during the bidding process.) However, the owner should make no exceptions, nor waive his or her right to enforce any formalities during the bidding process, if such action will affect the basic fairness of the procedure.

Validity

To be valid, each bid must conform to the requirements of the invitation and the instructions to bidders. Bids may be rejected if they do not conform to requirements regarding:

- The form of the bid; for example, if the form provided with the bidding documents was altered or not used
- The scope of work covered, if less than the full scope of work described in the contract documents was bid
- Time for performance

- Bonding requirements; for example, if the contractor is not bondable
- Terms of the proposed contract (Although contract terms between the owner and the selected bidder are often negotiable at the time the bids are due, the bidders must either be willing to accept the contract terms offered or decline to bid.)

If a bid does not conform to the applicable requirements, and the defect cannot be corrected or waived, the bid must be rejected.

Although, strictly speaking, bids are not invalid simply because they exceed the amount of funds available for the project, most owners retain the right to reject *all* bids if even the lowest exceeds the available funds.

Mistakes in Bids

If a bid is submitted before the deadline, and the bidder discovers an error in his or her quotation, the bid can be withdrawn without explanation and without penalty, and may be corrected and resubmitted—as long as this is done prior to the deadline.

After a bid has been accepted and the contract awarded, the selected bidder is generally responsible for *mistakes of judgment* that are reflected in his or her bid. For example, if a contractor thinks that a certain task will take one work crew three days to complete, the contractor will not be entitled to extra money beyond the bid price if it actually takes two crews three weeks to perform that same task. If a major mistake in judgment is made in a bid, and this mistake is discovered after the bid has been accepted, a bidder may either swallow it, so to speak, or withdraw the bid—with the subsequent loss of bid security, if any, and potential liability to the owner for damages. This means that the bidder would forfeit his or her bid bond, for example, and would be liable to the owner for the difference (if not covered by the bid bond) between his or her bid and the second lowest bid. (Refer to Chapter 16, "Insurance, Bonds, and Lien Laws," for more information about bonds.) Contractors who bid on projects are presumed to act knowledgeably, and must suffer the consequences of their own errors in judgment.

On the other hand, *mathematical and clerical mistakes* (such as misstating a subcontractor's bid) can be distinguished from judgment errors. For example, a bid may list a number of elements which together constitute the bid price. If each element is correct, but they are added incorrectly, the bidder will generally not be bound to that mathematical mistake. Similarly, an honest omission of a major element of work, which sometimes occurs in the frenzy of compiling a bid, is not considered a judgment error.

The law generally recognizes the basic unfairness of binding a bidder to a clerical mistake. In such instances, the bidder is usually allowed to withdraw the bid without penalty. Other bids are then considered without regard to the withdrawn bid. In short, the bidder is not made to suffer for his or her clerical mistakes, but neither is he or she allowed to profit from them.

Correcting Bids

Bidders are rarely, if ever, allowed to "correct" bids after they have been submitted and opened. The potential for favoritism and manipulation of the "corrections" is too high. Assume, for example, that a bidder submitted a bid of $740,000 because he thought that a price of that magnitude would be necessary to get the job. Once the bids were opened, he discovered that the next lowest bid was $800,000. The low bidder might be tempted to claim a mistake in the $50,000-to-$59,999 range so that he could still be the low bidder at approximately $790,000 and yet make the additional $50,000. To allow such a correction would be unfair to both the owner and all of the other bidders.

Sub-Bids for Public-Sector Projects

Some states and other governmental entities require that separate sub-bids (bids for the work of a single trade, like plumbing or electrical work) be submitted on each public-sector project. The number of categories for which sub-bids are required varies from state to state, and not all categories are applicable to every project. In Massachusetts, for example, there are at least 17 such categories. In addition to the sub-bids, a traditional general bid (intended to include all work not specifically covered by a category of sub-bid) is required.

Normally, the contract would be awarded to the general contractor who has submitted the lowest bid. In turn, that contractor would be required to subcontract with those trade contractors who have submitted the lowest sub-bid in each category. Such a procedure is intended to maximize competition and produce the lowest possible prices for the public's construction dollars.

This process can work well, but it often forces together general contractors and subcontractors who, for whatever reason, would not otherwise be willing to work with each other. Problems in coordination or communication among reluctant team members on a construction project may result in inefficiency and delays that can offset any savings that might have accrued as a result of the bidding process.

Evaluating Bids

The contract is not usually awarded at the bid opening. First, the contractors' qualifications may have to be validated if this has not already been done, and then the bids must be analyzed and evaluated. When evaluating bids, there is more to consider than the mere dollar amounts submitted by the bidders. For example, it is appropriate to compare the lowest bid to others to determine whether or not there are large disparities with respect to any given element of the overall price. Substantial differences may be an indication of a clerical mistake in the low bid, or may show that the low bidder simply did not understand the requirements for some part of the work. No one benefits if the owner blindly accepts a low bid that turns out to be substantially inaccurate.

Alternates

One of the elements of a bid that normally needs to be evaluated is the list of alternates or changes to specific parts of the construction drawings or specifications. Prices for alternates are provided by the bidders to enable an owner to tailor the bids to the available funds or to choose among different materials or equipment.

There is a basic package of construction work (the bid for which is called the *base bid*) to which alternates are added or subtracted. For example, in a renovation project, one alternate might be a new roof. Depending on the amount of the base bid, the cost of the alternate, and the total amount of money available, the owner may or may not elect to include the new roof in the work of the contractor.

It is possible for an alternate either to cover the addition of work to the basic package, or the deletion of work from it. For administrative simplicity, all alternates in any given bid proposal should be either additive or deductive – not both. For example, if there are three alternates to be bid, they all should be for items to be added to the basic work. None should involve deletion of any part of the basic work.

Generally speaking, owners of private-sector projects retain the right to accept *any* combination of alternates. Then, once the alternates are selected, the low bid is the combination of the base bid and bids for the alternates. For instance, if six alternates were bid, the owner could choose to include alternate numbers 1, 2, 5, and 6. It would not be necessary to agree upon alternate numbers 3 and 4 before 5 could be selected.

Regardless of the fact that the owner usually retains the unrestricted right to select among alternates, such selection should not be done to manipulate the bidding process. For example, if the base bid and alternate numbers 1, 3, and 5 are selected, Bidder A will be the low bidder; but if the base bid and alternate numbers 2, 4, and 5 are selected, Bidder B (the owner's brother-in-law) will be the low bidder. The owner should not chose alternate numbers 2, 4, and 5 just so that his brother-in-law will be awarded the contract.

By contrast, on public-sector projects, the alternates must normally be selected in order of listing. For example, alternate number 3 could not be selected unless alternate numbers 1 and 2 were selected as well.

Awarding the Contract

Once the owner and design professional have evaluated the bids and determined which alternates, if any, will be included, they are ready to award the contract to a selected bidder. Normally this is the lowest bidder. The contract must then be prepared in final form and signed by both parties before actual construction work may begin.

Notice to the Contractor

Once the owner has selected a contractor for the project, he or she should notify all of the other bidders that the contract has been awarded. The owner should state in the notice that, although a specific contractor has been selected, all bids remain available for acceptance for the period of time stated in the invitation to bid. That way, if the successful bidder does not sign the contract, for whatever reason, the owner can choose the second lowest bidder and that bidder's offer will still be open for acceptance. Normally, bid deposits are returned with the notice to all but the three lowest bidders. Deposits will be returned to those bidders once the actual contract has been signed.

List of Subcontractors and Suppliers

Once the contract is awarded, the contractor usually must submit a list of proposed subcontractors and suppliers for the major parts of the work. If the owner or design professional has reasonable objections to any entity on this list, the contractor must not subcontract with them on the project. The contractor may, however, propose a substitute subcontractor or supplier, and may adjust the bid price accordingly. If the owner does not accept the proposed substitute or the new price, the contractor may withdraw his or her bid without penalty.

Negotiating the Contract

Bidders should be entitled to rely on the integrity of the competitive bidding process. If contracts are awarded arbitrarily, there is no incentive for contractors to submit competitive bids.

Even so, minor changes that an owner may want to make to the contract documents can be negotiated with the low bidder before the contract is executed. Major changes that would require substantially revised contract bid documents, however, should not be accomplished through negotiation. Rather, changes should be made to the bid documents and the project then re-bid based on the new documents. Not only is this the most equitable approach for all bidders concerned, but it is usually in the owner's best interests as well. There is no guarantee that the original low bidder would have been the lowest had the revised documents been available to all interested bidders.

Conclusion

Competitive bidding remains the predominant method used to select a construction contractor for a project. The process is intended to identify the qualified contractor who will build the owner's proposed project, according to the owner's terms and conditions, for the least amount of money. If the bidding process is run fairly, and the bidders are provided with a reasonably complete set of construction documents describing the required work, that result should be achieved.

Owners in the private sector have considerably more flexibility than those in the public sector with respect to the bidding process. Otherwise, the process itself tends to be the same regardless of the type of funding available for the project and the identity of the owner. The information discussed in this chapter is an accurate representation of the actual bidding process, though particular details may vary from project to project.

CHAPTER 14 CONSTRUCTION CONTRACTS

Chapter 14 Construction Contracts

Introduction

There are a number of methods commonly used for contracting construction services. An owner may use one method for one project, and another method for a second project with different characteristics. Or, an approach to contracting that is appropriate for one owner may be inappropriate for another, even though both owners intend to build projects with similar characteristics.

This chapter describes common methods of construction contracting, the advantages and disadvantages of each method, and the legal considerations applicable to each. It is important for the owner to match the system of contracting to the characteristics of the project and the experience and expectations of his or her staff.

Construction Contracting

There are four major systems of construction contracting:

- Traditional
- Construction management
- Design/build
- Force account

Each of these contracting systems is discussed in the sections that follow.

Traditional

In the *traditional* method of construction contracting, the owner enters into a contract with a single *prime* (or *general*) *contractor* who assumes responsibility for the construction of the project. This relationship is shown in Figure 14.1.

Upon award of the construction contract by the owner, the general contractor normally subcontracts much of the work to specialist *trade contractors* (such as plumbers, electricians, masons, and others) who, collectively, perform as much as 80 percent of the total volume of construction. (Refer to Chapter 15, "Subcontracting," for additional information about subcontractors and subcontracting principles and procedures.)

The general contractor usually provides some services with his or her own forces. Carpentry and, perhaps, concrete construction work are commonly provided, plus other work of a general nature which might be impractical or inefficient to divide among the various trade contractors.

For example, the general contractor will frequently retain responsibility for developing and maintaining the project progress schedule, for providing and maintaining scaffolding, for clean-up, and similar matters. The general contractor also assumes responsibility for organizing and scheduling the sequence of the subcontractors' work.

One of the principal legal characteristics of traditional construction contracting is that the general contractor is responsible to the owner for the quality of all the work, whether performed by the general contractor's own forces or by subcontractors. This feature allows the owner to remove him or herself, if so desired, from the details of construction administration. This is advantageous for those owners who do not build frequently. It allows them to undertake projects of a scale or complexity beyond the capabilities of their own staff. It also allows them to hold a single entity responsible for the quality and timely performance of the construction work being performed by the various specialty trade contractors.

A disadvantage of the traditional method is that the contractor's superior knowledge and experience with construction projects may give him or her bargaining power that may result in a relationship in which the owner's interests are inadequately protected. To prevent this, the owner must either have the skills available on his or her staff to monitor the

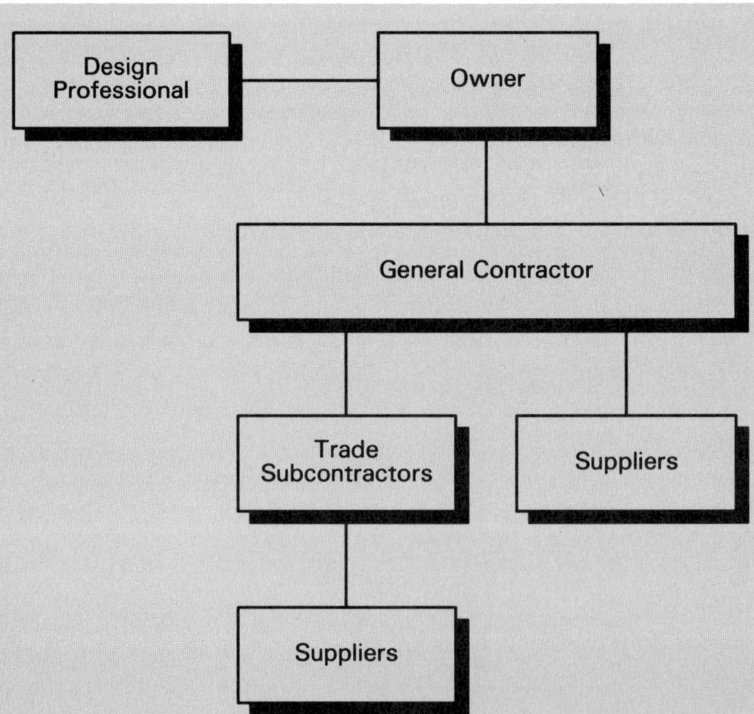

Traditional Construction Contracting

Figure 14.1

general contractor's performance, or the owner must hire design professionals and other consultants to do so.

Multiple Prime Contracts

A variation on the traditional approach is one involving *multiple prime contracts*. In this arrangement, the owner contracts directly with the major trade contractors, such as electrical, plumbing, and mechanical, and with a general contractor, who contracts for the balance of the required work including, possibly, coordination of all the contracts. This is shown in Figure 14.2.

There are a number of positive aspects of this method. The direct nature of the owner's relationship with the prime contractors often results in improved payment arrangements and working relationships. Each contractor has direct communication with the owner and knows that his or her payments will not be affected if another prime contractor fails to perform properly under a separate contract with the owner. Additionally, the size and duration for performance of each contract is relatively small and, therefore, not as risky for either the owner or contractor.

Further, contracts can be awarded closer to the time of their performance. This often results in more accurate cost estimates since the contractors do not have to speculate about the costs that they will incur months or years in the future. For example, a painting contractor in a traditionally structured project must submit prices for the painting work when the general contract is being bid, although the actual painting will not be performed until two years later. The painter must guess at the costs that will be incurred at that time and include money in the bid to

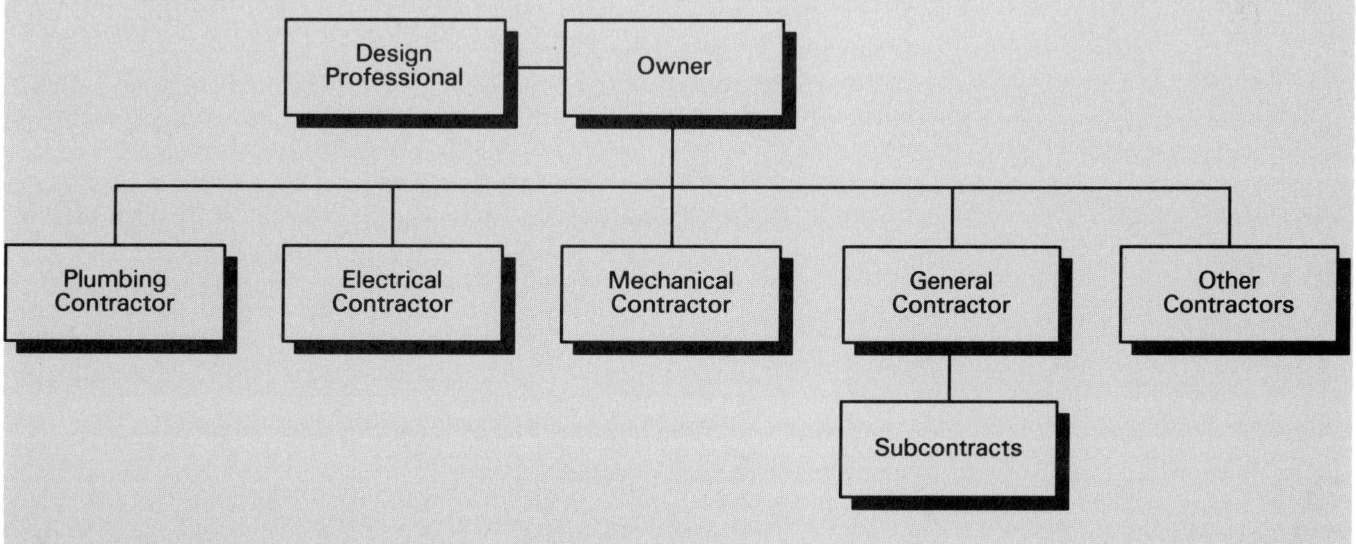

Multiple Prime Contracts

Figure 14.2

cover unexpected contingencies. Under the multiple prime contracting system, the painting contract could be awarded much closer to the time for performance, and the bid would not have to include so much money for contingencies.

Another positive aspect is that each contract may be written to reflect the technical characteristics of the specific work and to properly control its cost and the schedule for its performance. For example, a contract for excavation (which is subject to changes because of the indeterminate nature of subsurface conditions) may best be written with compensation determined as a function of costs incurred. On the other hand, a contract for painting (which is not as likely to change from unexpected conditions) may best be written with compensation stated as a fixed sum. By contrast, under the traditional system, the single prime contract would contain one price (or method of pricing) applicable to all the work.

Of course, there are also negative aspects to the multiple prime system of contracting for construction. First, the owner must establish a well-defined work package for each of the separate contractors. This involves additional work for the design professionals and may create gaps or overlapping among the work of the multiple prime contractors. Additionally, each of the prime contractors must be investigated and must prove that they are qualified to perform the required work. In addition to these tasks, it can be slow and expensive for an owner to solicit bids or negotiate for the separate parts of the construction work. Finally, the owner is responsible for scheduling and sequencing the work of the separate contractors, keeping in mind that delay or poor performance by any one of the contractors can adversely affect the performance of the others – to whom the owner is legally responsible.

Many owners want to be involved in project administration when there are only three or four prime contractors on the project, and the owner has the staff and experience to deal with them. Such owners are usually able to strike a balance between the positive aspects of this system and the increased legal responsibility that they assume for more than one prime contract.

Construction Management

Other owners recognize the benefits of a multiple prime contract system, but have neither the ability nor the inclination to become involved in managing even three or four contracts. For them, *construction management* is a viable alternative.

In this method of construction contracting, the owner hires a *construction manager*. This individual or entity possesses the skill and experience to manage all of the various trade contractors involved in the project.

There are two types of construction managers:

- Construction manager as general contractor
- Construction manager as independent professional consultant

These two types are discussed below.

Construction Manager as General Contractor

Some construction managers operate during the construction phase essentially like general contractors. This type of construction manager may:

- Perform some construction work with his or her own forces
- Contract directly with the trade contractors

- Provide a *guaranteed maximum price* (GMP)
- Incur profit or loss on the cost of construction labor and materials

This approach to construction management, shown in Figure 14.3, is very similar, legally, to the traditional method using a single general contractor. The owner contracts with one party who is responsible for the construction services. The principal distinction between this type of construction manager and a general contractor is that the construction manager performs services prior to the construction phase of the project.

During the preconstruction phases of a typical project, a general contractor-type construction manager consults with the design professionals to evaluate the proposed design in terms of construction feasibility, costs of materials, labor, equipment, and the sequencing of the trade contractors' work. Most likely, he or she would also develop a project time schedule (including activity sequences and durations) and a project construction budget (including preliminary estimates of construction costs).

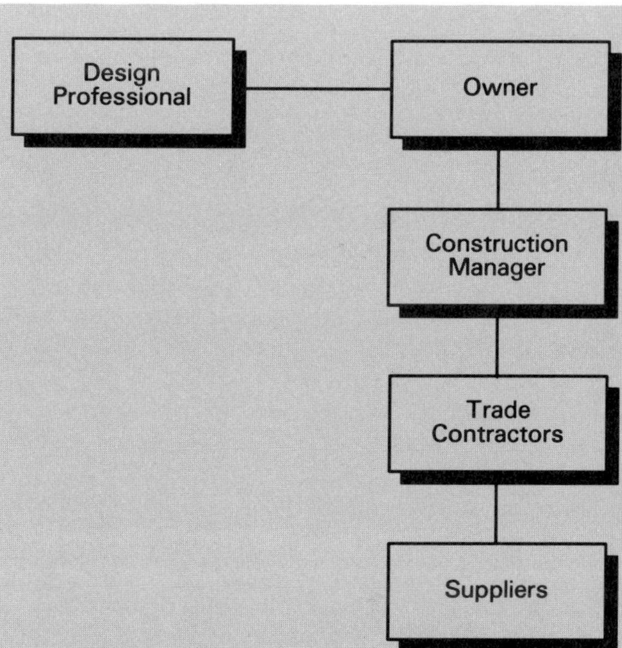

Construction Manager as General Contractor

Figure 14.3

During the construction phase, typical services might include:

- Monitoring and controlling the work of the trade contractors
- Providing on-site superintendence
- Providing on-site contract and construction administration
- Running project meetings
- Monitoring compliance with the progress schedule
- Monitoring the quality of the trade contractors' work
- Performing construction work not being performed by the owner or trade contractors
- Monitoring and controlling construction costs
- Reviewing and processing trade contractors' applications for payment (and making payment accordingly)

Construction Manager as Independent Professional Consultant

By contrast, construction managers who act as independent professional consultants do not perform any construction work with their own forces, nor do they contract directly with the trade contractors. These entities contract with the owner, and merely manage and administer the contracts between the owner and the trade contractors (see Figure 14.4). This consultant-type construction manager is generally paid a professional fee for his or her services, which does not vary with the cost of the construction labor and materials. Further, since a consultant-type construction manager has no contractual control over the trade contractors, he or she is generally not willing to provide a guaranteed maximum price.

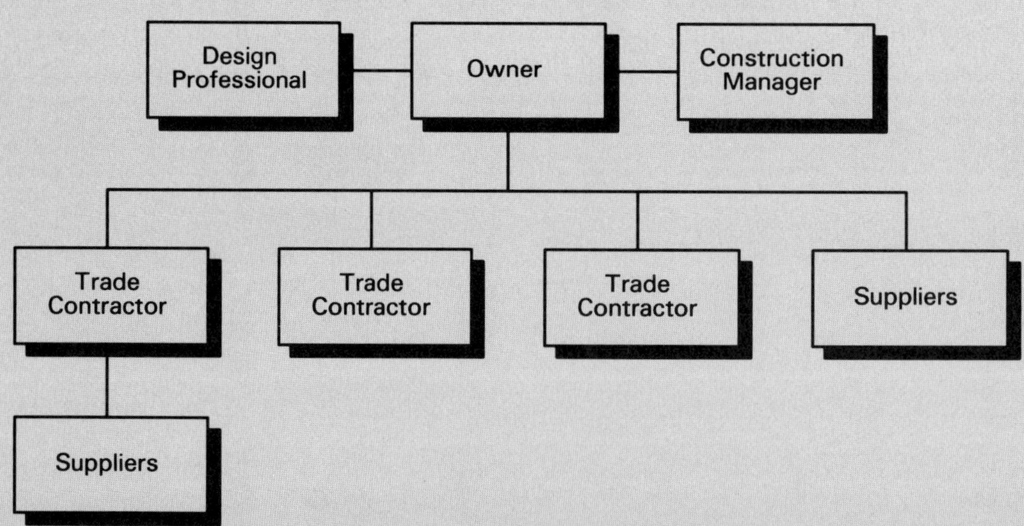

Construction Manager as Independent Professional Consultant

Figure 14.4

This approach to construction management is similar to the traditional system using the multiple prime contract method. The owner contracts with more than one contractor, each of whom is responsible to the owner for the construction services they perform. In turn, the owner is responsible to each prime contractor for the work of the other prime contractors. The consultant-type construction manager is able to provide management experience and capability that the owner does not have on his or her own staff.

The consultant-type construction manager also performs services during the preconstruction phase of a project. These are essentially the same services as a contractor-type construction manager. He or she also normally receives bids, recommends award of contracts or rejection of bids, and then assists in preparing construction contracts and reviewing the qualifications of proposed subcontractors and suppliers.

During the construction phase, a consultant-type construction manager:

- Coordinates all of the parties to the construction process
- Monitors and updates the project construction schedule
- Monitors and controls construction costs

Most other services that he or she performs are similar to services performed by design professionals, except that communication with the contractors is more frequent and more specific than that normally conducted by an architect or engineer. In many ways, the consultant-type construction manager acts as a buffer between the design professionals and the contractors.

Fast-track Construction

The multiple prime contract method, and construction management in particular, lend themselves to what is known as *fast-track construction*. In this procedure, construction contracts are awarded for parts of the project while other parts of the project are still being designed or bid by other contractors. For example, the site work, site utilities, and excavation might be contracted out and construction begun prior to completion of the foundation documents. Then the foundation documents would be completed and that work put under contract – prior to completion of the superstructure, mechanical, electrical, and plumbing documents.

Fast-track construction often shortens the overall duration of the construction phase of a project. This method may be economical for an owner because it decreases the total amount of financing charges to be paid on the project and enables the project to earn revenue earlier than what might otherwise be possible.

Design/Build

A third system for construction contracting is known as *design/build*. This is actually an ancient system based on the concept of the *master builder*, who was responsible for both the design and the construction of his projects. Its resurgence in the past decade seems to be a response to owners' desires to centralize responsibility for a project. Design/build largely removes the owner from detailed involvement with *both* design and construction disciplines. When problems arise, the owner does not have to discern whether it is design- or construction-related because the same person or entity is responsible for both aspects of the project. Figure 14.5 shows the relationship between the owner and the design/build firm.

There are other advantages to this system for construction contracting. Design/build brings the construction skill and experience of the contractor into the design phase, and can sometimes enable construction to begin while design work is being finished. (Design/build is similar to fast-track construction in this respect.) Furthermore, in most cases, design work can begin with relatively minimal criteria from the owner. In fact, even the design documents often show no more than "performance" criteria – merely describing the desired result without specifying in detail how that result is to be achieved. The typical amount of detail may never be included in the construction documents either, because the designer is also the builder and does not need the documentation.

Design/build is most effective for either specialized design work done for sophisticated owners with strong administrative staffs, or for very uncomplicated, straightforward designs, where the owner is quite flexible about the details of how the construction will achieve the project's requirements. In the first situation, the owner's staff is capable of evaluating and managing the services of the design/build contractor and is able to protect the owner's interests. In the second situation, the project is fairly simple and the details are relatively unimportant to the owner, who trusts the design/build contractor to perform properly. In

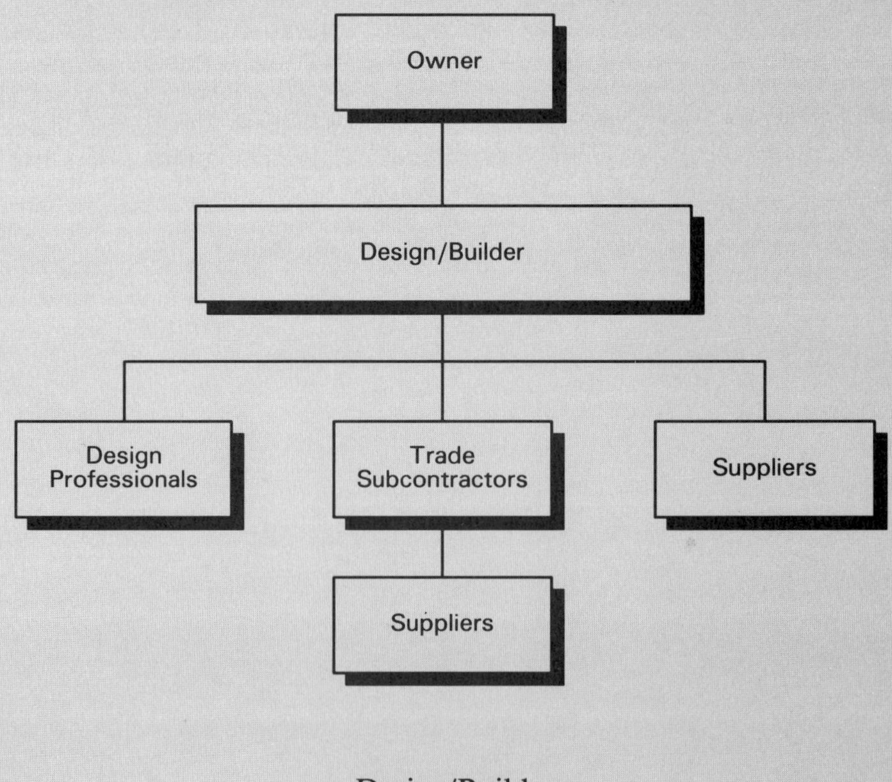

Design/Build

Figure 14.5

either case, to be successful, the owner should conduct an extensive prequalification screening of the design/build firm to be sure that it is experienced and trustworthy.

A design/build contractor assumes all of the usual liabilities associated with the performance of professional services (see Chapter 10, "Professional Services Agreements"), as well as those associated with construction services. (See Chapter 2, "Basic Contract Law," and Chapter 17, "Sources of Dispute.") Furthermore, a project designed and built by the same person or entity is similar, in some ways, to a "product," such as a television set or a washing machine. As such, there may be *implied warranties of merchantability* and *implied warranties for fitness for a particular purpose* with respect to the project (see Chapter 2 for more information about such warranties). Neither design professionals nor construction contractors are normally subject to these legal risks when performing services under separate contracts with the owner.

One frequent disadvantage of design/build is that there is no independent design professional working to protect the interests of the owner. In fact, if a design professional is involved in the project, he or she is normally on the staff of, or under contract to, the design/build contractor. Under these circumstances, the design professional's obligations are to the design/build contractor, not to the owner. The extent to which this is a cause for concern varies greatly from project to project, and depends on the type of building and the nature and capabilities of the parties involved.

Additionally, the quality of the design may suffer because the owner is not able to select from all of the available design talent. The owner must use the design professionals working on the staff of, or under contract to, the design/builder. Unfortunately, these individuals are not always the most qualified designers for the project.

Finally, it is normally quite difficult for an owner to know in advance precisely what the completed project will be like. This is because the design and construction documents normally lack the details that would have been included on documents used in competitive bidding. The owner may have a general idea of what the project will be like and how it will perform, but the details are often unavailable until the work is actually constructed.

Force Account

One last method of construction contracting which deserves mention is the *force account* system. This is really not a system of contracting at all. Rather, it is the classic "do-it-yourself" approach to construction. The owner actually hires the construction workers or has the work done by people already on his or her payroll. Forced account is typically used for small jobs such as repair work or interior modifications to office or plant facilities. (See Figure 14.6.)

An owner who intends to perform construction work on this basis must recognize that the costs of recruiting, training, and maintaining a qualified staff can severely limit available manpower. Additionally, the owner must maintain an inventory of construction equipment and supplies, and be willing to accept liability for construction-related tasks, like security from theft and the safety of the workers and the public in general.

Another more subtle disadvantage for many corporate owners who might consider this approach is that it sometimes generates adverse reaction in contractors who are seeking the opportunity to work for the corporate client. Any possible construction savings may well be offset by the loss of goodwill in the local construction community.

Construction Contract Provisions

Methods of Compensation

Construction contracts contain a number of provisions. Some of the most important concern the method for compensating the contractor for providing the labor and materials and performing the construction services required by the contract. Each individual construction contract, regardless of the chosen system for construction contracting, must be paired with at least one of the following methods of compensation:

- Fixed price
- Cost-based
- Unit pricing
- Cash allowances

This section discusses these methods of compensation, their advantages and disadvantages, and the legal risks associated with each.

Fixed Price

A *fixed price* (sometimes called a *stipulated sum* or *lump sum*) for a defined scope of construction work is still the predominant method of compensation for construction work in the United States. If the fixed price is to be meaningful to the owner, however, it must be based on comprehensive construction documents that leave no room for argument about the work that is to be provided for the agreed-upon price. The

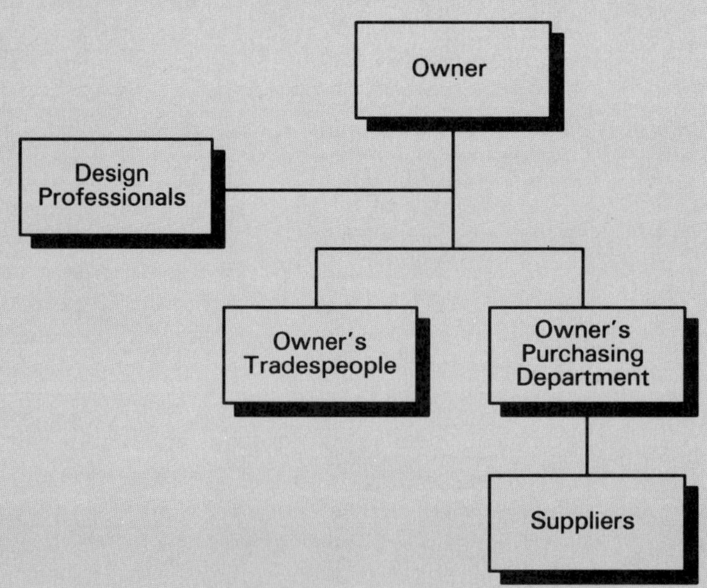

Force Account

Figure 14.6

effectiveness of this system also depends on the availability of experienced contractors willing to bid on the work. Without competition, the fixed price might not necessarily be the lowest.

Owners typically like fixed-price contracts because they offer relative certainty with regard to costs. That is, owners feel that they know, in advance, what the cost of the work will be. This can be a false sense of security, however, if the owner insists on substantial changes in the work after the contract is signed, if unanticipated conditions are discovered, or if the contract documents prepared by the design professional are not sufficiently complete and comprehensive. (See Chapter 17, "Sources of Dispute," for additional details on these matters.) In general, however, this method is the most reassuring for the majority of owners. Once the contract is signed, the owner need not be concerned with the contractor's records of costs. The owner is committed to paying the fixed contract price only, regardless of the costs the contractor actually incurs during the course of construction.

Fixed-price contracts are not well-suited to projects implementing experimental designs, new materials, or unusual construction methods. These variables breed uncertainty, which means that contractors have to include money in their price in anticipation of events that may never come to pass. It is likely that the owner will pay for work that need never be performed. Fixed prices are best used in contracts of relatively short duration, involving the least amount of guesswork for the contractor about future price movements. Such circumstances allow the contractor to minimize the amount of money included for contingencies so that the price is as low as possible.

One of the primary disadvantages of the fixed-price method is that there is an inherent conflict of interest between the owner and contractor. On the one hand, the contractor wants to do the least amount of work required to meet the demands of the contract, and as cheaply as possible. On the other hand, the owner expects the highest quality construction possible. To determine whether or not the contractor is meeting the expectations of the owner, the owner will frequently hire an architect or engineer to monitor the progress and quality of construction.

In a fixed-price contract, the contractor assumes legal responsibility for increases to the planned cost of performing the work. Performance cost increases are increases in the cost of labor and materials. They are unrelated to any change in the type or extent of work required by the contract. If the costs of labor and materials increase, and the contractor has not allowed for this possibility by including money in the fixed price to cover it, he or she may lose money on the project and may even have to abandon it. If the contractor does include money to cover such cost increases, and they do not occur, the owner will have paid more than necessary.

To minimize or share the risks of such increases, many fixed-price contracts include *escalation clauses*, which allow for time-related adjustments to the fixed price that result from labor and material cost increases. Additionally, *changes clauses* are often used to deal with changed circumstances which might affect prices. Usually such clauses cover subsurface conditions, unexpected or unusual conditions, delay, and acts of God (in legal terms, *force majeure*). (Refer to Chapter 17, "Sources of Dispute," for an extensive discussion of changes clauses and other related matters.)

Cost-based

An alternative to a fixed-price contract is a contract which includes *cost-based* compensation. In a cost-based contract, the contract price is derived from the contractor's actual cost of construction labor and materials, the amounts paid to subcontractors according to the subcontracts, plus an amount for overhead and profit. A cost-based system of compensation will be successful if there is a clear and comprehensive definition of what costs will be allowed plus a formula for calculating overhead and profit.

Cost-based contracts are most appropriate when construction documents are incomplete or sketchy, when the design is experimental, when new materials or unusual methods of construction will be used, or when construction is taking place in a remote, environmentally difficult, or hostile site. In all of those instances, there are an enormous number of unknowns. A contractor who is forced to quote a fixed price would necessarily have to include an amount to cover contingencies that may never occur, or that may occur on a greater magnitude than expected. In the long run, neither the owner nor the contractor can benefit from this kind of gambling.

One of the disadvantages of this system is that the owner does not know in advance what the construction work will ultimately cost. Many owners have limited funds and cannot operate with that uncertainty. Additionally, this type of contract puts a relatively large administrative burden on the owner and, usually, the owner's accountants and design professionals, to verify costs and monitor the contractor's record-keeping.

This type of contract also assures the contractor that he will recover all costs and will make a profit. There is little or no incentive, then, for the contractor to try to minimize project costs. Finally, when a cost-based compensation system is used, the owner bears the legal responsibility for cost increases, unknown conditions, and other such occurrences for which the contractor would probably be responsible under a fixed-price contract.

The cost-based contract described above is a "pure" system in which qualified costs are reimbursed without limit, and where the amount of overhead and profit varies in relation to the costs incurred. In practice, that system is very rarely used. Instead, there are a number of variations to the pure cost-based system that are more practical. These are:

- Cost plus a fixed fee
- Cost plus an award fee
- Cost plus a percentage of cost

These are discussed below.

Cost Plus a Fixed Fee: Here, the contractor's fee (for overhead and profit) is fixed at the time the contract is signed, and normally does not change unless there are substantial changes in the scope of work. The idea behind this type of contract is that the contractor should be able to determine the cost to administer the contract and establish the desired level of profit based on the nature and scope of the work effort, all without knowing precisely the actual cost of the construction labor, materials, and subcontracts. There is still no incentive, however, for the contractor to try to minimize costs, because they will be reimbursed without limit.

Cost Plus an Award Fee: Another variation on a pure cost-based approach is the *cost plus an award fee* contract. A base fee is established in the contract at the start of the project, then a *maximum* possible fee is determined at a level perhaps 2 or 3 times the amount of the base fee. Criteria are then established to judge whether or not the contractor should be awarded any amount of fee in excess of the base fee. Such criteria might include control and reduction of costs, quality of construction, and timely performance. At the end of the job, the owner evaluates the contractor's performance against the criteria and awards a fee somewhere in between the base and maximum amounts. This has the advantage of giving the contractor some incentive to do a good job without shifting all of the risk of unknown costs to the contractor.

Cost Plus a Percentage of Cost: Finally, there is the *cost plus a percentage of cost* system. Here, the contractor's fee is a fixed percentage of the cost of the work (for example, the fee may be 10% of costs). As a matter of policy, this system is not permitted in government contracting and is not as widely used in the private sector as it once was, because the contractor has an incentive to increase the cost of the work to increase the fee. Most owners consider this a conflict of interest.

Unit Pricing

Unit pricing is another method of compensation which is frequently used in the United States. It is sometimes used as the sole compensation method for a contract, but it is also used in combination with other methods. For example, in a fixed-price contract, unit prices might be included for use in pricing changes to the original scope of work. The units should be described clearly and be reasonably estimated. They are best used when the exact quantity of an item cannot be determined with certainty at the time the contract is awarded.

For instance, it is common for the owner and design professional to request prices to excavate each cubic yard of rock, to buy a lineal foot of electrical conduit of a specific diameter, or to buy a square yard of a specific type of carpeting. It must be clear whether or not the unit price includes the cost of transportation, handling, and installation, or whether it is just the purchase price.

When unit prices are requested, the contract sum is the unit price multiplied by the number of units installed. Clearly, it is in the owner's interest to carefully monitor the actual quantity of units supplied by the contractor. Frequently, the design professional can be of assistance in this regard. If there is a substantial overrun or underrun from the originally estimated quantities (something on the order of 15% or more), the contractor is normally entitled to an adjustment to the quoted unit prices. This adjustment is intended to reflect the economies of scale associated with establishing unit prices. In other words, there are always fixed costs to be divided among the total number of units sold. Therefore, if a contractor intends to buy 3,000 units of a particular item, he or she will likely be quoted a lower price per unit than if only 50 units were involved.

There is some risk to the owner of receiving "unbalanced bids" when a unit pricing system is used. This occurs when contractors price units of work that will be done *early* in a job at a rate that is *higher* than otherwise appropriate, and price units to be done *later* at a rate *lower* than otherwise appropriate. This is typically done to accelerate cash flow. Similarly, some contractors attempt to guess which units will have overruns, and then price them higher than is otherwise appropriate in order to obtain increased income if and when the overrun actually occurs.

Cash Allowances

Finally, *cash allowances* are often used as a method of compensation in construction contracts. Cash allowances are really a method of accounting for costs, and not a method for establishing such costs. When a cash allowance is used, an amount is included in the construction contract (often as part of a fixed-price contract) to account for the cost of a work item that will not be selected or fully defined until after the contract is awarded. Hardware and carpeting are often handled this way. The *extent* of the hardware or carpeting is known and described in the contract documents, but the *quality* may not be defined until later, when the specific brands and styles are selected.

In order for cash allowances to work effectively, the contract documents must clarify which elements of cost are covered by the allowance. When cash allowances are part of fixed-price contracts, usually only the direct cost of an item is charged against the allowance, while the cost of transportation and handling, taxes, labor, installation costs, overhead, and profit are considered to be part of the fixed contract price. Once the various allowance items are actually specified and purchased, their costs are totaled for comparison to the allowance. If the cost is greater than the allowance, the fixed contract price must be increased by the difference. Conversely, if the costs are less than the allowance, the owner would be entitled to a credit in the amount of the difference.

Other Provisions

Construction contracts typically contain a large number of provisions in addition to those dealing with compensation. Some of these include:

- **Administration of the Contract**: These provisions describe how the owner and design professional(s) will administer the construction phase of the project. See Chapter 10, "Professional Services Agreements," for more information on the role of the design professional.
- **Subcontractors**: These provisions establish policies and procedures for subcontracting by the general contractor. See Chapter 15, "Subcontracting," for more information on this topic.
- **Changes in the Work**: These provisions describe what constitutes changes in the work and how they are handled. See Chapter 17, "Sources of Dispute," for a discussion of changes clauses.
- **Time for Performance**: These provisions address the schedule, deadlines, and delays in the progress of the work. See Chapter 17 for an extensive discussion of delays and acceleration.
- **Protection of Persons and Property**: These provisions assign responsibility for safety. See Chapter 12, "Building Regulations," for a discussion of environmental regulations and building codes.
- **Insurance and Bonds**: These provisions establish requirements for insurance and bonding. See Chapter 16, "Insurance, Bonds, and Lien Laws," for more information on these topics.
- **Termination of the Contract:** These provisions describe procedures in the event of a breach of contract and other defaults. See Chapter 17, "Sources of Dispute," and Chapter 18, "Dispute Resolution," for more information.

Provisions in all of these categories are included in the standard form construction contracts published by the AIA, the EJCDC, and other organizations. These agreements are examined in the following section.

Standard Form Agreements

There are a number of standard form construction contracts available in the United States today. These forms are generally well prepared, and the ones published by the AIA are widely endorsed within the design and construction industry. Furthermore, the AIA forms have a long history of interpretation in the courts. It is unlikely that many owners or theri attorneys could match the depth and breadth of experience of the drafters of these standard forms. Furthermore, the provisions of these forms are well known in the industry and generally well understood. Wholesale modifications to the usual provisions are rarely justified.

On the other hand, the nationally marketed standard form agreements are necessarily general in nature. No single form can meet the precise needs of every project under the laws of every state and locality. Therefore, the standard forms must be modified intelligently and tailored to suit the individual circumstances of the users.

Copies of the following forms have been included in Appendix B:

AIA Documents:

- A101 Standard Form of Agreement between Owner and Contractor – Stipulated Sum, 1987 Edition
- A111 Standard Form of Agreement between Owner and Contractor – Cost of the Work Plus a Fee, 1987 Edition
- A201 General Conditions of the Contract for Construction, 1987 Edition

EJCDC Documents:

- 1910-8-A-1 Standard Form of Agreement Between Owner and Contractor on the Basis of a Stipulated Price, 1983 Edition
- 1910-8-A-2 Standard Form of Agreement Between Owner and Contractor on the Basis of Cost-Plus, 1983 Edition
- 1910-8 Standard General Conditions of the Construction Contract, 1983 Edition

Conclusion

The characteristics of the project as well as the needs and desires of the owner determine which approach to construction contracting is most appropriate. This chapter has discussed the advantages and disadvantages of the various systems of construction contracting and the various methods of compensating the contractor for his or her labor, materials, and services. Careful consideration of these factors, and of the ways in which available methods and procedures can be combined and adapted, will enable an owner to make the proper choice and start the construction phase of the project off right.

CHAPTER 15 SUBCONTRACTING

CHAPTER 15

SUBCONTRACTING

Introduction

The term "subcontractor" is used to describe individuals or entities based not on the specialized services they may provide, but on their relationship with another participant on a construction project. A subcontractor has a direct contractual relationship with the general, or prime, contractor to perform some of the work at the construction site. Often, he or she has a direct contractual relationship with other subcontractors as well. However, the subcontractor does *not* have a direct contract with either the owner or the design professional. An individual or business entity who has a contract with the owner is considered a prime contractor on that project.

Because of their position in the hierarchy of the project construction team, subcontractors often find themselves embroiled in a variety of legal matters, many of which arise from nonpayment by those to whom they are contractually bound. This chapter examines the various types of subcontracting arrangements, and the legal issues that typically arise under these arrangements.

Types of Subcontractors

There are three major categories of subcontractors:

- Trade contractors
- Suppliers
- Sub-subcontractors

Each of these is described in Figure 15.1, and in the following sections.

Trade Contractors

The majority of subcontractors are *trade contractors*, skilled in a specific type of construction work. For example, plumbers, carpenters, plasterers, masons, painters, and roofers perform work almost exclusively in their trade or discipline. They are referred to as subcontractors in circumstances in which they are hired by the general, or prime, contractor to perform some of the work that the general contractor has promised to perform for the owner.

Suppliers

Another type of subcontractor is the *supplier*. This individual or entity furnishes equipment, machinery, products, supplies, or materials directly to the general contractor, although he or she does not normally perform labor at the construction site. Like trade contractors, suppliers do not have a contract with the owner. Their contractual relationship is with the general contractor, frequently in the form of a purchase order (complete with various terms and conditions).

Sub-Subcontractors

Finally, *sub-subcontractors* are individuals or business entities who have contracts with other subcontractors. For example, a sub-subcontractor may be a supplier who is providing pipe and accessories to the plumber (the trade contractor). Or, a sub-subcontractor may be a firm of carpenters which has agreed to perform some of the carpentry work under contract to a much larger carpentry/drywall/painting firm which, in turn, has a contract with the general contractor for the installation of carpentry. In other words, a subcontractor (the large firm) may perform some of its contract obligations through sub-subcontractors (the smaller firm).

Advantages and Disadvantages of Subcontracting

There are a number of advantages and disadvantages of subcontracting portions of a project. One advantage is that subcontracting generally improves the overall efficiency of the project, since tasks are performed by laborers with specialized skills and equipment. Subcontracting also frequently reduces costs. General contractors who rely on subcontractors to perform specific types of work can compete for more projects than would be possible if they had to perform all of the work with their own forces. The greater number of competent firms there are to compete, the lower their prices will be in general. Similarly, competition among the subcontractors also tends to keep prices down.

Finally, subcontracting allows general contractors (as individuals or entities) to enter the construction field with relatively little capital. The subcontractors make the necessary investments in equipment and staff and pay for materials and labor before they are paid by the general contractor. As a result, the general contractor does not have to use his or her own capital for such purposes.

Types of Subcontractors

Trade Contractors	Suppliers	Sub-Subcontractors
Plumbers Carpenters Plasterers Electricians Roofers Masons Glaziers Painters Tile Setters etc.	Entities that furnish: equipment, machinery, supplies, etc. to prime contractors	Trade contractor or suppliers who have contracts with other subcontractors

Figure 15.1

There are disadvantages of subcontracting, however. Because the subcontractors have performed labor and purchased materials and equipment, they bear the greatest risk of delayed payment or nonpayment by the owner to the prime contractor. This is particularly troublesome when, as a result, a subcontractor cannot pay for labor and supplies purchased on credit. Of course, such cash flow problems increase with the percentage of work performed by subcontractors. This situation is exacerbated by the greater likelihood that nonperformance by one subcontractor will delay payment to others who have performed properly. Typically, the more subcontractors there are on the job, the less money the general contractor has tied up in the project, and the less he or she will be concerned about cash flow and payment from the owner. (Payment problems are discussed in more detail later in this chapter.)

Further, because the general contractor has a direct contractual relationship with the owner, he or she often has a substantial bargaining advantage over the subcontractors. This advantage is sometimes reflected in subcontracts that favor the general contractor. If the subcontractors don't like the arrangements, they simply won't be hired for the project.

Subcontracting also increases the potential for disputes over the requirements of the construction documents. General contractors commonly break up sets of construction documents and give subcontractors just those parts believed to pertain to their work. For example, the electrical subcontractor may only be provided with the appropriate section of the specifications pertaining to electrical work, along with the "E" drawings (the electrical drawings). This may result in confusion and disputes if some of the electrician's work is drawn or described in the "M" drawings (the mechanical drawings) and in specifications for the HVAC (heating, ventilating and air conditioning) work, which the electrical subcontractor would not have seen.

Other problems may arise over access to the site, accidents, scheduling, and so forth – all of which are the general contractor's responsibility. Such duties become increasingly difficult for the general contractor to manage as the number of subcontractors on a single project increases.

Finally, the owner is at risk when one or more subcontractors provide services for the project rather than just the general contractor. Those subcontractors may have a right to file a mechanic's lien in response to some payment dispute of which the owner may not be fully aware. For additional information about mechanics' liens, see Chapter 16.

Selecting a Subcontractor

Most often, the prime contract between the general contractor and owner does not deal expressly with the selection of subcontractors. The type of owner who chooses to use the single prime contract system normally does not want to be bothered with the details of arranging to have the work performed. The general contractor, therefore, usually has authority to determine which subcontractors will be used. Of course, this selection may be subject to the owner's approval. An owner may veto a general contractor's choice if the owner knows that the subcontractor is incompetent and unreliable. To a somewhat lesser degree, the owner may insist, after a contract has been signed, on removal of a subcontractor who is not performing in accordance with contract requirements.

Sometimes the prime contract contains clauses that deal expressly with the selection of subcontractors by requiring that they be approved in advance by the owner and/or design professional. The contract may also give the owner power to replace a subcontractor after the contract has been signed, without regard to the quality of the subcontractor's performance.

The *General Conditions* published by the AIA (Document A201, 1987 Edition) takes a compromise position. Under the terms of this document, the general contractor is not required to list the proposed subcontractors, either in the bid or during precontract negotiation. Once a general contractor is awarded the construction contract, however, a list of the principal subcontractors and suppliers must be provided. The owner can then demand, through the design professional, that a subcontractor be replaced if he or she does not meet with the owner's approval. If the substitute is more expensive than the original subcontractor, the owner can either pay the difference in cost, or allow the general contractor to withdraw his or her bid without penalty. The *General Conditions* published by the EJCDC (Document 1910-8, 1983 Edition) contain similar provisions. The AIA and EJCDC documents are reproduced in Appendix B.

The Subcontractor's Rights and Responsibilities

To determine the rights and responsibilities of a subcontractor, it is important to examine the contractual hierarchy established for the project. Typically, subcontracts are greatly affected by the contract that exists between the owner and the general contractor. In fact, it is not at all unusual for terms in the prime contract between the owner and general contractor to control subcontractual relationships.

For example, it is common to state in the subcontract that the subcontractor assumes toward the general contractor all of the duties and obligations that the general contractor assumes toward the owner (for example, the duty to indemnify in specific circumstances); and that the subcontractor is to benefit from those rights that the general has with respect to the owner (for example, the right to refuse to work in areas contaminated with asbestos or other hazardous materials). The terms in the prime contract can affect and control the subcontract, which can, in turn, affect sub-subcontracts. In common practice, each subcontract makes reference to, or incorporates, the prime contract into its terms.

In general, therefore, the terms of the prime contract are considered to take precedence over those in the subcontracts if they differ. This is because the owner and contractor have agreed on the terms and conditions under which construction services are to be provided. This agreement should not be compromised by the general contractor's decision to subcontract part of the work.

However, there are valid reasons for determining, in a specific case, that the terms of the subcontract take precedence over conflicting terms in the prime contract. For example, consider that:

- The terms of the subcontract are directly agreed to by both the general contractor and the subcontractor
- References in the subcontract to the prime contract are often vague or ambiguous
- Subcontractors generally have neither the time nor the ability to compare contracts and assess the ramifications of all the terms contained in the prime agreement

The following sections examine the rights and responsibilities of the subcontractor, general contractor, and owner when disputes arise and one of the parties defaults on its contract.

The Subcontractor's Default

When there are difficulties with the work on a project, the owner looks principally to the general contractor to remedy the situation. The general contractor, of course, provides the majority of the work required by the contract through subcontractors. Therefore, the general contractor is responsible to the owner for the subcontractors' defaults (failures to perform in accordance with contract terms), as long as the owner has given the general contractor the right to manage the job without unreasonable interference.

On the other hand, the general contractor is not responsible for subcontractors' work if the owner takes away the power to manage and control such work. To have such effect, the owner must remove substantial power from the general contractor. Merely retaining the right to approve of a subcontractor for the project is not sufficient to relieve the general contractor of responsibility.

The General Contractor's Recourse

Defaults often result in disputes. Common sources of dispute between owners and general contractors are discussed in Chapter 17, "Sources of Dispute." Since the relationship between owner and general contractor parallels that between general contractor and subcontractor, that chapter should be consulted as to the general contractor's recourse when a subcontractor fails to perform properly under its subcontract. Also refer to Chapter 2, "Basic Contract Law," for additional information on breaches of contract in general.

The Owner's Recourse

Although the general contractor is primarily responsible to the owner for the work of the subcontractors, an owner usually may assert some ancillary claims directly against subcontractors. Such claims may be brought under one of three principal legal theories:

- Third-party beneficiaries
- Negligence
- Agency

Third-Party Beneficiaries: According to this theory, the owner is the intended beneficiary of the contract between the general contractor and the subcontractor. Therefore, the owner has the power to enforce the terms of that contract. (Refer to Chapter 2, "Basic Contract Law," for more information about this legal theory.)

Negligence: This theory is based on the allegation that the subcontractor's performance did not measure up to what would normally be expected of subcontractors in similar circumstances. (See Chapter 4, "Basic Tort Law," for additional information on negligence.)

Agency: Under this theory, the owner contends that he or she was the principal of the general contractor who, as the owner's agent, contracted with the subcontractor. (See Chapter 3, "Basic Agency Law," for more information about this legal theory.) Of the three theories mentioned, this one is the least applicable in light of the fact that most project construction teams are not really organized this way.

The General Contractor's Default

Sometimes general contractors default on their contracts with subcontractors. This may be the result of a default by the owner. For example, if the owner fails to pay the general contractor in a timely manner, the subcontractor may not be paid in a timely manner either. In other cases, a default by the general contractor may not be anyone else's fault. For example, the general contractor may fail to provide scaffolding for a masonry subcontractor, or may fail to coordinate the work of the various subcontractors, causing delays for the other workers.

The Subcontractor's Recourse

If the general contractor defaults in his or her obligations to pay for work that has been properly completed, subcontractors will normally be able to file claims for nonpayment. To assess and resolve such a claim, it is important to look to the prime contract and the subcontract to see which of the two common methods of payment apply.

One method treats each subcontract as a separate line item in the general contractor's requisition to the owner for payment. When that approach is used, it is easier for the subcontractor to justify full payment upon full performance. The percentage of each item of work that has been performed, as well as the related payment history, is clear. This method tends to be favorable to subcontractors who can demonstrate that any failure by the owner to pay the general contractor is not a result of problems with their work.

The other method establishes conditions which must be met before payment is due to the subcontractors. This method is favorable to the general contractor in those cases in which payment to the general contractor by the owner must be made before the general contractor is obligated to pay the subcontractors. In other words, if the general contractor does not have to pay the subcontractors unless and until the general contractor is paid by the owner, it is largely irrelevant that the subcontractors have properly completed their work.

There is an obvious unfairness, from the subcontractor's point of view, in using the latter method. In fact, recent court cases have generally ruled that it is reasonable to assume that parties who perform properly should be paid for their work. Therefore, if a subcontractor is *not* going to be paid for actual, proper performance of contract work, the contract language must clearly specify that the subcontractor knowingly assumed such risk of nonpayment. The rationale behind this ruling is based on a number of points:

- The subcontractor has little choice about the language used in the contract between the owner and the general contractor. Further, subcontracts often incorporate provisions of the prime agreement, about which the subcontractors have had no say. Also, as previously noted, the general contractor has quite a bit of leverage on the typical project, and can often impose his or her version of the subcontract on the subcontractor.
- The subcontractor usually has little or no chance to evaluate the owner's solvency. Therefore, it is illogical to assume that the subcontractor would implicitly base his or her right to payment on an owner's unevaluated ability to pay.
- It is unfair for the subcontractor to bear the full risk of nonpayment when the general contractor is in the best position to evaluate the owner's solvency and make provisions to assure payment.

Finally, when assessing whether or not the subcontractor is entitled to payment from the general contractor, the reasons for nonpayment are important. If the general contractor has not been paid because the subcontractor has not performed properly, then it is highly unlikely that the subcontractor can force the contractor to pay. On the other hand, if the general contractor is the one who performed improperly or failed to make reasonable attempts to be paid, or another subcontractor performed improperly, then the subcontractor normally is entitled to payment from the general contractor.

The Owner's Default

Subcontractors are not paid directly by the owner in the normal course of affairs, since their contract is with the general contractor. They are often not paid at all *unless and until* the general contractor is paid by the owner. Clearly this can be a problem for the subcontractor, who has little or no control over the owner-contractor relationship, or over the quality of the work of the contractor and the other subcontractors. The subcontractor may suffer as a result of failures in any of those areas which may affect payment by the owner.

The Subcontractor's Recourse

If the owner does not pay the general contractor, and neither the general contractor nor the subcontractor are at fault, the general contractor may have to pay the subcontractor despite not having been paid by the owner. Similarly, if the owner becomes insolvent or bankrupt and *cannot* pay, the general contractor normally bears the risk because he or she was the one who entered the contract with the owner and was in best position to judge the owner's financial capability.

A subcontractor may normally make a claim for nonpayment against almost anyone in the contract chain of which the subcontract is a part. Obviously, the subcontractor may sue the general contractor for breach of contract. Perhaps not so obvious is the fact that the subcontractor may sue the owner based on a theory of *unjust enrichment*. That simply means that it is unjust for the owner to obtain or retain benefits without paying a fair price for them. (See the discussion on quasi-contracts in Chapter 2, "Basic Contract Law.") Although this theory appears to be fair, the subcontractor could be denied the right to be paid because either:

- The owner can show that he or she has paid someone (usually the general contractor) and has not received anything for which he or she has not paid
- Retaining the benefits of the subcontractor's work is not unjust because the subcontractor can protect himself by using statutory remedies. The next section examines the types of remedies available to subcontractors on projects that have gone bad.

For a summary of the recourses of each of the participants, see Figure 15.2

Statutory Remedies

The legal rights typically available to subcontractors are not always of practical assistance to them. For example, one can say that a general contractor is legally responsible to pay for labor and materials supplied by a subcontractor but, realistically speaking, the subcontractor will not be paid if that general contractor is bankrupt and out of business. In recognition of such circumstances, state and federal laws have been passed which provide relief. These are called *statutory remedies*, and include the following:

- Mechanics' liens
- Stop notice statutes
- Trust fund statutes
- Compulsory bonding laws

Each of these is discussed below.

Mechanics' Liens

Mechanics' liens are created by statute based on the idea that an owner should not be enriched by a subcontractor's labor and materials without an enforceable obligation to pay for them. The lien gives the subcontractor a claim against the owner's property which, if unpaid, can result in a sale of that property (with the claim being paid out of the sale proceeds). (See Chapter 16, "Insurance, Bonds, and Lien Laws," for more details about mechanics' liens.)

Subcontractors' lien rights may arise when:

- The subcontractor's work has improved the owner's property
- The right to a lien is based on the subcontractor's legal right to payment from the general contractor. However, if the subcontract contained a payment condition, as described earlier, the subcontractor may lose his or her right to a mechanics' lien, because there would be no right to payment.

Unfortunately for subcontractors, mechanics' liens cannot be placed on public buildings. Further, the lien process can be quite complicated – exacerbated by the fact that most state lien statutes require strict compliance with the procedural rules if a lien is to be effective. Finally, a lien claim may be worthless because prior security holders (such as a mortgage lender or other contractors) have foreclosed and there are no funds left with which to pay the subcontractor's claim. (Refer to Chapter 9, "Financing," for additional information about foreclosures.)

Stop Notice Statutes

Stop notice statutes have been enacted in only about ten states. According to these statutes, a surety may advise the owner to stop paying the general contractor on a project that is behind schedule, over

Matrix of Subcontracting Remedies

Defaulting Party	Recourses		
	Subcontractor	**General Contractor**	**Owner**
Subcontractor		See contract with subcontractor	• Third-Party Beneficiary • Negligence • Principal – Agent
General Contractor	See contract with subcontractor		See contract documents
Owner	• Mechanics' lien • Stop notice statutes • Trust fund statutes • Compulsory bonding laws	See contract documents	

Figure 15.2

budget, or otherwise in trouble. Thereafter, money is to be held by the owner or paid to the surety so that the surety can protect its interests and its obligation to complete the work. This benefits the subcontractors indirectly because it tends to increase the money available for payments to them. If the owner continues to pay the general contractor after having received a stop notice, the surety will be released from its obligations to the extent of such further payments. Such action by the owner would typically strengthen potential claims by the subcontractors against the owner, if they are not paid by the general contractor or the surety.

Trust Fund Statutes

Trust fund statutes are similar to stop notice statutes. They require funds that the general contractor receives from the owner for work performed by subcontractors to be placed into a trust account. The general contractor, as trustee, must pay the subcontractors from the trust account funds and may not use the funds for general purpose spending. The value of this arrangement is that the money paid into the fund on account of the subcontractors' work is more likely to actually reach the appropriate subcontractors.

Compulsory Bonding Laws

Finally, in many states there are compulsory bonding laws. These compensate for the unavailability of lien remedies against public structures. The laws require that on public projects, the general contractor must post payment bonds to protect subcontractors against nonpayment by the general contractor. (For more information on bonds, see Chapter 16.)

Conclusion

Subcontractors are in an unenviable position in the legal scheme of a typical project. They normally do not have an opportunity to deal directly with the owner, yet their legal rights and obligations are substantially determined by the owner and passed down through the general contractor. Although they often have little negotiating leverage, they invest their money and effort and depend on the general contractor to obtain payment from the owner so that they, in turn, may be paid.

If they do not perform in accordance with their subcontracts, payments to them will be jeopardized. Of course this may be so even if they do perform properly. Unfortunately, the usual legal remedies may not be of much practical value. There are special statutory remedies that may be helpful, however, if a subcontractor is vigilant in asserting his or her rights.

CHAPTER 16

INSURANCE, BONDS, AND LIEN LAWS

CHAPTER 16 INSURANCE, BONDS, AND LIEN LAWS

Introduction

Individuals and business entities in the design and construction industry face a number of risks throughout the construction process. In order to protect themselves, they must first identify and determine the magnitude of the risk, and then either diminish, transfer, or insure against loss because of these risks.

Some risks can be diminished by contractual agreements that limit the amount of liability a party must assume. Occasionally it is possible to disclaim or avoid responsibility for specific damages that may occur. It is sometimes possible to transfer risk from one person or entity to another through indemnification agreements and other such devices (see Chapter 4, "Basic Tort Law," for more information about indemnification). This chapter discusses the three basic ways of managing risks in the design and construction industry:

- Insurance
- Bonds
- Lien laws

Insurance

Insurance is a type of contract in which one party agrees to indemnify (reimburse) another party for a specific loss. The party who agrees to pay is the insurance company (sometimes called the *underwriter*). The party who is protected by the insurance is referred to as the *insured*. The actual agreement between the insurance company and the insured is the *policy*, and the amount paid by the insured to the insurance company for the policy is called the *premium*.

There are four basic types of insurance coverages available to parties involved in construction:

- Business Insurance
- Workers' Compensation Insurance
- Professional Liability Insurance
- Property Insurance

Individuals and business entities involved in construction should consult with their insurance and legal advisors to determine which coverages are required, and in what amounts. More than the basic coverage may be required, depending on a project's nature and risks. Each of these four types of insurance is discussed below.

Business Insurance

Contractors and design professionals normally purchase insurance to protect against losses from damages arising out of their general business operations. Usual coverage includes:

- Comprehensive General Liability Insurance
- Contractual Liability Insurance
- Comprehensive Automobile Liability Insurance

Comprehensive General Liability Insurance

A standard form of general liability insurance, referred to as Comprehensive General Liability (CGL) Insurance, was developed by a national trade organization known as the Insurance Services Office. CGL policies insure against property damage and bodily injury. Typically, every business entity purchases this type of insurance.

Types of CGL Policies: There are two types of CGL policies: *occurrence policies* and *claims-made policies*. Originally, CGL insurance was written on an occurrence basis. This meant that the insurance policy that was in effect at the time the injury or damage allegedly occurred would cover the risk of loss, even if that policy had expired by the time the claim was filed. For example, if property damage allegedly occurred in 1980, the CGL policy that was in effect at that time would pay for resulting losses, regardless of whether or not the policy was still in force in 1989 when the claim was made.

In 1983, the Insurance Services Office developed a claims-made policy that could be used instead of the occurrence policy. This policy covers only those claims that are made while the policy is in effect. It allows the insurance company to select a *retroactive date*, and to deny coverage for bodily injury or property damage that occurs prior to the retroactive date. For example, the retroactive date may be the date when the insured first purchased CGL coverage, say, 1980. If the insured's CGL coverage is in force in 1989, it will cover a claim made in 1989 for damages that allegedly occurred in 1985. If the retroactive date was 1986, however, that claim would not be covered. The intent of the claims-made form is to decrease the insurance company's risk, which will tend to make policies less expensive. Both occurrence policies and claims-made policies are available, but not all states permit the use of the more restrictive claims-made policies.

In 1986 *Commercial* General Liability Insurance was introduced. This policy is similar to the *Comprehensive* General Liability policy, with some changes. Policy language was simplified and new restrictions were added. For example, pollutants that are emitted during the regular course of an insured's business are excluded from coverage under the Commercial policy. The policy does, however, cover pollution which is emitted suddenly and accidentally. (Refer to Chapter 12, "Building Regulations," for information about regulations on pollutants and hazardous wastes.) Depending on the individual policy, there may be other exclusions. The new policy can be written on either a claims-made basis or an occurrence basis.

Limits of Liability: The new Commercial General Liability policy limits the amount payable for each bodily injury or property damage claim. This policy also limits the total amount payable per year for all bodily injury and claims. The standard Comprehensive General Liability policy also establishes a limit per occurrence, but does not limit the total amount payable in a policy period (normally a year). The owner normally specifies the limits that the contractor must maintain for the project. Such limits should be determined based on the advice of the owner's insurance advisor.

Coverage: The standard Comprehensive General Liability policy protects the insured from losses due to damage to property owned by other persons or entities. For example, the CGL policy would cover damage to a fence on land that is adjacent to the project site if the damage was caused by a contractor's worker. It is important to note that the property must be *physically* harmed to qualify for coverage. The standard CGL policy does not cover loss of profits, loss of reputation, or loss of future business.

Claims for bodily injury are also covered by the CGL policy. *Bodily* injury refers to actual harm to a person's body. CGL policies do not pay for emotional distress or libel, which are considered *personal* injuries. (Refer to Chapter 4, "Basic Tort Law," for additional information about personal injuries.) Personal injury protection is a part of the Commercial General Liability policy, but must be added by endorsement to the older Comprehensive General Liability policies.

Exclusions: Although the standard Comprehensive General Liability policy insures against property damage, there are exclusions that substantially limit its scope. For example, the policy excludes coverage for damage to property "in the care, custody, or control of the insured or as to which the insured is for any purpose exercising physical control." This means that a contractor's CGL policy does not cover structures that he or she is building, because they are within his or her physical control. (Other types of insurance, particularly Property Insurance, is available to protect against damage to the actual construction work.) CGL policies also usually exclude coverage for the contractor's faulty workmanship, which would be covered by a performance bond or, perhaps, Property Insurance.

Finally, the standard CGL policy excludes coverage for explosion, collapse, and underground (XCU) risks. This exclusion relates to property damage arising out of blasting or explosion, damage to other structures adjacent to the contractor's work, and damage to buried pipe, conduits, and the like. Often it is important on projects in urban areas to have this exclusion removed by endorsement to the policy so that these risks will be insured.

Contractual Liability Insurance

The standard Comprehensive General Liability policy only insures against damages for which the insured is legally responsible. It does not cover liability that the insured voluntarily agrees to assume. Such liability is known as *contractual liability* because it exists only by virtue of having been agreed to in a contract. Insurance that protects against contractually-assumed risks is called Contractual Liability Insurance, and is another type of business insurance.

The standard form owner-contractor agreements published by the AIA and the EJCDC contain provisions that require the general contractor to indemnify the owner and design professional for claims, damages, losses, and expenses arising out of the contractor's negligent acts, or the negligent acts of anyone for whom the contractor is legally responsible (such as subcontractors). (See Chapter 4, "Basic Tort Law," for additional information on negligence and indemnification.) Contractors who agree to such indemnification clauses are not automatically covered by the normal CGL policy. They must obtain a contractual liability *endorsement*. Such an endorsement modifies the standard CGL policy to provide the required insurance coverage for the project.

Comprehensive Automobile Liability Insurance

A third type of business coverage is Comprehensive Automobile Liability Insurance. This insurance covers the insured for bodily injury and property damage caused by vehicles that are owned by the insured and used in the insured's business. The basic business auto policy can be modified by endorsement to include coverage for rented and leased vehicles, and other vehicles (such as employees' personal automobiles used in connection with the business) that are not owned by the insured.

Workers' Compensation Insurance

Most states have workers' (or workmen's) compensation statutes, under which employers are legally responsible for the costs of injuries that their employees sustain in the course of their employment. Employers are typically required to obtain Workers' Compensation Insurance to pay such expenses if they occur. The Workers' Compensation policy will pay, on the employer's behalf, whatever benefits are required under applicable state law. Generally, contractors, design professionals, and other business entities who hire employees are required to purchase Workers' Compensation Insurance.

The benefit of this insurance is that injured workers do not have to sue their employers to prove that their injuries resulted from the employer's negligence. Once a worker demonstrates that his or her injury occurred on the job, benefits are automatically paid. On the other hand, the statutes provide that injured workers who are covered by Workers' Compensation Insurance are *not* entitled to sue their employers for more money than was paid by the Workers' Compensation Insurance.

The idea behind Workers' Compensation Insurance is to ensure that benefits will be paid in a timely manner to all workers injured during the course of their employment. Without such insurance, an injured worker could be faced with delay in compensation, incur additional expenses, and suffer the uncertainty of the outcome of a lawsuit against the employer. In return for such reassurance and timely payments, workers must forego the potentially larger amount of money they might win if the matter went to court.

Sole proprietors and partners of a partnership are normally considered owners or employers, *not* employees. Therefore, they usually are not covered under state Workers' Compensation statutes. They must obtain accident and health insurance to protect themselves from the costs of work-related injuries. The officers of a corporation *are* employees of the corporation, however, and are protected by Workers' Compensation laws. Consequently, officers are covered under the corporation's Workers' Compensation Insurance policy for injuries sustained during the course of their employment.

Employer's Liability Insurance

Employer's Liability Insurance coverage is provided in Workers' Compensation Insurance policies. This insurance defends an employer against lawsuits based on negligence and other legal theories brought by employees who suffer job-related injuries or who contract a disease as a result of the working conditions of their job.

As stated above, if an injured worker collects benefits under Workers' Compensation Insurance, he may not sue his employer for additional damages, if any. Some states, however, allow employees (prior to an injury) to waive benefits available under Workers' Compensation statutes. The employee would then have the right to sue the employer if injured during the course of employment. This may be an attractive alternative in states where the maximum amount paid under Workers' Compensation Insurance is unlikely to compensate fully for an injury. The employer's insurance company would defend the employer and pay the employee the amount for which the employer is obligated.

Even in those cases in which the employee is covered by Workers' Compensation Insurance and is precluded from suing the employer for job-related injuries, the employee's spouse or child(ren) may still sue the employer, claiming that they suffered damages (such as loss of the emotional support normally provide by the employee) as a result of the employee's injury. Employer's Liability Insurance would also compensate such a person if they are legally entitled to compensation, on behalf of the employer.

Professional Liability Insurance

Professional Liability Insurance protects against losses and expenses associated with liability for performing or failing to perform professional services. The scope of coverage depends on the terms of the individual policy and the information that the design professional provides in his or her application for insurance. Many insurance policies make the completed application an integral part of the policy.

In the application, the design professional provides information about his or her practice, including the types of services performed and the nature of the projects with which he or she is associated. This information allows the insurer to evaluate the potential risks and establish the proper annual charges (the *premiums*). For example, a structural engineer is likely to pay proportionately higher premiums than an electrical engineer because structural engineering services have historically been associated with a higher frequency and magnitude of claims than those of electrical engineers. Similarly, a firm whose practice is heavily involved with the design of condominium developments will probably face relatively high insurance costs due to the vast amount of litigation associated with such projects.

If the design professional performs a service that was not described on the application, the insurer could refuse to cover claims arising from that service. Further, the insurer could deny coverage because information in the application was inaccurate or omitted. Finally, the design professional must include in the application information about pending claims to which he or she is a party, if any, as well as circumstances that may lead to a claim in the future.

Design professionals normally carry Comprehensive General Liability Insurance (CGL insurance). However, the CGL policy does not include Professional Liability, which is written on a separate policy sometimes referred to as an *errors and omissions*, or *malpractice*, policy. CGL and Professional Liability Insurance policies are complementary and do not overlap.

Exclusions

Like all insurance policies, Professional Liability Insurance contains some exclusions from coverage. For example, losses that are covered under other insurance policies, such as CGL, Workers' Compensation, and

Employers' Liability, are excluded from coverage under normal Professional Liability policies. These exclusions prevent overlapping coverage. Other areas commonly excluded from coverage are services that design professionals in general do not ordinarily perform, and services that the insured in particular does not ordinarily perform.

Intentional Acts: Professional Liability Insurance policies normally will not provide coverage for a design professional who intentionally causes harm or damage to another person or entity, whether by fraudulent, criminal, dishonest, malicious, or other types of actions.

Contractual Liability: Design professionals should not agree to assume more liability than they would typically be responsible for under normal legal principles in the usual performance of professional services. Generally, design professionals are liable for their own negligent practices as well as for those of persons and entities for whom they are vicariously liable. (See Chapter 3, "Basic Agency Law," and Chapter 4, "Basic Tort Law," for additional information on negligence and vicarious liability.) A design professional who agrees to accept liability for the acts or omissions of other individuals or entities, or who promises in their contract to be more careful and skillful than could reasonably be expected of other design professionals practicing in similar circumstances, will not be covered by Professional Liability Insurance.

Warranties and Guarantees: Warranties and guarantees are also excluded from normal Professional Liability Insurance coverage. For example, if an architect tells his client that the basement of his new building will not leak, the architect will not be covered for claims arising out of this guarantee if the basement does leak. If the basement leaked due to the architect's negligent failure to specify waterproofing, but the architect made no guarantee that it wouldn't, he would be covered by the Professional Liability policy.

Promises to meet specific deadlines and estimates of construction cost are similar to warranties and guarantees and are also excluded from Professional Liability Insurance coverage.

Joint Ventures: Another common exclusion is for liability resulting from acts as a member of a joint venture. (See Chapter 1, "Basic Business Structures," for more information on joint ventures.) Insurance coverage for joint ventures can normally be obtained by endorsement to the Professional Liability policy if all joint venture members are insured and the insurance company is advised of the details of the joint venture's activities. In that case, the insurance company has an opportunity to evaluate the risks and may impose a proper charge for such an endorsement.

Hazardous Materials: Claims arising from a design professional's involvement with asbestos, PCBs, pollution, and other hazardous and toxic wastes are excluded from coverage under normal Professional Liability Insurance policies. (See Chapter 12, "Building Regulations," for additional information about environmental regulations with respect to toxic or hazardous substances.)

Limits of Liability

Every Professional Liability Insurance policy has a limit to the amount the insurance company will pay on an individual claim, and to the total amount of all claims paid in a policy period (usually one year). Design professionals should consult their insurance brokers and attorneys before deciding on the limit to request. Factors that should be considered when determining the appropriate policy limit are the firm's size and volume of business, and the types of projects with which the firm is associated. In general, the higher the limit, the higher the annual premium.

If a design professional carries $250,000 of Professional Liability Insurance, and there is a claim that results in a judgment against the design professional for $300,000, the insurance company will only pay a maximum of $250,000. The design professional must pay whatever *deductible* is provided for in the policy – for instance, $5,000 (deductibles are discussed in the next section). Furthermore, the fees of the defense attorneys are subtracted from the policy limits. Thus, if the attorneys' fees are $60,000, only $190,000 out of the $250,000 policy limit will be available to pay the injured party. In total, the design professional would have to pay $115,000 on this claim: $300,000 (the amount of the claim) – $190,000 (the amount available after attorneys fees) + $5,000 (the deductible), plus the cost of the insurance premiums for that year.

Deductible

Every policy contains an amount that the insured is responsible for paying, if necessary, on each claim. This is called the *deductible*. Assume, for example, that an engineer has a policy with a deductible of $5,000, and a claim is made alleging negligence. If it can be settled for $3,500 plus $500 in attorneys' fees, the engineer would have to pay the full $4,000 (since it was less than the $5,000 deductible), and the insurance company would pay nothing. After the engineer pays the deductible amount, the insurance company will pay additional costs and expenses up to the policy limits. The deductible applies to every claim, no matter how many are made in a year.

Policy Forms

Professional Liability Insurance is presently issued only on a claims-made basis. This means that the policy covers only those claims that are made against the insured during the policy period. This can be quite risky for design professionals. Assume that an engineer was insured from 1980 through 1986, but was unable to renew his policy in 1987 or 1988 because the insurance company went out of business and no other company was writing new policies. If a claim is made in 1988 for damages resulting from negligent services performed in 1985, the engineer will *not* be covered. Although the engineer was insured when the services were performed, he was not insured when the claim was made and therefore has no coverage. To be safe, design professionals must maintain insurance as long as they practice, plus the period of time thereafter in which claims may still legally be made.

Handling Claims

Design professionals are required to notify the insurance company as soon as they receive notice of a claim. A claim is typically defined as a demand made upon the insured for money or services. For example, a claim may be a letter from a client to an architect alleging negligence or improper performance of services for which the client demands money or free services. A claim does not have to be in writing; verbal demands may be sufficient. Certainly, commencement of an arbitration proceeding or a lawsuit against the design professional is a claim under the policy. To be safe, design professionals should notify the insurance company even if there is doubt as to whether or not certain circumstances constitute a claim. Delay in reporting a claim can hinder the design professional's legal defense and could cause the insurance company to refuse coverage.

After the insurance company is notified of a claim, it arranges for a meeting with the insured and conducts an investigation. The insurance company assigns a defense attorney to handle the design professional's defense of the claim. Some policies provide that an insurance company cannot settle a claim without the insured's consent. Insurance companies tend to look at situations in terms of dollars and cents. However, design professionals have their reputations to think of as well. It is possible that the insurance company will offer to settle a claim to avoid the potential for a larger judgment, while the design professional may want to vindicate his or her good name. If a design professional does not agree to a settlement, and the insurance company is forced to continue to defend the claim, the insurance company will not be responsible for paying the amount, if any, by which the final judgment exceeds the amount for which the case could have been settled. For example, if the case could have been settled for a payment of $25,000, but the design professional refuses and ultimately is found liable for $40,000, the insurance company will only pay $25,000 out of the $40,000 judgment. In addition, the design professional must pay the costs and expenses incurred in continuing the defense beyond the date when the case could have been settled.

Property Insurance

The fourth type of insurance commonly used in construction is Property Insurance. Property Insurance policies insure against physical damage to a structure under construction caused by such perils as fire, windstorms, lightning, and explosions. The Property Insurance required under the standard form construction contracts published by the AIA and the EJCDC is also referred to as *Builder's Risk Insurance*. Although the terms of Builder's Risk policies vary, they generally cover damage to the permanent structures under construction, including additions, attachments and permanent fixtures, as well as temporary structures, materials, equipment, and supplies which are used on the project. The contractor's machinery and tools, if they are located on or within 100 feet of the construction site, are also covered under Builders' Risk Insurance.

The Builder's Risk policy becomes effective either at the start of construction or at a time specified in the policy. Coverage ceases upon occupancy of the structure or completion of construction, whichever occurs first.

Who Purchases Property Insurance?

The AIA and EJCDC forms require the *owner* to purchase Property Insurance. Under these forms, legal title to the materials and equipment used in the project passes to the owner upon payment or incorporation into the project, whichever is earlier. (Refer to Chapter 5, "Basic Real Property Law," for information about fixtures and ownership of real property.) Therefore, it is proper for the owner to insure his or her own property.

It is possible to modify the standard forms to require the contractor to purchase the necessary property insurance, with the cost of the policy to be included in the contract price. This decision is normally based on whether the owner or the contractor can purchase the required insurance more economically.

Policy Forms

Property Insurance may be written on either a *named-peril* policy or an *all-risk* policy. A named-peril policy covers only those risks that are specifically named in the policy. An all-risk policy covers damage from all perils except those that are specifically excluded. The all-risk policy offers somewhat broader coverage than the named-perils policy, and is much more common. Damage from fire, smoke, lightning, windstorms, hail, and explosions is usually covered under an all-risk policy, whereas damage from catastrophic natural occurrences, such as earthquakes and volcanoes, is normally excluded.

Regardless of which policy form is used, Property, or Builder's Risk, Insurance is normally written on a *completed value* basis. That means that the estimated value of the project upon completion is the limit of the potential payments under the policy. The insured should take care to adjust the initial policy amount if subsequent changes in the scope or quality of the project increase the insurable value.

Exclusions

Normal all-risk policies do not cover losses resulting from design defects or substandard construction workmanship. Such losses must be insured under the architect's or engineer's Professional Liability Insurance policies, the contractor's CGL policy, or the performance bond.

Deductibles

The AIA and EJCDC standard form contracts specify that the contractor is responsible for the normal deductible that must be paid when claims are made. The amount of the deductible (usually about $500) should be stated in the contract documents. If the owner chooses a higher deductible than is mandatory, in an attempt to reduce premium costs, the contractor should not be expected to pay the higher amount unless notified of that fact in the contract documents.

For a summary of the coverage and exclusions of each type of insurance described above, see Figure 16.1

Subrogation

Frequently, an individual or business entity has a legal right to be reimbursed for losses incurred because of another's wrongdoing. (Refer to Chapter 4, "Basic Tort Law," for a discussion of these rights.) If insurance is available to reimburse the injured party for their loss, that person or entity need not take legal action against the wrongdoer. The insurance company, however, has the right to substitute itself for the injured party, who it insured and paid, and sue the wrongdoer to recover its costs and expenses. This substitution of rights is referred to as *subrogation*.

Parties to a contract may waive their subrogation rights. Such a waiver is contained in the standard form documents published by the AIA and the EJCDC. Under these documents, the owner, design professionals, and contractor all waive rights against each other and their respective subcontractors and agents for damages covered by Property Insurance. The effect of such a waiver is that once the insurance company compensates the insured for its damages, it may not sue the party who caused such damages. The idea behind the waiver is to prevent extended litigation. The insurance company was paid a premium to cover such losses, and that should be the end of it.

However, parties who intend to agree to a waiver of subrogation should be sure that their insurance policies allow such an action. Some all-risk Property Insurance policies stipulate that the policy will be ineffective if the insured agrees to a waiver of subrogation.

Bonds

A *bond* is a legal instrument under which one party agrees to answer to another party for the debt, default, or failure to perform of a third party. The purpose of bonds, also referred to as *surety bonds*, is to minimize or eliminate financial loss to the owner if the general contractor does not enter into a contract, does not perform the work properly, or does not

Comparing Insurance Coverages

Policy	Coverage	Exclusions
Comprehensive General Liability Insurance	• Damage to property owned by others • Bodily injury	• Personal injury* • Property in the care, custody, or control of the insured • Faulty workmanship • Coverage for explosion, collapse, and underground risks
Contractual Liability Insurance	• Voluntarily assumed liability	
Comprehensive Automobile Liability Insurance	• Bodily injury and property damage caused by vehicles owned by the insured	• Rented or leased vehicles*
Workers' Compensation Insurance	• Injury to employees sustained during the course of their employment	
Professional Liability Insurance	• Liability for performing or failing to perform professional services	• Coverage under other policies • Services the insured does not ordinarily perform • Intentional acts • Contractual liability • Warranties and guarantees • Joint ventures • Hazardous materials • Others
Property Insurance (Builder's Risk)	• Damage by fire, windstorm, lightning, or explosion to structures under construction	• Damage from earthquakes and volcanoes • Losses resulting from design defects or substandard construction workmanship

**Can be added by endorsement.*

Figure 16.1

pay his or her subcontractors or suppliers. Bonds are purchased from commercial bonding companies called *sureties*.

Sometimes the general contractor will require his or her subcontractors to obtain bonds. These are called *subcontractor bonds*. They offer to the general contractor the same security with respect to the acts and omissions of the subcontractors that the general contractor's bonds offer to the owner.

Types of Bonds

There are three major types of bonds commonly required on construction projects:

- Bid bonds
- Performance bonds
- Payment bonds

Each of these is defined in Figure 16.2 and examined below.

Bid Bonds

A bid bond is a contract between the bidding contractor, the owner (or awarding authority), and the surety. The surety stands behind the bidder's promise that, if awarded the contract, he or she will actually enter into a formal contract with the owner for the amount bid. If the successful bidder fails to enter into the contract, the owner will be paid the difference in cost between the low bid and the second lowest bid, plus costs associated with awarding the contract to the second lowest qualified bidder. Such payments may not exceed the face amount of the bond (called the *penal sum*). The request for, and amount of, the bid bond is normally stated in the invitation to bid. This information is available to all parties interested in bidding on the project.

Assume, for example, that Bidder A was the low bidder with a bid of $495,000, and that Bidder B was the second lowest bidder at $510,000. Each bidder was required to provide a bid bond in the amount of 5% of their bid. Bidder A was awarded the contract but, prior to signing the

Types of Bonds	
Bid Bond	The surety promises to pay the difference between the contractor's bid and the next lowest bid if the contractor refuses to enter into a formal agreement with the owner.
Performance Bond	The surety promises to perform, or to pay the cost of having another perform, the construction contract as required, if the contractor fails to do so.
Payment Bond	The surety promises to pay for labor and material used in the performance of the construction contract if the contractor fails to do so.

Figure 16.2

formal agreement, he discovered a major error in his bid. He then refused to sign what he knew would be an unprofitable contract.

The owner proceeded to sign a contract with Bidder B for the $510,000 that she bid. The owner can make a claim against Bidder A's bid bond for the $15,000 difference between the lowest and second lowest bids. If Bidder A provides a bond with a face amount of, say, $24,750, the owner can collect the $15,000 and begin the project in the same financial position he would have been in had Bidder A not backed away from his original bid price.

Performance Bonds

Under a performance bond, the surety guarantees that the general contractor will perform all of the terms in the construction contract. Frequently, there is the misconception that the surety only stands behind the obligations to perform construction work. In fact, the surety is backing the full performance of all of the general contractor's obligations under the construction contract. This includes warranty obligations, obligations to provide insurance, indemnification obligations, and other such matters contained in the construction contract.

Performance bonds are written for either 50% or, more often, 100% of the contract price. The amount is important to know because the face amount of the bond is the limit of the surety's liability under it. For example, assume that a performance bond was obtained in the face amount of $300,000, which was the full contract price. If the contractor defaults after being paid $200,000, and it costs an additional $125,000 to finish the project, the surety would be required to pay $25,000, which is the difference between the contract price and the final cost of construction.

The owner must always pay the amount that he or she contracted to pay, but is not normally expected to pay more. In the example above, the owner had to pay the full $300,000 contract price: $200,000 to the original contractor and $100,000 to the surety. The surety paid the second contractor $125,000 to finish the project.

In practical terms, if there is a 100% performance bond on a project, the ultimate cost of a project on which the contractor defaults could be as much as 200% of the contract price before the owner will suffer any financial loss as a result of the contractor's failure to perform.

It is usually not practical to write a bond for more than the contract price. Even a contractor who defaults early in the project typically has accomplished something of value, and the project can usually be completed for the face amount of the bond (the original contract sum) or less.

Payment Bonds

A payment bond, sometimes referred to as a *labor and materials payment bond*, is the surety's guarantee to the owner that the contractor will pay for labor and materials used in the performance of the contract. This type of bond is typically written in conjunction with the performance bond, and is issued at the same time (often on the same form) without any additional premium.

If the contractor on a private-sector construction project fails to make payments, the subcontractors and suppliers may look to the surety for payment. This is usually a more attractive solution for both parties than placing and defending liens on the owner's property. (Liens will be discussed in greater detail later in this chapter.)

The Surety's Role

The surety's role is defined in the bond itself. One widely used standard bond form, which illustrates the typical role of the surety, is AIA Document A312, *Performance Bond and Payment Bond*, 1984 Edition. This form was developed in conjunction with other professional organizations, contractors' organizations, and the surety industry.

Based on the provisions of this document, if the contractor defaults (assuming that the owner has not defaulted), the surety must take one of the following actions:

- Arrange for the original contractor to complete the contract. This would only be done with the consent of the owner, and in a situation where the contractor was doing work of acceptable quality but was having cash flow problems that affected subcontractor relations and schedules. Once the surety steps in to solve the cash flow crisis, the contractor may be perfectly capable of completing the project.
- Take over and complete the project, either by itself or through agents or independent contractors. The owner would be required to pay the remaining balance due on the original construction contract, as described above.
- Find a successor to the defaulted contractor, through competitive bidding or other methods. The surety would then have to arrange for a new contract between the successor and the owner, with appropriate new surety bonds.
- Pay the amount for which it is liable under the bond (after due investigation). This may include paying for the cost of correcting defective work, paying legal and design professionals' fees resulting from the contractor's default, paying delay costs resulting from the contractor's default, and paying actual damages (or liquidated damages, if any) caused by the contractor's nonperformance or delayed performance of the construction contract. Keep in mind that these obligations are normally limited to the face amount of the bond. The owner is required to pay the balance of the contract price to mitigate his or her damages.
- Deny liability in whole or in part.

Bonding Requirements

Owners in the private sector may decide for themselves whether or not to require bonds from contractors. Owners in the public sector do not have that flexibility; certain bonds are required of contractors by law. The bonding requirements on private- and public-sector projects are discussed below.

Private-Sector Projects

On privately funded projects, the owner and the owner's advisors (such as design professionals, legal counsel, and insurance counsel), determine whether or not bonds are needed, the types of bonds, and the qualifications of the surety(ies) chosen to provide them. Typically, the owner states in the invitation to bid whether or not the general contractor must obtain bonds, the types of bonds required, and the amounts of each. Thereafter, the owner pays indirectly for the cost of the premiums for the bonds, since bidders include such costs in their bids. This is only fair because, although the contractor is the one who secures the bonds, the owner is the one who benefits from them – in that the surety promises to perform for the owner if the contractor fails to do so.

The following steps may be taken to minimize the need for a bond(s):

- Carefully selecting the general contractor reduces the likelihood that the contractor will fail to perform his or her obligations.
- A carefully managed payment plan may decrease the risk that subcontractors and suppliers will not be paid. Such a plan may call for *dual payee checks* (checks made out to both the contractor and the subcontractor/supplier, which cannot be cashed by the contractor alone), or *lien waivers* with each payment application. (Lien waivers are discussed in more detail later in this chapter.)

An owner who does not require a bond(s) for his or her building program may eventually have to pay for some losses if a contractor defaults on one of the projects. Over a long period, however, such losses, if any, may be less than the total cost of the bond premiums.

Public-Sector Projects

The *Miller Act*, enacted by Congress in 1934, requires performance and payment bonds to protect subcontractors and suppliers on federally funded construction projects, since statutory mechanics' liens are not available against public buildings. (Refer to Chapter 7, "Basic Labor Law," for a more detailed discussion of the Miller Act.) These bonds provide subcontractors and suppliers with additional security (in lieu of liens) for payment of their claims.

Changes in the Construction Contract

Traditionally, the law has treated any change in the basic obligations of the principal on a bond (normally the general contractor) as grounds for releasing the surety from its obligations. This was based on the assumption that sureties were volunteers, such as friends and relatives of the principal, not commercial operations charging fees for their services. Because of the personal and uncompensated nature of their commitment, the courts felt it was unfair to hold the surety to his or her promises if the terms of the underlying arrangement changed in any way that might increase the surety's risks.

Over the years, this assumption proved to be incorrect. Today, sureties are usually profit-oriented commercial enterprises, and the attitude of the courts has changed so that now most performance bonds state that the surety "waives notice of any alteration or extension of time made by the owner" and of other changes in the construction contract which may affect its risk. The courts reason that because of the size and complexity of today's construction projects, it would be a considerable inconvenience if the owner or contractor was required to notify the surety and obtain its consent for every change that occurs during project delivery.

Filing Claims

Statutes determine who can file claims on bonds required for publicly funded construction projects. On privately funded projects, the terms of the individual bonds specify who may file claims under them. In general, however, the following is true.

The owner is the beneficiary of the general contractor's performance bond, and may file suit under it if the contractor does not perform fully according to the terms of the contract. In those cases in which the general contractor has, in turn, required subcontractors or suppliers to obtain performance bonds, the subcontractor bonds benefit the general contractor, and the surety promises to pay the contractor in the event that the subcontractors or suppliers do not perform fully under their contracts.

If a payment bond is provided by the general contractor to the owner, the subcontractors and suppliers can file suit directly under it if they are listed as *claimants* (those people or entities who have a direct contract either with the contractor or with a subcontractor on the bond. In that event, they may enforce the surety's obligations under the bond if the general contractor does not pay them. If a payment bond is provided by a subcontractor to the general contractor, the situation is similar with respect to claims of sub-subcontractors.

Time Limitations

It is common for performance bonds to require that legal action be commenced within two years after 1) the contractor defaults, 2) the contractor ceases working on the project, or 3) the surety refuses or fails to perform under the bond.

Typical payment bond provisions require claimants to give certain notices of their claims, as described below. Subcontractors (those with a direct contract with the general contractor) must notify both the surety and the owner before filing suit. Those who do not have a direct contract with the general contractor (sub-subcontractors) must notify both the general contractor and the owner within 90 days after last having performed labor or supplied materials or equipment for the project.

If claimants give the proper notices and are still not paid, they have one year from the date of the notice, or the last date on which they performed labor or provided materials, whichever is earlier, in which to bring a claim against the bond.

Defenses

If a claimant does not comply with the bond requirements, the surety may deny liability. For example, if the claimant did not perform its contractual obligations fully and properly, if proper notice was not given, or if a lawsuit was commenced after the time established in the bond, the surety could refuse to pay the claimant. Courts often interpret bond requirements in favor of claimants, however, by requiring the surety to show if, and how, it was adversely affected by the claimant's noncompliance. For instance, formal notice may not have been delivered on time, but if it was only a day or two late, or if the surety actually knew from the contractor or otherwise that the claimant had not been paid, it is unlikely that the delay had any detrimental effect on the surety. Since the surety was paid for issuing the bond and was not adversely affected by such noncompliance, it should not be allowed to escape its obligations.

Because the surety takes the place of the contractor upon default, most of the defenses that the contractor would have against claims of the owner (such as, interference, failure to pay, defective construction documents, and so forth) are equally available to the surety. Similarly, defenses that the general contractor could assert against a claimant (such as delay, failure to comply with the requirements of the contract documents, and the like) can usually be asserted by the surety, as well.

Reimbursement

After the surety company pays the owner or claimant, it usually will seek reimbursement from the defaulting general contractor. Unfortunately, it is often impossible for a general contractor who has defaulted to reimburse the surety. After all, if the contractor had the money, he or she probably wouldn't be in default. To protect itself, the surety requires that, upon default of the general contractor, the owner send to the surety all future payments, if any, that would ordinarily have gone to the contractor.

Additionally, in case of default by the contractor, the surety usually demands that *retainage* held by the owner at the end of the job be paid to it. Such funds belong jointly to the general contractor and the surety, and should *never* be paid to the general contractor without the surety's consent. If they are, the surety may be relieved of its obligations in later claims on the theory that its risks were substantially increased without its consent. In other words, the retained funds, which would have been available to the surety to pay claims, were paid instead to the contractor by the owner, to the detriment of the surety.

Lien Laws

Lien laws have been enacted in each state to secure payment for labor and materials used to improve real property. Commonly referred to as *mechanics' liens*, these laws were first enacted in the early years of this country's history to encourage artisans, craftsmen, and others to work on the buildings which were required by this nation to grow and prosper. Such workers did not otherwise have effective remedies available to them if their clients failed to pay. They were neither interested in instituting, nor able to afford, lengthy lawsuits to collect their money. Lien laws provided a practical solution.

Lawmakers recognized that once labor had been performed and materials provided and incorporated into some improvement to real property, such labor and materials could not readily be repossessed or reclaimed if the property owner failed to pay for them. Accordingly, lien statutes were conceived based on notions of restitution and the prevention of unjust enrichment. (Refer to Chapter 2, "Basic Contract Law," for additional information about these concepts.) In short, the legislators decided that an owner should be required to pay for the labor and materials that have improved his or her property.

Interestingly, there are certain aspects of lien laws which tend to work against each other. On the one hand, they need to be interpreted liberally in order to achieve their purpose. This means that the courts will try not to allow technicalities to preclude workers from establishing valid lien claims. On the other hand, since lien laws are purely statutory and have no basis in the common law, they are to be strictly construed, meaning that a party seeking the benefit of the law must comply strictly with its requirements.

Courts normally resolve these competing considerations by demanding that the requirements prescribed by the legislature in each jurisdiction for establishing a lien be met precisely. Once the lien is established, however, the courts will normally do their best to produce equitable (fair) results for the lienholder. If the owner does not pay the lienholder, the owner's property to which the lien has been attached will be sold, with the proceeds of the sale used to satisfy the lien.

Who Can Claim a Lien?

Most statutes allow any person who performs specific work or provides materials with respect to *improvements* to real property to claim a lien. For example, general contractors, subcontractors, sub-subcontractors, construction managers, material suppliers, and design professionals are among those types of people or entities who can claim a mechanics' lien in most states.

Improvements

The term *improvement* generally refers to buildings or structures, although courts have been rather liberal in defining the term. For example, roadways and utility lines are now considered to be improvements, too. A properly filed mechanics' lien will attach to the portion of the improvement on which work was actually done, and it may extend (depending on the particular state law) to the whole building site and as much of the adjacent property as is reasonably critical for the use of the improvement.

Not all improvements are subject to liens, however. For example, public property is generally not lienable. (See Chapter 7, "Basic Labor Law," for information about the Miller Act.) Private property is generally lienable if the owner either authorizes or knowingly allows the work to be performed. If the owner hires the contractor, authorization is reasonably clear. If a tenant orders improvements, the landlord's property may be subject to liens if the landlord consented to the work being done. In some jurisdictions, it is enough that the tenant improvements are made with the "knowledge" of the landlord. In other jurisdictions, it is necessary to demonstrate express approval (such as written permission) by the landlord before his or her property may be subject to a mechanics' lien.

Services, Materials, and Equipment

When design professionals perform services for a project that never proceeds to the construction phase, it is questionable whether or not their services, in and of themselves, constitute an improvement. If the construction activity to which the drawings and specifications relate has at least begun, it is easier to prove that there has been an improvement to which the lien may be attached.

It is necessary to look at the language of each applicable lien statute to determine what types of services, materials, and equipment are considered improvements and, therefore, are lienable. Generally, the materials must actually become a part of the improvement, and the services must contribute to the creation of the improvement, if they are to qualify for a lien.

Usually it is reasonably clear whether or not materials and equipment qualify for a lien. Depending on the specific statute, the following categories of work are typically lienable:

- Repairs
- Landscaping and excavation
- Materials delivered to the site but not incorporated into the work
- Specially fabricated materials and other materials stored off the site (The applicability of this category varies from state to state, often depending on whether or not the materials are usable for other projects.)

The following categories are usually *not* lienable:

- Fuel and consumable supplies
- Demolition

In some states, essentially all services performed by the architects, engineers, and other professionals working on a project are lienable. In other states, merely preparing drawings and specifications is not enough to establish a lien claim, whether or not the project proceeds into the construction phase. The reasoning is that the services do not constitute visible work on the premises. In many states, on-site observations of construction by architects and engineers, sometimes inaccurately called "site supervision," are considered lienable services because they contribute to the successful completion of the project.

Establishing a Lien

Normally there are several steps in the process of establishing a mechanics' lien. Each step must be completed properly and within the applicable time limits if the lien is to be effective. Those steps and the usual time limits are described in the following sections.

Pre-Claim Notices

Those who intend to claim a mechanics' lien are typically required by law to notify the owner prior to filing the lien. However, those who deal *directly* with the owner, such as general contractors, are usually not required to give formal notice of the fact that they have supplied labor and materials on the project. The theory is that the owner knows of their existence as contractors on the project and the scope of their potential claims. Others, like subcontractors, who have a contract with the general contractor only, and who normally do not deal directly with the owner, typically must file pre-claim notices to alert the owner to the fact that they have supplied labor and/or materials to the project and to the potential scope of their lien claims.

The information to include in the notice is normally specified in the statute itself. In general, the following items are required:

- An accurate description of the property sufficient to identify it clearly
- A description of the work performed and/or the materials supplied
- The last date on which work was performed or materials furnished
- The net amount claimed

It is important to comply with all notice requirements that appear in statutes for filing notices and lien claims. Otherwise, it is possible that lien rights will be forfeited.

Statement of Claim

Once the requirements for any pre-claim notices have been met (including, perhaps, recording in the public records), the actual lien claim may be prepared and filed. This is called the *statement of claim*. Lien claims are generally required to be verified (notarized by a notary public) and recorded with the appropriate public documents. In this way, the owner is notified, as are third parties, that a claimant has attached a lien of a specific amount to the property. Failing to file the lien claim in a timely manner according to the requirements of the applicable statute will probably result in forfeiture of such lien rights.

Enforcing Liens

Each statute defines the limited time period during which mechanics' liens are valid. The lienholder must enforce the lien prior to the expiration date. Liens are typically enforced by suing the owner. The claimant should also sue all who have an interest in the property, including all other lien claimants. Legal procedure requires that all controversies arising out of a single situation be resolved in one proceeding, if possible. If all interested parties are included in the lawsuit, a final judgment may be issued.

Waiving Lien Rights

It is usually permissible for a contractor, subcontractor, or other party to waive its right to file a mechanics' lien. The waiver can be made at nearly any point during the course of a construction project. By definition, to waive something is to intentionally relinquish a right or claim that you know is yours. Therefore, an effective lien waiver must be based on a party knowingly giving up their right to file a lien. In other words, it must be done voluntarily. A lien waiver that is fraudulently induced is not enforceable. To be considered fraudulent, the owner must have given information to the contractor with the intent to deceive. The contractor must then have relied, to his or her detriment, on that information, and without knowledge of the facts.

There are *total* lien wavers and *partial* lien wavers. In other words, the waiver may cover all, or part, of the work performed or materials supplied by the potential claimant. Normally, partial waivers cover a period of time, such as "up to the present requisition for payment." In any case, what appears or is presented as a partial waiver of liens is sometimes far more extensive and may, in fact, be a total waiver. It is important to read the language of every lien waiver carefully to avoid waiving more rights than intended.

Sometimes owners include lien waivers in the owner-contractor agreement itself. If the language of such a waiver is clear, it will usually be upheld. Some states prohibit these "no-lien" contracts, however, because they believe that they contradict the general public policy of the state. Subcontractors should be wary of any lien waiver in the general contractor's agreement with the owner. The subcontractor may well be bound to the same waiver due to the fact that the owner-contractor agreement is frequently incorporated into the subcontract by reference.

Releasing Liens

Unlike a lien waiver, which states that no lien claims will be made in the future, a *release* is made after the claim has already been established. If a lien has been recorded and the claimant is subsequently paid, the lien may be removed from the property via a release. Since a release is based on satisfaction of the lien claim, once it is given, the lien cannot normally be revived.

Conclusion

It is impossible to eliminate risk, particularly in construction. This chapter has described three basic ways of *managing* many of the risks that arise out of design and construction activities.

The materials in this chapter are not part of the day-to-day experience of most design and construction professionals. Insurance policies and construction bonds are complex documents with which few people are well versed. Similarly, lien laws are very detailed, they differ substantially from state to state, and they must be followed precisely by those who would take advantage of their provisions. Most design and construction professionals turn to insurance and legal advisors for help in untangling what can often be a complicated and precarious legal and financial situation.

However, it is important for all design and construction professionals to have a basic understanding of the risk management devices discussed in this chapter. Those who ignore the risks they face in business are living on borrowed time. Those who understand such risks, and the basic ways in which they can minimize or compensate for them, will improve their business' chances to grow and prosper.

CHAPTER 17

SOURCES OF DISPUTE

CHAPTER 17 SOURCES OF DISPUTE

Introduction

A construction site is fertile ground for disputes. They seem to sprout and grow like weeds, with the potential to choke the life from a project. Project participants who are aware of the sources of dispute are better able to evaluate the legal implications of their situation, and the options available for addressing the conflicts. Four common sources of dispute during the construction phase of a project are:

- Contract changes
- Differing site conditions
- Time for performance
- Payment

Each of these is examined in the sections that follow.

Contract Changes

In construction, a change is a deviation from the scope of work defined in the drawings and specifications, which were the basis of the original contract. There are many reasons why owners request changes in the construction work. It may be as simple a matter as changing his or her mind about some part of the project; or perhaps an error or omission is discovered in the documents; new materials or construction techniques may become available; or maybe unanticipated site conditions demand that the changes be made. One thing is certain, however: There will be changes made during the construction phase of every project. If the associated modifications to the contract documents are made improperly, there could be serious legal repercussions for the parties involved.

In general, changes or modifications to a contract require mutual consent and legal consideration in order to be binding. In other words, both the owner and the contractor have to agree on the precise nature and extent of the changed work, the cost of such work, and the time required to perform the work. (See Chapter 2, "Basic Contract Law," for information about required elements of an enforceable contract.)

Unfortunately, coming to an agreement on these matters is not always easy. Because of this, owners typically include in their construction contracts clauses giving them the right *unilaterally* (*without* first obtaining the contractor's consent) to order a contractor to make changes within the general scope of the contract and pursuant to all of its terms and conditions. These clauses are called *changes clauses*.

Changes Clauses

The driving force behind changes clauses is the owner's desire to maintain design flexibility, to control costs, and to minimize delays in the construction schedule. Changes clauses give the owner the power to order work to be performed even if the contractor disagrees on the price and required time for performance of such work. This power is not unlimited, however. The changes must be *within the general scope of the work* being performed under the contract.

For example, if the contract work includes a 10,000-square-foot roof, and the owner wants to expand it to a total of 12,000 square feet, this work appears to be within the general scope of the contract. In such a case, the owner may instruct the contractor to perform the changed work, *according to all the terms and conditions of the original contract*, and the contractor must do so. This provision is important because without it, the contractor would have enormous economic leverage, able to negotiate a favorable deal every time the owner wanted or needed to make a change from the original contract documents. This leverage comes from the fact that the contractor is practically the only one in a position to accomplish the changed work in the middle of the project. In many instances, the contractor must do the work ordered by the owner regardless of whether or not the appropriate price and time changes are then agreed upon. Just in case the parties cannot agree, the contract must contain a mechanism for calculating adjustments to the contract price and time for performance.

Types of Contract Changes

It is important to distinguish three specific types of contract changes from normal, or generic, contract changes. They are *cardinal changes*, *minor changes*, and *constructive changes*. They illustrate the legal significance of the changes clause in the administration of the construction phase of a project. Each of these is shown in Figure 17.1 and described below.

Types of Contract Changes

Change	Description
Cardinal Changes	• Beyond the general scope of the contract
Minor Changes	• Consistent with contract documents • Involves no additional time • Involves no additional cost
Constructive Changes	• Where circumstances of the project warrant an adjustment to time and price, even though no change was ordered

Figure 17.1

Cardinal Changes

Changes clauses attempt to determine whether certain work is considered an "extra," or whether the work is part of the original contract requirements. Changes that are *beyond* the general scope of the contract are called *cardinal changes.* Whether or not specific work is a cardinal change is something that must be decided on a case-by-case basis. Contractors cannot be ordered, without their consent, to perform cardinal changes. This is because the pricing, financial, and risk allocation provisions of the original contract that apply to normal change order work (work that is within the general scope of the contract) do not necessarily apply to cardinal changes.

Minor Changes

Construction contracts often give the owner the right to order, through the design professional or another representative, *minor changes* in the contract work. Minor changes:

- Are consistent with the intent of the contract documents
- Involve no additional time for completion
- Involve no additional cost to the owner or contractor

Examples of minor changes include, perhaps, changing from one standard paint color to another, or relocating a door along the same section of wall.

Constructive Changes

Where the circumstances of the project entitle the contractor to an adjustment in contract price or time, such adjustment should be made, even though no change was explicitly ordered. This is called a *constructive change.* Such circumstances frequently arise when:

- A contractor is required to perform work different from, or faster than, the work or speed required by the contract documents
- A change order should have been issued, but was not

For example, a contractor may request a change order for a time extension due to changes in the work which the contractor considers to be extra, and which will take an additional two weeks to perform. If the owner and design professional delay in approving the change order, the contractor may, as a practical matter, be forced to accelerate his work to meet the original scheduled completion date (on the chance that the change order will not be approved). In this case, the contractor would probably be entitled to be paid for the costs associated with accelerating the work, even if the time extension is ultimately granted. Although neither the owner nor design professional ordered the contractor to accelerate, the circumstances of the delayed approval created a situation in which the contractor had to do so, or risk breaching his contract by not completing the work within the contract time. A court will usually assume that the contractor did what was appropriate under the circumstances and will enforce a change in the contract price and time.

Authorizing Contract Changes

Who has the authority to order changes to the construction contract? Clearly, the owner has authority to do so. The owner's agents or representatives, including the architect or engineer, often have authority to order changes as well. There are different kinds of authority, however, as discussed below.

Apparent v. Actual Authority

Sometimes representatives who have *apparent* authority to order changes do not have *actual* authority to do so. Agents have apparent authority when the acts of the owner lead the contractor to believe that the agent has authority to order changes in the work. For example, a design professional has apparent authority when the owner knows he or she is executing change orders and makes no objection; or when the owner pays the contractor for changed work based on change orders issued by the design professional. To avoid disagreements or unpleasant surprises, contractors are well advised to determine who has the actual authority to order changes. (See Chapter 3, "Basic Agency Law," and Chapter 10, "Professional Services Agreements," for more information on these topics.)

The Change Order Process

A *change order* is a legal document presented to a contractor directing that a change be made from the original drawings, specifications, or other contract documents. The change order process is usually quite formal. There are several practical rules that must be followed: 1) prior notice must be given; 2) the change order must be in writing; and 3) change orders must be signed by the owner, design professional, and, perhaps, the contractor. Each of these formalities is discussed below.

Prior Notice

It is important for the owner to know if the contractor perceives requested changes as changes in the scope of the work which require more time or money. Therefore, most construction contracts require prior notice from the contractor if he or she intends to claim additional time or money on account of changes in the work. Given proper notice, the owner may decide that the work is not needed after all; the owner or design professional may offer a cheaper alternative for accomplishing such work; or the owner may defer or delete other work to stay within the budget originally established for the project.

In general, such notice must be given *prior* to starting work on the changes, and within a specific period (usually 21 days) after the conditions giving rise to the claim occur or are first discovered. The exception to this is a situation in which the safety of people or property is at risk. In an emergency, prior notice is usually not necessary for the contractor to be paid for extra work.

Writing Requirement

It is important that change orders be in writing. Most contracts state that no changes or modifications to the contract documents will be valid unless they are in writing. Regardless of whether or not such a provision is part of the contract, it makes good sense to document all agreements–particularly those that address a volatile and dispute-prone aspect of construction contracting.

Signature Requirement

Under the standard form AIA and EJCDC documents, both the owner *and* the design professional are required to sign change orders. The design professional's involvement is necessary to verify that the changes do not violate codes or other building regulations, and do not in some other way threaten the integrity of the original design. The design professional also tends to act as a "flywheel," keeping emotions at a steady level and advising the owner against unwise or precipitous actions.

The contractor should also sign the change order if he or she agrees with the proposed change in the contract price and time for performance. Of course, if the contractor does not agree, he or she should not sign. The disagreement over price or time will need to be resolved at some future date in accordance with the other terms of the contract. However, regardless of whether or not the contractor signs the change order, the work must be performed by the contractor, as ordered by the owner.

Waiving Change Order Formalities

Unfortunately, the formalities so carefully established and documented in the construction contract are often ignored in practice. When this happens, the question is whether or not such formalities have been *waived*. This is important in cases in which a written change order was not signed by the owner but alleged extra work was done by the contractor. In such cases, the contractor might be entitled to the reasonable value of the work performed, according to quasi-contractual principles. (Refer to Chapter 2, "Basic Contract Law," for additional information on quasi-contracts.) Furthermore, it is important to determine whether or not there was a waiver in circumstances in which no formal notice was given, but where the design professional was well aware of the circumstances that the notice would have described.

In public-sector construction contracts, formalities usually may *not* be waived, and the courts are willing to assume that they have not been waived. In private-sector work, the answer is not so clear. Typically, only those with the authority to order changes have the authority to waive formalities. Of course, the owner has the authority to waive provisions of the contract which are in his or her favor, as does the contractor. Those problems that do arise often result from oral orders given by the design professional or the owner's project representative (presumably as the owner's agent) which are later disavowed by the owner.

Courts are reluctant to enforce formalities if the purpose for which they were established has been achieved and it would be inequitable to enforce the letter of the contract. Assume, for example, that the engineer and owner's representative are at the construction site when rock is discovered. The contractor's superintendent points out the condition and states that extra time and money will be required to remove the rock. If the engineer and owner's representative instruct the contractor to proceed, it is likely that the contractor could recover his or her costs and some profit, despite the fact that prior notice was not given and there was no written authorization from the owner to proceed.

Each case must be examined individually. The courts are aware, however, of the tendency of construction professionals in the field to concentrate on the physical problems of building the project, not on the paperwork. Courts, therefore, are frequently willing to waive formalities, either by ruling that the parties had established an agreed-upon course of conduct, or by using some other legal rationale. If the owner's rights have not been violated, the contractor, in most cases, would be entitled to be compensated fairly for changes in the work.

Pricing Work Changes

Assuming that a proper procedure is followed for implementing a change order, the changed work is generally priced by the following methods:

- By a lump sum price agreed upon by both parties
- By unit prices (preestablished in the contract or agreed upon at the time) for each unit of changed work
- By applying a fixed or percentage fee to the actual cost to the contractor of the changed work, and then by adding cost and fee together

- By the design professional or, possibly, the construction manager, based on the contractor's itemized costs, plus an allowance for overhead and profit. When a single change order includes both additions and deductions of work that is of a substantially similar nature, it may not be clear how to calculate overhead and profit. One method is to add together the pluses and minuses, and apply overhead and profit to the net amount only.

Ideally, the parties would agree upon a lump sum price for the changed work and a specific time adjustment would be established before the work was begun. This is not always possible, however, since the scope of changed work is not always known at the beginning. In those instances, unit prices or time and materials charges plus overhead and profit often work well. The last method described above is considered "fail safe" in that there is an established pricing procedure even when no other agreement is possible.

If proper procedures are not followed, or if additional work is beyond the general scope of the contract, claims for payment for such extra work are frequently based on *quantum meruit* ("the value of the thing"). To be paid on that basis, a contractor must demonstrate to the court that:

- The work for which payment is sought was outside the scope of work required by the contract
- The owner asked for, or demanded, the extra work or services
- The owner agreed, either by words or conduct, to pay for the extra work or services
- The extra work or services were not furnished gratuitously
- The extra work or services were not required due to any fault of the contractor

The contractor must clearly demonstrate that the owner ordered or consented to the extra work or services. The owner has a right to control his or her financial obligations and should not have to pay for extra work that was neither wanted nor requested.

Time Changes

In addition to changes in price, changes to the work also frequently result in changes to the allowable time for performance of the contract (to account for the fact that more or less work than originally planned must be accomplished). Most standard form construction contracts state that the contract time may be altered by change order only.

The problem of delay is addressed later in this chapter. It is sufficiently serious and involves such potentially substantial amounts of money that it merits a discussion all its own. For now, readers should understand that each change order should resolve all time-related issues associated with changes in the contract work.

Differing Site Conditions

Contractors base their prices and schedules on the information provided by the owner in the documents that are part of the construction contract. Some of that information relates to the site conditions that the contractor can expect to encounter. If the actual site conditions *differ* in any substantial way from what the contractor could reasonably expect based on his or her knowledge and the information given in the construction documents, changes will probably be required to the contract price and time for performance.

There is potential for dispute with respect to: 1) the conditions which should have been expected; 2) whether or not the differences, if any, are substantial; 3) how much time and money are due the owner or contractor as a result of the differences; and 4) the legal implications of the provisions in the contract which allocate the risks in these matters.

There are three broad categories of site conditions changes. They are:

- Subsurface conditions
- Conditions within an existing structure
- Acts of God

Each of these is discussed below.

Subsurface Conditions

In the construction process, there is always a risk that unexpected, preexisting subsurface or subsoil conditions will be discovered. When this happens, the contractor's costs and time required for performance of the agreed-upon work can increase dramatically.

A subsurface investigation program, including test borings, can reduce the possibility of surprise to some extent. However, test borings of a single area, or borings performed at spaced intervals across a larger area, do not always give a true indication of the subsoil condition of the entire construction site. Furthermore, because of their expense, only a limited number of borings can usually be made for any given project. Unfortunately, the fewer dollars spent in investigation, the greater the risk of discovering unanticipated conditions.

The owner normally provides survey and soil information to the bidders on each project. However, as part of the bidding and construction documents, the owner may grant them permission to conduct independent testing of the subsoil conditions. Unfortunately, bidders normally have neither the time nor the capital to test the soil for each project on which they intend to submit a bid. Therefore, they often bid on a project knowing little more about the actual conditions than what is stated in the information provided by the owner.

Allocating Risk

Given all this, how does the law allocate the risk associated with the discovery of unanticipated subsurface conditions? Generally, unless the owner provides information about subsurface conditions *upon which the contractor is entitled to rely*, a contractor working under a fixed-price contract bears the risks associated with unexpected subsurface conditions. The contractor must perform all increases in the amount of work and bear all of the expenses associated with the unexpected conditions.

The contractor may be entitled to relief in the following extreme cases:

- *Commercial impracticability*: where performance by the contractor is only possible at a cost that is so high it would not be commercially viable to proceed
- *Mutual mistake*: where both parties thought they knew what the circumstances were and contracted accordingly, only to find out they *both* were wrong.

In short, if the contract documents are *silent* about subsurface conditions, and the contractor encounters a condition that does not change the work as shown in the drawings and specifications, but merely makes the job more expensive or difficult to do, then it is likely the contractor *cannot* recover additional costs from the owner. On the other hand, if the unanticipated condition makes it necessary to change the drawings or specifications, then a change order should be issued and the contractor should be paid for the change and, if appropriate, granted a time extension. This is illustrated in the following paragraph.

Assume that the contract documents on a particular project do not address subsurface conditions at all or, at least, to the extent that the contractor may reasonably rely on the accuracy and completeness of the information given. The design of the project requires excavation for a basement. If the contractor encounters rock where no rock was anticipated, and if the contractor must remove the rock in order to proceed, he would probably not be entitled to extra time or money associated with the rock removal, since no changes were necessary to the original drawings and specifications. On the other hand, if the contractor encountered soil whose load-bearing value required a change in the foundation design, the contractor would have a valid claim for extra time and money spent as a result of the finding.

Can the contractor enforce a promise by the owner to pay additional compensation for an unforeseen condition? Probably not. Usually, an owner's promise to pay a contractor additional compensation for the same work that was required in the original construction documents is unenforceable. This is based on the principle that the contractor had a preexisting duty under the original contract to perform the work, and thus gave no consideration for the owner's promise to pay. (Refer to Chapter 2, "Basic Contract Law," for additional information on the legal consideration in a contract.)

In cases involving subsurface conditions, however, the courts will often enforce such a promise as long as the resulting modification to the contract is fair and equitable in view of circumstances that were unanticipated by the parties when the contract was made.

Disclaimers

Disclaimers and *exculpatory language* (provisions designed to shift responsibility or avoid liability in a particular circumstance), are often included in the contract documents to prevent the contractor from relying on whatever information is provided by the owner, and to shift the risk and the consequences of differing site conditions to the contractor, as described in the previous section.

To determine whether or not disclaimers or exculpatory language will protect the owner, consider the following factors. Silence alone is generally not actionable. This means that if the owner provides absolutely no information on the subsurface conditions, the contractor will be responsible for whatever conditions are discovered, as previously described.

It is critical to distinguish between *silence*, *partial disclosure*, and *misrepresentation*, however. If the owner did not provide accurate information, it is important to know whether or not the inaccuracies were a result of innocence, negligence, or intentional conduct, and whether or not there was a reasonable opportunity for the contractor to conduct independent tests of the site.

If the information provided by the owner was accurate but incomplete, and the contractor had the opportunity to make an independent investigation, the contractor will normally bear the risk of discovering unexpected conditions. On the other hand, a contractor has the right to terminate the contract and to be reimbursed for any losses he or she may have incurred, if the owner intentionally misled the bidders with inaccurate or deceptive information. A disclaimer cannot shield an owner from fraud or bad faith.

Of course, many owners believe that it is in their best interest to allow the contractor an equitable adjustment in price and time for performance if the subsurface conditions are substantially different from what was anticipated. They reason that if contractors must build contingencies into their bids to cover subsurface conditions that may or may not occur, then the owner will pay for such conditions whether or not they ever occur. In the long run, it is cheaper for the owner to pay only for those conditions that actually do change.

Conditions Within an Existing Structure

Sometimes hidden and unanticipated conditions are encountered during renovations of, or connections to, existing structures. For example, it might be normal to expect that concrete of a specific density would have been used for a type of structure and that it would have been only lightly reinforced with steel. If part of a contractor's work is to make openings in such concrete, he or she will probably raise the issue of changed conditions when it becomes apparent that the concrete is twice as dense as expected, and heavily reinforced.

When conditions *within* an existing structure or facility are found to be different than expected, the analysis is basically the same as that for changes in subsurface conditions. Such changed conditions create the same potential for dispute, have the same effect on the contractor's price and time for performance, and are subject to the same legal analysis as unexpected subsoil conditions.

Acts of God

Finally, there are those events that are beyond human control. These are commonly referred to as *acts of God* (or *force majeure*). Examples of so-called acts of God include:

- Fire
- Flood
- Tornados
- Adverse weather conditions that could not be reasonably anticipated (such as 10 inches of rainfall in a month that has historically received an average of 6 inches)

These events are no one's fault, and so both the owner and the contractor must bear their own burdens. Typically, this means that the contractor will be allotted extra time for completion, if required; and he or she will be compensated for direct costs incurred as a result of the *force majeure*, but will not receive any damages for delay.

Time for Performance

Time is an important factor in every construction project. Construction contracts commonly specify the time available for the contractor to perform the work, and the consequences of failing to do so. Many disagreements arise over questions about time extensions and the costs associated with them. The following material addresses disputes over changes in the time for performance and the legal implications of such changes.

Commencement of Work

The construction phase of a project usually *begins* either upon signing the construction contract or upon the contractor's receipt of a *notice to proceed.* A notice to proceed is often used to mark the start of the construction phase if the construction site is not available for some reason at the time the contract is signed, if the owner needs additional time to file financing documents, or for some similar reason. A notice to proceed allows the owner some flexibility with respect to the starting date – but not without limits. The contract will normally include a definite period within which the notice must be issued. Without it, the contractor's cost estimates might become outdated by a delay in the start of the project.

Completion of Work

The construction phase *ends* at completion of the work specified in the contract. However, in calculating the time for performance, "completion" typically means "substantial completion." According to the AIA standard form of *General Conditions* (A201, 1987 Edition), substantial completion is achieved on the date "when construction is sufficiently complete, in accordance with the Contract Documents, so that the Owner can occupy or utilize the Work or designated portion thereof for the use for which it is intended." (Refer to Appendix B for a sample copy of this document.) This definition reflects common practice. Under the widely used standard form documents, the design professional determines when substantial completion has been achieved.

Many people equate substantial completion with "beneficial occupancy" and the issuance from a local government authority (usually the building inspector) of a *certificate of occupancy*. However, a certificate of occupancy may not have any bearing on whether or not a construction contract is substantially complete. In many cases the minimum amount of construction that must be in place to satisfy the local officials that the structure is fit for human occupancy may be far less than the amount of work required under the construction contract.

The time span allotted for construction work may be expressed in two ways. It is possible to state a precise date by which the work must be completed. This requires either a fairly accurate knowledge of when the contract will be signed, or a willingness to accept a certain amount of flexibility with regard to construction *duration*, as long as the work is completed by the specified date.

More commonly, the construction time is stated as a specific number of days available to complete the contract work. When a number of days is stated, the contract should specify whether time will be measured in *calendar days* or *working days*. AIA and EJCDC standard form documents specify calendar days; and it is generally agreed that calendar days are easier to use than working days. Among other reasons, they prevent arguments about which holidays are working days, whether or not a rain day is considered a working day, how Saturdays are counted, and so forth.

Clearly, durations or completion dates are listed for a reason. The owner desires to have the work completed by the stated deadline, beyond which he or she will probably incur some inconvenience and additional costs. In response to this, it is common to see the clause "time is of the essence of the contract" in construction contracts. Those are magic words that elevate *any* delay in completion to a material breach of the contract. If not for the inclusion of this clause, minor delays in completion would not be considered substantial enough to justify termination of the contract, nor would they entitle the owner to reimbursement of costs or other damages due to the delay.

Progress Schedules

Given that the construction contract has a definite starting date and completion date, how is time managed between these dates? The answer is that, absent any other provisions, contractors are required to work at a reasonable pace and to complete the project by the appointed date. In reality, it may be quite difficult to determine what constitutes a "reasonable pace." On all but the simplest projects, contractors are required to prepare *progress schedules*.

Types of Progress Schedules

The two most widely used methods of scheduling the progress of a project are:

- Bar charts
- Critical path method (CPM) schedules

Bar Charts: Bar charts show the duration of the various construction activities plotted against a time line. Unfortunately, other than when they are scheduled to occur in time, no relationship is apparent among the activities depicted in the bar chart. It is impossible to determine with any certainty whether the start of one activity depends upon the completion of any other activity(ies). As a result, it is nearly impossible to determine the effect that a delay to any specific activity will have on the project as a whole, using this method of scheduling. Figure 17.2 illustrates a bar chart.

Critical Path Method (CPM): In part because of the limited usefulness of bar charts, critical path method (CPM) schedules have been developed to document the relationship between the activities (which activity must be completed before a subsequent activity may begin), and the planned duration of each activity. From that information, the "critical path" of events through this network can be calculated. The critical path is the shortest route (time-wise) through the network, and shows which activities must be completed on schedule if the overall completion date of the project is to be met. Figure 17.3 illustrates a typical CPM schedule.

CPM schedules are usually stored on computers and updated frequently to reflect actual events. Today it is possible, even on personal computers, to monitor manpower and equipment requirements, as well as the cash flow associated with each event.

Review and Approval of Progress Schedules

The owner and/or design professional may reserve in the contract the right to review or approve of the contractor's proposed schedule. Their aim is not to perform the contractor's job or to relieve him or her of responsibility for managing the project, but to verify that the contractor has developed a realistic and workable schedule. Design professionals are frequently leery about becoming involved in scheduling matters for fear of liability exposure if the contractor does not meet the schedule, or if it proves to have been inaccurate or ill-conceived. They worry that the owner will hold them partly responsible for the contractor's performance under the schedule, since they reviewed it and agreed that it was reasonable. Therefore, disclaimers and exculpatory clauses abound with respect to progress schedules.

Interim Dates

However established or approved, the interim dates on a progress schedule usually have little contractual significance. Although important milestone dates (by which certain aspects of the work are to be completed) may be established and enforced, it is the overall completion date that is normally most important. Contractors typically need flexibility in dealing with the interim schedule to respond to the actual unfolding of events.

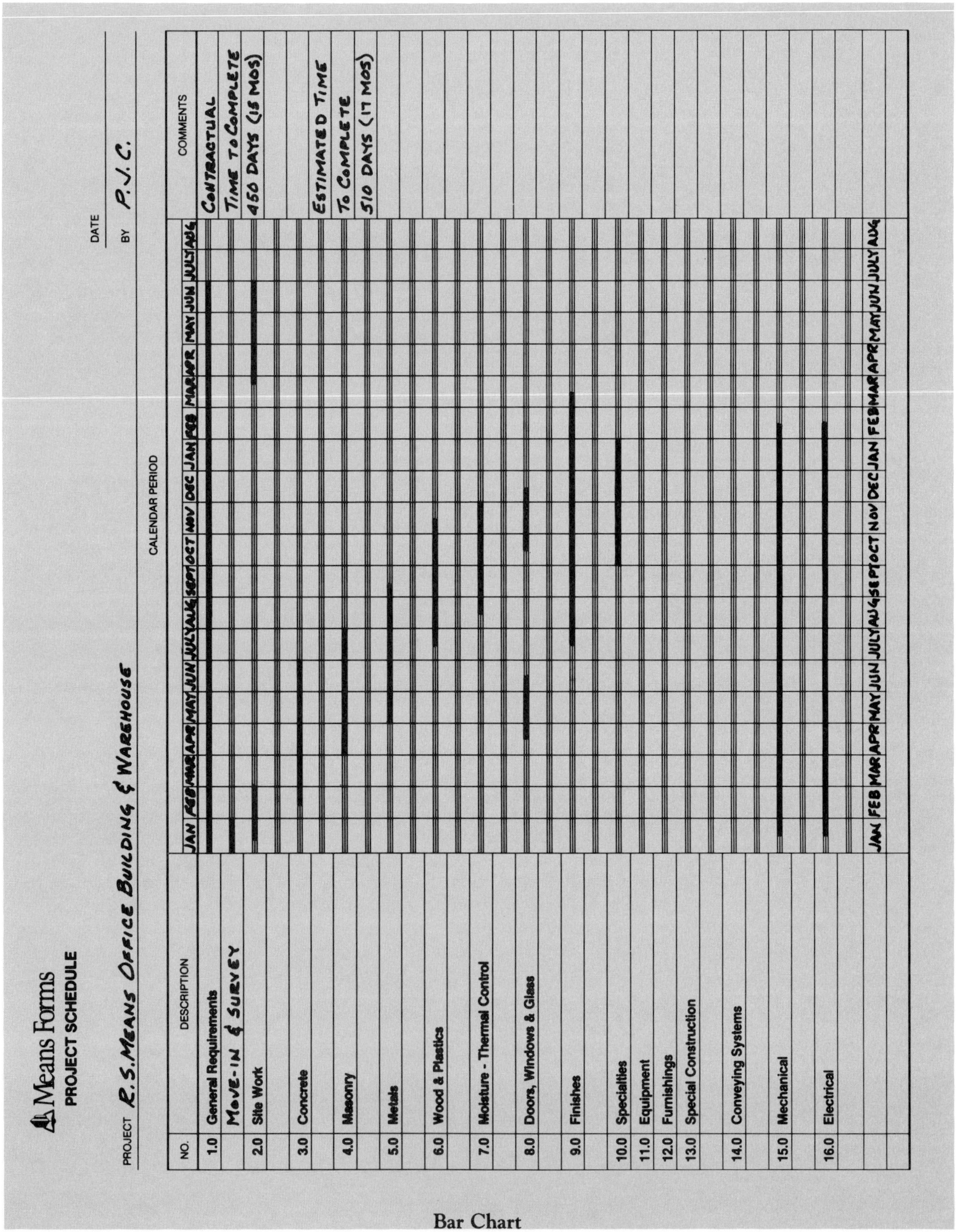

Bar Chart

Figure 17.2

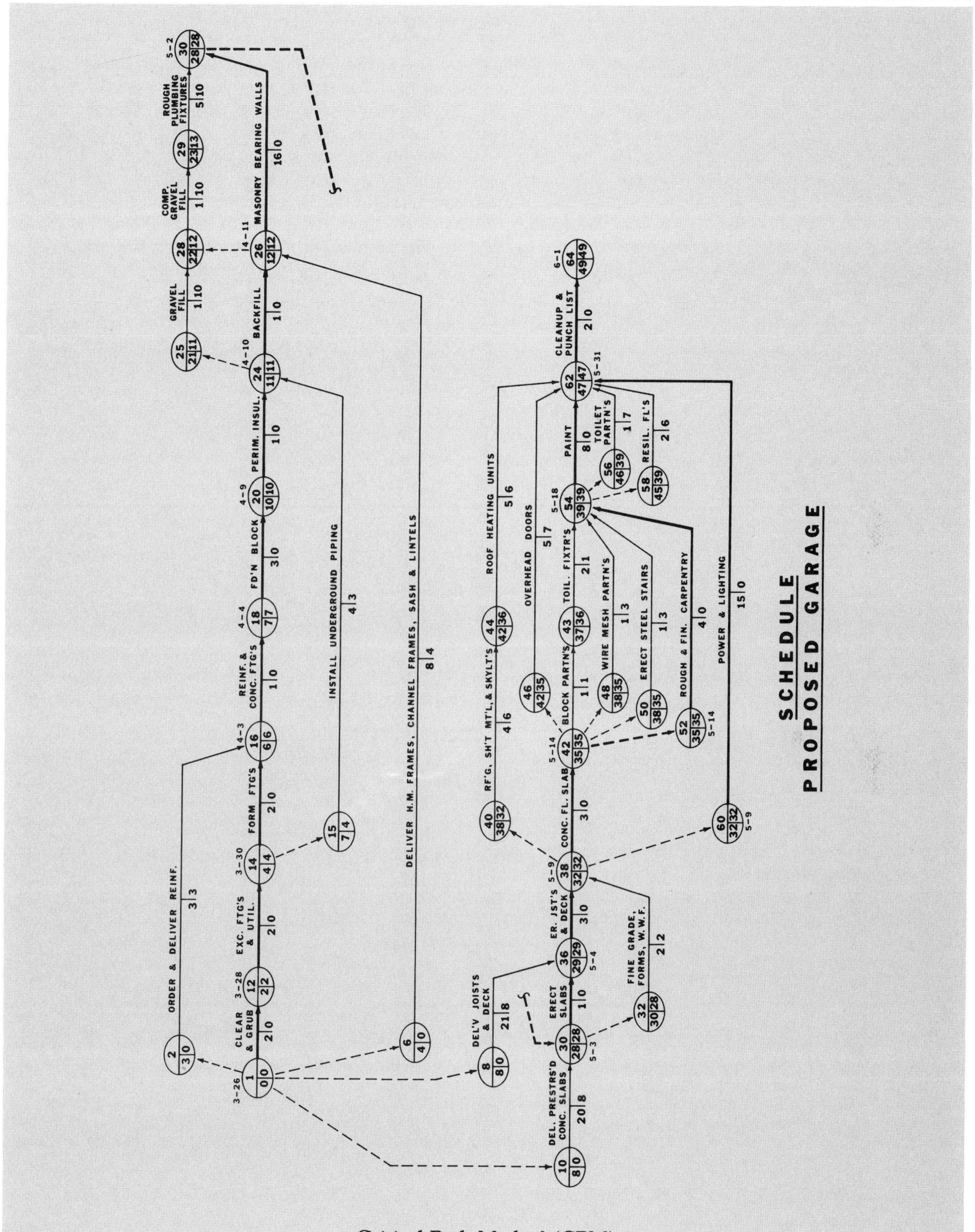

Critical Path Method (CPM)

Figure 17.3

Schedule Float

There is a continuing debate over who has control of the *float* in a schedule. On a CPM schedule, certain activities are "critical," in that a delay in completing any one of them will delay the overall project. The other activities are "noncritical," meaning they contain a certain amount of float. The float is the difference between the time noncritical activities actually take to complete, and the time available to complete them. For example, a concrete foundation may take 5 days to pour. If that work can be accomplished any time within the next 10 days without delaying subsequent activities, it contains 5 days of float. The work could start on day 0 and finish on day 5, or it could start on day 5 and finish on day 10, or any combination in between. See Figure 17.4.

The law varies on who "owns" the float. As a general rule, the contractor decides how it is used. This may change, however. The 1987 edition of AIA's *General Conditions* (AIA Document A201, reproduced in Appendix B) provides that the owner has the right to suspend all or

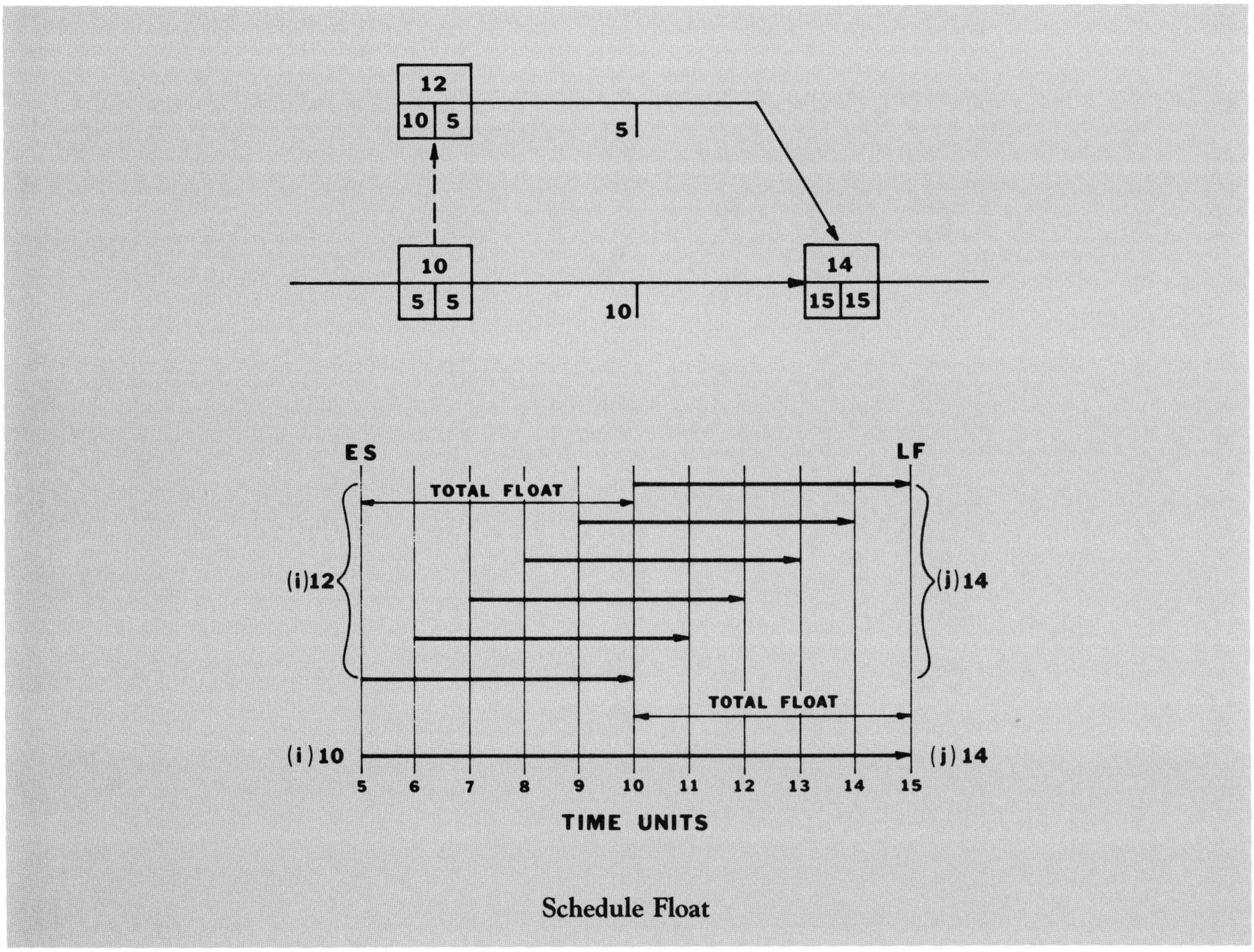

Schedule Float

Figure 17.4

part of the work at any time, with or without cause, as long as he or she is willing to pay the costs associated with such action. It appears therefore, that the owner has the power to dictate the use of the float in a schedule if he or she so desires.

Acceleration

Generally, an owner has the right to order a contractor to accelerate the pace at which the work is being performed. Acceleration is usually considered to be a "change" under the changes clause and would entitle the contractor to an equitable adjustment to the contract price. Of course, if the acceleration is necessary to get the project back on schedule after delays for which the contractor was responsible, then there would be no additional compensation.

If the owner orders an acceleration that is so extreme it is beyond the scope of the original contract requirements, the request should be treated as a cardinal change. This means that the contractor would have the right to refuse to accelerate the pace without an agreement with the owner for additional compensation.

Delays and Disruptions

Delays are treated differently than most other breaches of contract. Time is generally not considered to be equivalent to labor or materials. Normally, there is no bargained-for exchange of time for money. Therefore, a mere delay in the construction schedule usually does not automatically terminate the delaying party's right to continue performance. It's a good thing, too, because delays are a part of almost every construction project.

Types of Delays

Delays can be characterized in three basic ways:

- Excusable delays
- Inexcusable delays
- Compensable delays

Excusable Delays: The first category of delay is one for which the delaying party (frequently the contractor) has a valid excuse. An excusable delay is caused by 1) a reasonably unforeseen future event that is 2) beyond the contractor's control (the contractor could not have prevented it or continued working because of it) and which 3) occurred without the fault or negligence of the contractor. Generally, an excusable delay entitles the contractor to an extension of time to complete the contract, although the contractor may or may not be entitled to additional compensation as a result of the time extension.

The following events are normally valid excuses for delay:

- Acts of God (*force majeure*)
- Labor disputes
- Unusual delays in transportation

Inexcusable Delays: An inexcusable delay is one for which the delaying party (the contractor, for example) has no legally recognized excuse. The contractor accepted the risk that the events which caused the delay might occur. When they did occur, the contractor had no one else to blame.

Since not all delays entitle a contractor to additional time or money, it is necessary to distinguish between those risks that the contractor accepted under the contract, and those that he or she did not accept. Examples of risks that are normally assumed by the contractor include:

- Shortage of capital
- A subcontractor's breach of contract
- Failure to staff the project properly
- Failure to order materials on time
- Failure to coordinate subcontractors properly

Delays caused by such events are generally considered to be inexcusable.

Either party may have the right to terminate the contract for the substantial, inexcusable delay of the other. In many cases, a party may be entitled to reimbursement for reasonably foreseeable losses if they were caused by another party's inexcusable delay.

Compensable Delays: A delay to either party may be compensable. Inexcusable delays are never compensable. Some excusable delays are compensable. Compensable delays for a contractor are generally caused by the owner or someone for whom the owner is legally responsible.

For example, compensable delays may be caused by:

- The owner's failure to provide access to the site when required
- Correcting work based on faulty design or incorrect construction documents
- Waiting an inordinately long time for shop drawings to be reviewed and approved
- Changes made to the work

Compensable delays entitle the contractor to compensation in the form of additional time and additional money.

Allocating the Risks of Delay

To determine whether or not specific delays are excusable, it is important to determine how the risk of the delay was allocated under the contract. In general, if the contract is silent on this issue, the contractor is considered to have assumed most of the risk. Contracts usually do not completely ignore the issue of risk allocation. Contract language can define the specific risks assumed by each of the contracting parties. An example of this is the *"no damages for delay" clause* described next.

"No Damages for Delay" Clause: This type of clause is intended to place the financial risk of delay on the contractor. It denies a contractor any compensation for delay other than an extension of the time available to complete the work. The following is an example of such a clause:

> *The contractor agrees that delays caused by any act or failure to act by the owner or any of the owner's representatives will be fully compensated by an extension of the contract time, and in no event will the contractor be entitled to money damages due to such delays.*

This type of clause is generally enforceable, but only in strict accordance with its terms. In other words, the courts will not strain to make this type of clause work. It must apply explicitly to the events that actually occur or it will not be upheld.

Many courts have taken the position that only delays that were "reasonably within the contemplation of the parties" at the time the contract was signed are covered by these "no damages" clauses. For instance, courts in some jurisdictions have held that an owner's "active interference" could not reasonably be expected, and so is not covered by these clauses. In other words, the contractor will not be prevented from recovering damages for such delays, despite the existence of a "no damages" clause.

Not all courts have taken this approach, however. One case that caused contractors in New York State considerable dismay was *Kalish-Jarcho, Inc. v. City of New York*. (This is described in Figure 17.5.)

"No damages for delay" clauses can be expensive for an owner. They encourage contractors to include in their bids an amount to cover the costs of delays, even though the delays may never occur. Current standard form documents published by both the AIA and the EJCDC neither provide for, nor prohibit, delay damages.

Actual Damages Caused By Delay

If a contract allows for, or at least does not preclude, a party from recovering actual damages they incurred as a result of delays, what might those damages be?

For the owner, actual damages might include:

- Loss of use of the facility
- Loss of revenue
- Additional financing costs
- Costs of additional services required of design professionals

A contractor who has been delayed might be able to demonstrate actual damages for the following reasons:

- Idle or under-employed equipment and labor
- Inefficiency (loss of productivity)
- Labor and material price escalations
- Demobilization/remobilization costs (from stopping the job and then restarting)
- Increased maintenance, including winter protection (such as heat, temporary enclosures, and so forth)
- Shift premiums (higher wages associated with non-daytime shifts)
- Extended supervision (salaries of superintendents which must be paid for longer than anticipated)
- Equipment/machinery rentals
- Extended overhead (supporting the project in the home office for an extended period of time)
- Travel

Kalish-Jarcho, Inc. v. City of New York, 58 NY 2d 377, 448 NE 2d 413, 461 NYS 2d 746 (1983)

The mechanical contractor for the New York City Police Headquarters Building sued the city for damages caused by delays allegedly due to hundreds of drawing revisions, failure of the city to coordinate the work of the separate contractors, and interference with the contractors' schedules. The court finally held that even "active interference" by the owner *was covered* by the "no damages" clause, and that to recover under this type of clause, a contractor had to prove "bad faith" or "malice" on the part of the owner.

Figure 17.5

- Increased insurance premiums
- Interest

Many of these items are difficult to prove with reasonable certainty – particularly the proper amount of unrecovered, extended home office overhead. Where a contractor has not kept detailed records, a court will either deny recovery of damages because the contractor has not proved its losses with reasonable certainty, or it will establish a rough basis for approximating the contractor's losses.

Calculating Actual Damages

There are two formulas that courts may allow to approximate the actual damages owed to a contractor as a result of delays. They are:

- Eichleay formula
- Total cost method

Eichleay Formula: One of the most widely used methods to approximate a contractor's damages for extended home office overhead is the *Eichleay Formula,* so called because it came into use in a case involving a contractor by that name. The Eichleay Formula applies only to extended home office overhead, and not to the other elements of contractor delay damages. This formula computes the daily overhead rate by taking the contractor's total overhead for the contract period, multiplying it by the ratio of job billings to total billings for the contract period, and dividing the result by the actual days of contract performance. The daily rate is then multiplied by the number of days of compensable delay to compute the reasonable extended overhead amount for which the contractor has not otherwise been compensated. An example is shown in Figure 17.6.

Total Cost Method: Another method for calculating damages due to delays is known as the total cost method. This method is used to approximate the costs of all delay-related costs itemized above. Here, the contractor claims that his or her damages equal the total cost of performance minus the preperformance estimate of costs. In other words, the contractor says, "This is what I thought it would cost, and this is what it actually cost. The difference is the damages due to delay."

The Eichleay Formula

$$\frac{\text{Contract billings}}{\text{Total billings}} \times \frac{\text{Total overhead for contract period}}{\text{Actual days of contract performance}} = \text{Daily overhead}$$

$$\frac{\$100{,}000}{\$1{,}000{,}000} \times \frac{150{,}000}{300} = \$50$$

Daily overhead × Compensable delay days = Unabsorbed home office overhead

$50 × 30 (days delay) = $1,500

Figure 17.6

This method is rarely allowed by the courts, because it presumes that 1) the original estimate was appropriate, 2) the contractor's work was performed efficiently, 3) all costs incurred were reasonable, and 4) the contractor had no part in causing the delay. Rarely can all of these factors be proven.

Liquidated Damages For Delay

In many cases it is difficult, if not impossible, for an owner to calculate the actual damages caused by delays. For this reason, the parties may agree in the contract that an unexcused delay by the contractor will result in a specific amount of damages to the owner. In other words, the amount of damages is liquidated at a specific figure (preestablished in the agreement).

Originally, liquidated damages clauses were not viewed favorably by the courts. The tendency was to require proof of actual damages. Today, liquidated damages clauses are typically enforceable. Factors that are important in determining enforceability include:

- The clause should have been bargained for between two relatively equal contracting parties.
- The clause should not be used to impose a penalty, that is, it should not be set much higher than the probable actual damages in an attempt to coerce timely performance. If the clause imposes a penalty, it will not be enforced.
- The clause should be based on a good faith preestimate of delay damages.

Other practical considerations include the following:

- The clause should give the owner the right to deduct liquidated damages from amounts retained from the contractor's periodic payments (or other amounts due the contractor) so the owner won't have to sue to get his or her money.
- The clause should apply even when the contractor abandons the project, or when the owner terminates the contract for just cause.

Sometimes liquidated damages clauses are joined with *bonus clauses*. Under bonus clauses, the contractor gains a daily bonus if he completes the project before the scheduled completion date. It is acceptable to combine bonus clauses with liquidated damages clauses if the owner so chooses. However, *penalties* are never enforceable – even when combined with bonuses.

Time Extensions

It is often difficult to decide exactly how much of an extension is required for any particular delay. An updated CPM schedule can be a valuable tool in this endeavor, because it tracks the precise effects of each delay at the time it occurs by determining whether or not the delayed event is on the critical path. If the event is on the critical path, the delay will affect the ultimate completion date of the project. If it isn't, the completion date will not be affected. Generally, only delays on the critical path justify an extension to the contract time for performance or are compensable.

Most contracts require that prior notice be given to claim that an event will require a time extension. Such notice is similar to that required under the changes clause discussed earlier.

The usual deadline for giving such notice is 21 days after the event that caused the delay, as specified in the AIA standard form documents (20 days in the construction contracts used by the federal government). Notice enables the owner and design professional to attempt to remedy the situation or to minimize applicable damages in the manner they see fit.

Separate Contracts

When a separate contract system (multiple prime contracts) is established for a project, the process for settling dispute claims can be a nightmare. This is because each independent contractor may have a separate claim against the owner if they are delayed by the actions or failure to act of any of the other separate contractors working on the project.

To help alleviate the owner's burden, one of the contractors is usually assigned responsibility for scheduling and coordinating the work of *all* of the contractors on the project. The AIA standard form documents, for example, recognize this problem and require that if one contractor accuses another of causing delay, the accused contractor must defend the suit and indemnify the owner (hold the owner harmless) for damages, if any, incurred by the delayed contractor. Some courts have even allowed the delay*ed* contractor to sue the delay*ing* contractor *directly* (leaving the owner out of the picture if he or she has no responsibility).

Payment

Finally, a fourth major source of disputes is the process of making or withholding payments to the contractor. Under old English law, contracted work had to be fully completed before payment was due. American law, which is derived from English law, generally follows this premise. Unless the contract provides otherwise, payment is not due until performance is "complete." Because of this, most construction contracts expressly provide for *progress payments* tied in some manner to the progress of the services and work to be provided under the contract.

Schedule of Values

Payments and work are most commonly connected through use of a *schedule of values*. This device divides the total contract price among the various categories of work so that the owner and design professional can assess the progress of each element of the project. Often, the schedule of values is based on the 16-division format developed by the Construction Specifications Institute (CSI) to categorize construction products and materials. Each division heading is used as a line item for payment purposes. On major projects, the schedule of values can be an extensive and detailed listing of work categories. (See Figure 17.7)

The schedule of values is the basis of progress payments under stipulated sum contracts. Each scheduled amount is multiplied by the achieved percentage of completion for that line item. The product equals the gross progress payment less *retainage*, if any. (Retainage is discussed later in this chapter.) For cost-based contracts in which the final price is indeterminate, the schedule represents a budget allocation, listing each element of work with its planned cost.

To accelerate cash flow, contractors sometimes "front load" this schedule by putting a disproportionately high value on items of work which will be completed early in the project's life. This is something that owners and design professionals should be aware of, to avoid overpaying in the early stages of construction, leaving insufficient funds available to complete the work.

Application for Payment

Generally, the contractor submits his or her application for payment prior to the time when payment is due so that the owner can evaluate the application, usually with the design professional's help, and make proper payment. The owner and design professional may also require the contractor to submit documentation such as receipts and delivery tickets, to verify the contractor's right to payment. Furthermore, some contracts require contractors to state under oath that all subcontractors and material suppliers have been paid properly out of money advanced on

County School Schedule of Values	
BOND	$ 12,050.00
PERMIT/TEMP. FAC.	$ 4,060.00
SITEWORK	$ 35,254.00
ASPHALT, EXT. CURBS	$ 27,221.00
CONCRETE	$ 70,542.00
MASONRY	$ 10,114.00
METAL FAB.	$ 5,605.00
ROUGH CARPENTRY	$ 66,437.00
TRUSS JOIST	$ 28,000.00
GLU LAMS	$ 5,600.00
FINISH CARPENTRY	$ 7,328.00
ARCH. WOODWORK	$ 61,000.00
BUILDING INSULATION	$ 7,785.00
ROOF MEMBRANE	$ 20,282.00
FLASHING AND SHEET METAL	$ 26,590.00
STEEL DRS, WD. DRS, HDWARE	$ 15,655.00
ALUM. STOREFRONT, ALUM. WIND.	$ 31,713.00
GYPBOARD	$ 40,916.00
ACOUSTICAL	$ 14,904.00
DEMOUNTABLE PARTITIONS	$ 18,093.00
EXTERIOR FINISHES	$ 43,161.00
FLOOR FINISHES	$ 39,965.00
PAINTING	$ 6,545.00
CHALK, TACK, CUBICLES, SIGNS, LOCKERS, TOILET ACCESS, APPLIANCES	$ 4,500.00
PLUMBING	$ 48,319.00
HVAC	$ 73,295.00
ELECTRICAL	$ 70,727.07
MECHANICAL REVISIONS	$ 1,946.92
TOTAL CONSTRUCTION COST:	$797,607.99

Figure 17.7

account of prior applications for payment, or to submit partial releases of liens from these entities. (See Chapter 16 for additional information on liens.)

The typical application for payment provides the following information:

- The schedule of values
- The percentage of the work in each category that is complete as of the date of the application
- The total value of the work and materials as of the date of the application
- A credit for previous payments
- The amount then due
- A signature by an authorized official of the contractor's business entity

Certification for Payment

Under fixed-price contracts, design professionals generally visit the project site to verify that the percentage of completion stated by the contractor in the application for payment for each item of work is accurate. This percentage is multiplied by the scheduled value, then previous payments are deducted to arrive at the gross payment to be certified for that item.

On cost-based contracts, the design professional verifies that cost receipts have been presented for the amounts claimed, and then certifies to the owner that payment for these receipts is due to the contractor. The AIA and EJCDC standard form documents require that materials be incorporated in the project or delivered to, and properly stored on, the site before they are paid for by the owner. Contractors may be paid for materials and equipment stored off-site only if special provisions are made to do so.

Most standard form contracts allow the design professional to issue a "partial" certificate for payment (one in an amount less than the contractor requested) if the design professional disagrees, in whole or in part, with the content of the contractor's application. AIA and EJCDC documents also allow the design professional to withhold a certificate or to revoke a previous one for sufficient cause. The design professional may do this if:

- Defective work by the contractor is discovered
- Subcontractors have not been paid by the contractor after the contractor has been paid with respect to the subcontractors' work
- Legal claims are made or threatened against the owner with respect to any of the work on the project

A contractor involved in a dispute with an owner may allege that the fact that payments were made for certain work indicated that such work was properly performed and acceptable to the owner. However, progress payments do not usually constitute a waiver of defects in the work, nor do they constitute final acceptance of the work. In fact, contracts often specifically state that progress payments do not, by implication, waive defects in the work.

Similarly, contracts should make clear that a contractor who accepts one delayed payment has not necessarily waived the right to insist on timely payments in the future. In general, courts do not look kindly on purported waivers of rights unless they are:

- Express
- Related to a relatively minor matter
- Related to a relatively technical matter

However, none of these elements would normally apply to a purported waiver by a contractor of the right to insist on timely payments by the owner.

Furthermore, progress payments generally do not make construction contracts divisible or severable with regard to the work covered by the various payments. This means that parts of the work for which payment has been made will not be considered fully performed, solely because of such payment. The owner maintains the right to sue for breach of contract if the remaining work is not finished.

Generally, progress payments are made to the general contractor only; rarely are progress payments ever made directly to the subcontractors. Joint payments to the contractor and subcontractor are sometimes made, however. This might be done by making a check out to both the general contractor and the subcontractor, and giving it to the general contractor. Neither party can cash the check without the other's signature. The idea is that the general contractor would determine whether or not payment was due to the subcontractor and, if so, would sign the check and give it to the subcontractor to cash. If not, the contractor cannot cash the check on his own. This method allows the contractor to maintain control over the flow of cash, while protecting the owner's interest and allowing subcontractors to take an active role in their own destiny. In fact, many subcontractors' trade organizations are working to promote a system of *direct disbursement* of payments from the owner to the subcontractors. They want to ensure that money paid by the owner for work performed by a subcontractor will actually reach that subcontractor.

The Surety's Interest

Since the surety (the bonding company) may be responsible for paying subcontractors and suppliers if the general contractor fails to do so, the surety will want to protect its interests by asserting control over the money paid by the owner. Sometimes, the surety will request that an owner stop paying a general contractor who is behind in his or her payments to subcontractors and suppliers. They do this to mitigate the damages for which they might be responsible if the contractor defaults on his or her contract with the owner. After such a request, the owner either pays the money to the surety or places the money into an *escrow account* (a bank account administered by a neutral party acting under specific instructions for dealing with the money) so that it can be properly applied on the project.

Remedies for Nonpayment by the Owner

The contract itself typically specifies the remedies available to the general contractor in the event of late payment or nonpayment by the owner. Three types of remedies are available to contractors:

- Interest payments
- Suspension and termination of work
- Damages

Each of these is discussed below.

Interest Payments

In addition to other remedies, contractors may be entitled to interest on late payments. If the amount of the payment due is reasonably clear, interest will begin to run on the due date. The rate of interest must normally be specified in the contract. AIA standard form documents state that if the rate is not specified, it will be the "legal rate" of interest in the state where the project is located. Such a rate is set by law and is between 4 and 12 percent in most states.

Suspension and Termination of Work

Unless there are provisions in the contract which state otherwise, contractors usually do not have the right to *suspend* work because of late payments. To do so could be considered a breach of contract by the contractor. However, if the payments involved a large amount of money, and if the period of nonpayment was prolonged, the owner has committed a substantial breach of the contract. Under these circumstances, the contractor normally has the right to *terminate* the agreement.

Damages

If the contract is terminated because of a substantial breach of contract by the owner, the contractor may ask for reimbursement for his or her damages. In most cases, the contractor would be entitled to the *benefit of the bargain* that was made with the owner. Damages calculated on this basis put the contractor in the position he or she would have been in had the contract been fully performed. Usually this means that the contractor would have been paid the full contract price, with the difference between that amount and the cost to perform the work as the contractor's profit. Under this theory, therefore, the contractor would be entitled to his or her expected profit, if any, as damages for the owner's default.

An alternative is an award of damages based on *restitution*. Damages calculated on this basis would put the contractor in the position he or she was in before the contract was signed. This is particularly attractive to a contractor who made a poor bargain to begin with and stands to lose money performing under the contract. It would entitle the contractor to the reasonable value of the labor performed on, and materials supplied to, the project. (See Chapter 2, "Basic Contract Law," for additional information on remedies.)

Remedies for Contractor Default

During the payment process, the owner obtains some protection against the contractor's potential default by being careful not to pay in advance for work that has not yet been performed. Of course, the owner and design professional check carefully to determine that the payment requested by the contractor each month is appropriate for the amount of work properly completed. This process is inexact, however. Therefore, for additional protection, owners normally *retain* some of the contractor's money, that is, they do not pay 100% of the amount that appears to be due.

Retainage

Retainage is the name given to the process whereby the owner withholds, or retains, a percentage of the amount of each progress payment that is otherwise due to the contractor. This process is intended to help the owner secure proper performance of the contractor by, in effect, creating a fund out of which the owner can be reimbursed for the cost of potential claims it may have against the contractor. Retainage is not usually considered a liquidated damages figure, nor is it the limit of the damages that may be claimed by the owner if the contractor breaches the agreement.

There are many systems for administering retainage on a construction contract. One system withholds a fixed percentage (such as 10%) of each progress payment and pays the balance (90%) to the contractor. Another withholds a specific percentage of each line item in the schedule of values. These percentages may vary, or they may be consistent for all line items. A third creates a sliding scale of retained amounts. For example, the contract could provide for retainage of "10% until the contract is 50% complete and then no further retention." On that basis, at completion, the total retained amount equals 5% of the original contract sum.

Substantial Completion

As stated earlier, a contractor must fully or substantially complete performance of his or her contractual obligations before he or she has a legal claim for *any* money. Once substantial completion is achieved, however, the contractor will be entitled to the full contract price minus damages to the owner, if any. Such damages may be one or both of the following:

- The cost of completing or correcting the incomplete or incorrect work
- The amount by which the value of the project is diminished by the work that was incomplete or incorrect

When the contractor believes that the work is substantially complete, he or she prepares a list of incomplete or incorrect work. This is referred to as the *punch list*. Upon receipt of the punch list, the design professional inspects the work and may add to the punch list.

Under each of the retainage systems described above, the owner normally reduces the amount retained by paying it to the contractor upon substantial completion of the work. The owner normally holds approximately 150% of the estimated cost of performing the work on the punch list, to cover the cost of finding and paying another contractor to complete such work – in the event the original contractor neglects or otherwise fails to do so.

The punch list of incomplete or defective work is generally not taken to be an exclusive or exhaustive list of defects in the work. If formerly accepted work is damaged, or if additional defective or incomplete work is discovered, they may be dealt with at the time of discovery.

Final Completion and Final Payment

When the contractor believes that the work is complete, and the contract has been fully performed, he or she then makes an application for final payment, and the design professional makes a final inspection of the work. Assuming that the work is completed, the final payment process begins.

Many of the provisions related to final payment have to do with reassuring the owner that subcontractors and suppliers will not file mechanics' liens against the project. Making joint payments to the general contractor and the subcontractors, as well as providing the owner with signed lien waivers, are often useful in that regard, as described earlier.

Final payment is normally seen as a waiver of most, or all, past and future claims by both parties. The AIA and EJCDC standard form documents have provisions addressing this. They state that by *accepting* final payment, the contractor waives all claims against the owner except those previously made in writing and identified by the contractor as unsettled as of the date of the final application for payment.

Similarly, by *making* final payment, the owner waives all claims except those arising from:

- Unsettled liens
- Defective work that is discovered after substantial completion
- Work that does not comply with the requirements of the contract documents
- Terms of special warranties applicable to the work

The contract documents normally require the contractor to warrant that the work will be of good quality and as specified. Additionally, there may be warranties from manufacturers and suppliers which apply to parts of the work. None of these warranties is normally waived by making final payment. (Refer to Chapter 2, "Basic Contract Law," for more information about warranties and the Uniform Commercial Code.)

If the contract does not contain provisions similar to those described above, final payment may or may not be considered a waiver of claims, depending on the nature of the defect to which the claim relates. If the defect is *patent*, that is, known or observable at that time, final payment will probably be taken as a waiver of claims. On the other hand, if the defects are *latent* (unknown or not reasonably discoverable by inspection or observation), final payment will probably not be viewed as a waiver of claims related to the defects.

Conclusion

This chapter has addressed four aspects of the construction phase of a project which are prone to dispute. Three of these (contract changes, differing site conditions, and time for performance) relate to, or arise from, various changes to the factual basis on which the contract was initially made. The fourth, payment problems, is endemic to every large and expensive endeavor.

Frequently, disputes in these areas arise and grow because one or both of the parties involved does not fully understand their legal rights and responsibilities. There is a tendency to fight battles that should not be fought. A thorough review of the material in this chapter should enable owners, contractors, and others to minimize the friction and disputes that can mar this most challenging phase of project delivery.

CHAPTER 18
DISPUTE RESOLUTION

CHAPTER 18 DISPUTE RESOLUTION

Introduction

Disputes are common in the construction industry. There are many factors that can lead to a dispute, such as a delay in the construction schedule, the discovery of unanticipated site conditions, allegedly inadequate construction documents, and ambiguous contract language. There are also a number of factors that can exacerbate misunderstandings. Consider, for example, the fact that construction projects usually depend on the cooperation of a number of individuals and business entities who normally have no ongoing relationship with one another, and that projects typically involve large amounts of money. Parties to design and construction contracts who cannot come to an agreement among themselves have basically two options:

- Arbitration
- Litigation

Arbitration and litigation are the predominate methods of dispute resolution when informal discussions and settlement negotiations have failed. This chapter examines these two processes as they apply to the construction industry.

Arbitration v. Litigation

Arbitration is a method of settling disputes whereby parties to a contract agree to present their grievances to one or more impartial parties, called *arbitrators*, and to accept the decision of the arbitrator(s) on the issues. *Litigation* is the process by which disputes are resolved through formal legal proceedings (before a judge and/or jury in a court of law).

There are a number of differences between the arbitration and litigation processes. First, arbitration is entirely voluntary. It is possible only if both parties have agreed to arbitrate. Such an agreement may be reached either before a specific dispute develops or during the course of a dispute. Litigation does not depend on the mutual consent of the parties. Either party may *compel* the other to litigate simply by filing a complaint in court. (The process for doing so is discussed later in this chapter.)

In arbitration proceedings, under commonly used rules, the parties are not required to be represented by a lawyer. For relatively small claims, this can reduce expenses substantially. However, arbitration is relatively expensive to initiate, with the minimum fee to file a demand for arbitration starting at $300, and escalating in proportion to the amount of the claim. Arbitrators often serve for free for one day, but thereafter (and sometimes right from the beginning), they are paid by the parties roughly $300 to $500 per day. The filing fees for litigation are considerably less than arbitration filing fees because the court system is supported by tax money, but parties in a court proceeding typically spend a larger amount of money on lawyers' fees.

Arbitration hearings are informal compared to the judicial process. Arbitrators conduct hearings with dignity and fairness, but they are not bound by the formal rules of procedure which, by law, dictate how court cases are conducted and how alleged facts must be proven. In most cases, arbitrators tend to allow all evidence about alleged facts to be heard, if relevant, giving it the weight and credibility they feel it deserves. The formal rules that apply in litigation have been developed over the years to promote *due process of law* (essential fairness with respect to procedure) and to protect individual rights. Although these rules work very well, they can be cumbersome and tend to slow down the pace of the proceedings by focusing more on formal procedures than on a substantive search for the truth.

Another positive feature of arbitration hearings is that they are private. The parties normally have the right to exclude all non-interested people from the hearings and to prohibit any public disclosure about the proceedings. Judicial proceedings, on the other hand, are open to the public. This may be an important distinction for design professionals in particular. Often, disputes call into question an individual's professional skill and care – which is not the kind of publicity that professionals typically seek.

For many construction professionals, the primary advantage of arbitration is that the arbitrator(s) has construction experience. Many of the issues that arise during the course of dispute resolution are based on technical factors with which laypeople are unfamiliar. It is reassuring to the parties involved to know that a person(s) who is well versed in the construction industry will be making the decision on their dispute. Although many judges are familiar with the common issues and legal principles applicable to construction cases, it is likely that they have had no formal training in the construction industry. By the same token, however, arbitrators are not required to be lawyers, and their unfamiliarity with legal principles may be a disadvantage, depending on the precise nature of the dispute.

The court system usually permits several parties, each with different claims, to be joined into one lawsuit so that one trial will resolve all of the issues at one time. Consolidating claims can be beneficial to some, but rarely all, of the parties. There are normally too many issues involved for all parties to feel satisfied that they will receive a fair hearing if their claims are combined into one proceeding. For that reason and others, a *joinder of claims and parties* is usually not invoked in arbitration proceedings, which are strictly consensual.

The standard form contracts published by the AIA and the EJCDC call for arbitration of disputes under the Construction Industry Rules of the American Arbitration Association (AAA). The AAA is a private corporation that administers arbitration proceedings in the construction industry, in labor disputes, and in many other areas, on a national basis. These rules allow for more than two parties to participate in a single arbitration proceeding. However, the forms of agreement between the architect or engineer and the owner published by the AIA and the

EJCDC contain a *non-joinder clause* that prohibits a person who is not a party to the agreement from participating in arbitration proceedings. The non-joinder clause may pose a problem for the owner who does not know whether the design professional, the contractor, or both, were at fault in a given situation. If the owner proceeds to arbitrate, he or she will most likely have to endure two separate arbitration hearings: one involving the design professional and one involving the contractor. Participation in two arbitration hearings is time-consuming and costly. The owner could, conceivably, lose in both matters. If this same case were filed in court, the owner could sue both the design professional and contractor at the same time, in the same proceeding, and expect a final resolution of the matter.

Non-joinder clauses mainly benefit design professionals, who are legally required to perform services without negligence. If they have not performed negligently, they generally are not liable – even if an error or omission of some type is found in the design work. Contractors, on the other hand, generally are judged on whether or not they have performed as promised in the contract. It is sometimes difficult for an arbitrator to make the distinction between a contract claim and a negligence claim. There is a tendency to say that if something is wrong with the building, either the design professional who designed it or the contractor who built it *must* be responsible. In fact, that is not always true. The design professional, for example, may have performed without negligence, but designed something that does not work well. The contractor may have properly performed all of the work required by the contract. Neither would be liable under the law. Unfortunately, if both the contractor and design professional are part of the same proceeding, an arbitrator(s) is unlikely to exonerate *both* of them. There is a tendency to spread the blame, to some degree, among all available parties. Experience has shown that the design professional is often the one who suffers in such cases, and is most likely to benefit from the non-joinder provision.

Speed is another factor to consider. Arbitration tends to be a faster process than litigation. This can be a plus or minus, depending on the nature of the case and one's position in it. A contractor who is trying to recover money from the owner (who has not paid because of alleged defects in the work, for example) would want a speedy hearing to facilitate payment. That same owner would not be in such a hurry if it meant having to make those payments sooner. Similarly, a design professional facing a large claim for professional negligence – a claim that could put him or her out of business if lost – would probably also not be in a hurry to reach what might be a potentially disastrous conclusion to the case. It is a different matter, however, for the design professional who is trying to collect unpaid fees.

One last difference between arbitration and litigation is that an arbitration decision is final and *binding* on the parties, with little or no opportunity for appeal to a higher authority. A court decision may generally be appealed at least once to another court. Again, this can be a two-edged sword. For substantial matters that threaten an individual's livelihood, most individuals would want the right to appeal an unfavorable decision in the hopes that the decision will be reversed. In a more run-of-the-mill case, however, it might be a blessing to settle the issue permanently so that the individuals involved can get back to business. See Figure 18.1 for a summary of the features of each dispute resolution method.

The Arbitration Process

Arbitration is only available to parties who agree to arbitrate. The parties may give their consent at the time a dispute arises, or they may agree in their written contract to arbitrate all disputes that may arise in the future with respect to the contract. This statement is called an *arbitration clause*. An arbitration clause should specify the rules under which the hearings will be conducted. Most construction arbitrations are administered according to the Construction Industry Rules of the American Arbitration Association (AAA).

Initiating the Arbitration Process

There are a number of formal steps that must be taken to initiate the arbitration process according to the rules of the AAA. These include:

- Filing a demand
- Filing an answer
- Selecting an arbitrator
- Conducting an administrative conference

These are described below.

Filing a Demand

To initiate an arbitration proceeding, one party must file a *demand* with both the AAA and the party with whom he or she has a dispute. A

Dispute Resolution Methods	
Arbitration	• Voluntary • Relatively expensive to initiate • Informal • Private • Faster process • Joinder of parties not normal • Arbitrator has construction experience • Binding
Litigation	• Either party may compel the other to litigate • Open to public • Slow process • Joinder of claims and parties allowed • Governed by formal rules • Process can be expensive • Decision can be appealed

Figure 18.1

demand is a notice of a party's intent to arbitrate. The demand should describe the nature of the dispute, the amount involved, and the relief or remedy sought for the harm or damages allegedly caused by the other party. The party initiating the arbitration proceeding should also file with the AAA copies of the arbitration provisions of the contract, along with the specified filing fee.

Filing an Answer

The party on whom demand for arbitration is made has seven days to *answer* the allegations in the demand. If some of the allegations are true and can be admitted, the area of dispute can be narrowed. If that party does not file an answer, it is assumed that the claim is denied. Parties who receive arbitration demands frequently reply by denying the claim of the initiating party and alleging claims of their own (called *counterclaims*) against that party.

Selecting an Arbitrator

If the parties' agreement does not specify the number of arbitrators who will hear the claim, the dispute will be heard by one arbitrator. The AAA has the authority to appoint more than one arbitrator, however, and will normally do so in disputes involving approximately $100,000 or more. A panel of three arbitrators, for example, a design professional, a contractor, and an attorney, is more appropriate for such cases because it includes more diverse expertise and represents greater experience in the construction industry.

If the parties have already named an arbitrator in their contract, that person will be appointed. Similarly, if the contract provides a method for appointing an arbitrator, that method will be followed. Otherwise, the AAA sends a list of potential arbitrators from which the parties can choose. The list includes the educational background and relevant experience of each arbitrator. The parties cross out those names on the list to which they object. For example, a design professional may have had a bad experience with an attorney at some previous time and does not want any attorneys on the panel. Or a contractor may feel that all engineers are subconsciously biased against contractors, and in favor of owners. He, therefore, wants to strike the names of all engineers.

The AAA then attempts to appoint an arbitrator selected from those names to which neither party objected. If the parties do not agree to any of the persons named, the AAA will make the appointment from among other members of the *National Construction Industry Arbitration Panel* (the AAA's data bank containing the names and backgrounds of qualified and potentially available arbitrators). The chosen arbitrator is expected to be neutral and to disclose any actual and potential conflicts of interest.

Conducting an Administrative Conference

At the parties' request, or at the AAA's discretion, there may be an administrative conference with an AAA staff administrator, the arbitrator, and the parties to exchange information and to document agreement on uncontested facts.

The Hearing

The next step in the arbitration process is the *hearing*. At the hearing, the party who initiated the arbitration proceedings presents his or her case by introducing witnesses who are able to testify to what they know about the matters in dispute. He or she will also present documents, photographs, charts, graphs, and other evidence in an attempt to prove the truth of the alleged matters. The witnesses are questioned by each party. The defending party then presents his or her witnesses and evidence in a similar manner. The arbitrator has the authority to vary this process at his or her discretion, as long as the procedure is fair. The rules governing the type of the evidence that will be received by the arbitrator and the procedures for submitting it are not as formal as those in a court of law. Most evidence that is relevant and substantial should be heard and given the weight and credibility it deserves.

Frequently, the parties' presentations are completed in one day, but they may take considerably longer depending on the nature of the case. The parties may only communicate with the arbitrator while they are at the hearings. All other oral or written communications to the arbitrator must be made through the AAA.

Closing the Hearing(s)

After the parties have presented their evidence, the arbitrator declares the hearing(s) closed. Unless the parties have agreed otherwise, the award will be made no later than 30 days from the closing date of the hearing(s). The arbitrator may reopen the hearings at the request of one of the parties, if such request is made prior to the time the decision is rendered. This will only be done in extraordinary circumstances, for example, if new evidence is uncovered.

The Award

The arbitrator has the power to award any remedy that is fair and within the terms of the parties' contract. For example, an arbitrator can order a party to pay interest on late progress payments only if the contract provides for such interest. Additionally, the arbitrator may order the parties to split the AAA's and the arbitrator's fees and expenses, or may order one party to pay the entire amount. If the parties have voluntarily reached a settlement on some or all of the issues during the course of the arbitration proceedings, the arbitrator should put the terms of the settlement in the award. If there is more than one arbitrator on the panel, all decisions, including the award, must be made by a majority of the panel.

The Litigation Process

The judicial system in this country is available to all citizens, and is primarily supported by tax revenues. It is possible to sue for a wide variety of reasons. In fact, outrageous tales of "frivolous" suits abound. However, just because a party sues and loses does not mean that the suit was frivolous. Judges are reluctant to dismiss suits before the aggrieved party has had a chance to present its case. The following materials elaborate on the litigation process and the procedures involved in pursuing a lawsuit in court.

Weighing the Risks of Litigation

Lawsuits are usually expensive to prosecute, with costs often exceeding the final judgment. Before filing a lawsuit, therefore, the estimated costs of litigation should be carefully weighed against the probable outcome. The party who initiates the lawsuit should understand that there is always the risk of losing. A losing party not only fails to recover any money, but is normally responsible for court costs and lawyers' fees.

Counterclaims

The party who files the lawsuit should also recognize the risks of a *counterclaim* by his or her opponent. A counterclaim is made by the party originally being sued, against the initiating party, generally with respect to related matters on the project. For example, assume a design professional is under contract with a homeowner to design an addition to the homeowner's house. The design professional completes the plans and sends the homeowner an invoice for design services. The homeowner pays the design professional all but $5,000 of the invoice. The design professional sues for the remaining balance. The homeowner then files a counterclaim against the design professional, alleging that the design professional failed to perform services properly under the contract. The counterclaim seeks a judgment against the design professional for $50,000, which is the amount the homeowner claims it will cost to have the plans corrected and for other damages. At this point, the design professional wants to drop the lawsuit because he realizes that it is going to cost far more than $5,000 to litigate his claim and to defend himself against the homeowner's claim. The design professional cannot extricate himself from the lawsuit, however, unless the homeowner agrees to drop the counterclaim. Clearly, it is better to think through the possibilities prior to instituting a lawsuit to collect fees.

Selecting the Appropriate Court

A lawsuit cannot be initiated in just any court. The court must have jurisdiction over the matters at hand. This means that the court must have the authority to deal with the particular kind of claim that is brought before it. The court must also have the right to make its ruling binding on the persons (called personal jurisdiction) or property (*in rem* jurisdiction) involved in the case. Such authority is granted to the courts by applicable state or federal law.

Federal and State Court

There are two essentially parallel court systems in the United States: the *federal* and the *state court systems*. To determine whether a lawsuit should be initiated in federal or state court, both the rules of legal procedure and case law must be consulted. In both the state and federal judicial systems, there are various levels of courts. The lower courts handle cases that involve small amounts of money. Cases involving larger amounts go directly into a higher level court. Usually, a lower court decision that a party believes is legally incorrect may be appealed to the next higher court. For example, a *district court* decision may be appealed to the *superior court* and then, perhaps, to the *Supreme Court*. On appeal, the judge (called the *appellate judge*) assumes that the facts of the case were correctly stated. He or she reviews and rules only on the correctness of the application of the law to the facts of the case.

Initiating the Lawsuit

The *plaintiff* is the party on whose behalf the lawsuit is filed. The plaintiff files the suit against the *defendant*, who is the party the plaintiff accuses of wrongdoing. The defendant must defend him or herself against the plaintiff's claims, or probably lose the case.

Filing a Complaint

To initiate a lawsuit, the plaintiff's attorney files a *complaint* in court. The complaint describes what the defendant has allegedly done wrong and how the plaintiff has been injured as a result. A plaintiff cannot initiate a lawsuit simply out of retribution or spite. The plaintiff's complaint must allege or claim that he or she was harmed and must usually request a specific, concrete remedy, such as an award of money damages. (Refer to Chapter 2, "Basic Contract Law," and Chapter 4, "Basic Tort Law," for additional information on the elements of common claims.)

Serving the Summons

The plaintiff's lawyer *serves* (delivers) the complaint to the defendant via a *summons*. The summons gives the defendant notice of the claim, and requires the defendant to file in court an *answer* to the complaint. The summons states the amount of time the defendant has to file an answer (typically 20 days). In the answer, the defendant either admits, denies, or claims a lack of knowledge about, the items alleged in each paragraph of the complaint. If the defendant does not file an answer in the allotted time, he or she is in default, and the court may automatically decide the matter in the plaintiff's favor. The defendant may also assert in its answer defenses and counterclaims that it may have against the plaintiff.

Request for Dismissal

If the defendant believes that there is no legal basis for the plaintiff's claims, he or she can ask the court to dismiss the case. If the court denies this request, the case will proceed as follows.

Discovery

Once the initial *pleadings* (complaint and answer) have been filed, the parties begin to uncover the facts that are important to the case. This stage is called *discovery*. The general rule is that each party has the right to discover the information about the case which the other party possesses. There are some exceptions to this rule, however. For example, a party does not have a right to privileged information, including the substance of the other party's conversations with his or her lawyer made while pursuing or defending the lawsuit.

In construction matters, the design professional and the contractor normally have on hand more data about the design and construction process than does the owner. In that respect, discovery may be more important for the owner than either of the other two entities. It is important to remember that, with the exception noted above, an individual's or a firm's files become an "open book" in litigation. It is therefore unwise to write memos or letters during the course of a project if those documents contain conclusions drawn from incomplete information. Such conclusions often are incorrect and can be used against the writer later on.

The exchange of information that takes place in the discovery stage sometimes leads to settlement. It also enables each party to avoid surprises prior to the trial. Perry Mason-type trials, in which surprise evidence or witnesses are introduced at the last moment to save the day, are not typical.

Discovery is an important phase of litigation. It can take years, however, depending on the complexity of the issues and the number of people and documents involved. There are four tools commonly used to discover information:

- Interrogatories
- Depositions

- Requests to produce documents
- Requests for admissions

Interrogatories

Interrogatories are written questions that one party sends to the other to answer in writing. In most states there is a limit to the number of questions that can be asked in a set of interrogatories (typically 30). An example of a design professional's interrogatory to an owner is the following:

> *Please state with particularity the facts upon which you predicate the allegation in paragraph 7 of the Complaint that [Engineer] breached its obligations under the contract and the implied covenant of good faith and fair dealing by failing to perform work in a proper professional manner; failing to properly design and engineer the job; failing to properly supervise; failing to make decisions in a timely manner; failing to use cost-effective designs; and by submitting bills for work in excess of the agreed hourly charges.*

Depositions

Depositions are also used to acquire information. They are questions asked orally to another party or to a witness. The party or witness is questioned about their knowledge of the facts involved in the dispute, and must answer the questions under oath. The questions and answers are recorded and later transcribed. Depositions are especially helpful if it is likely that a witness will not be able to attend the trial; for instance, if the witness is terminally ill and may die before the trial. Depositions also allow the questioner to explore issues that are raised in the answers to other questions.

Requests to Produce Documents

Each party in a lawsuit may request that the other party(ies) *produce documents* about the case. The party who requested the documents reviews them and makes copies of those that are relevant or helpful to its case or its defenses. An example of part of such a request for production is as follows:

11. *All progress schedules, including bar charts or network diagrams, updatings, and accompanying narrative reports prepared in connection with the Project.*
12. *All minutes, memoranda or notes of preconstruction meetings or conversations in connection with the Project, whether made by [Owner] or received by [Owner] from others.*
13. *All bid tabulations, comparisons, and evaluations.*
14. *All correspondence, memoranda, and notes which document communication between [Owner] and consultants other than [Architect] regarding [Architect's] drawings, including copies of drawings submitted to such consultants.*

Requests for Admissions

Each party may send a *request for admissions* to the other party(ies). The party receiving the request is asked to either admit or deny that specific facts are true. The purpose of requests for admissions is to find out which allegations are not disputed. The parties may then focus their trial preparation on those elements of the case that are disputed. For example, the owner might be willing to admit that a specific amount of money had been paid to the contractor, while still contesting that more than that was due. Once the parties have conducted as much discovery as they deem necessary, the case is ready for trial.

Trial

The trial may either be a *jury trial* or a *bench trial* (one before the judge only). The plaintiff is entitled to request a jury trial in most matters, but may choose to waive its right. Usually, that would be done for a specific strategic reason. For example, a blue-collar jury might not be very sympathetic to a wealthy real estate developer, regardless of the legalities involved. In such an event, the case would be heard by a judge.

The trial usually begins with the plaintiff's and defendant's opening statements (their summary of the case as they see it unfolding). Then witnesses are called by the plaintiff and questioned through *direct examination*. The defendant then questions the witnesses through *cross examination*. After the plaintiff has presented all of its evidence, it rests its case.

At this point, the defendant may ask the judge to rule in its favor on the basis that it is entitled to win as a matter of law. In other words, even if everything that the plaintiff has claimed is true, the law says that he is not entitled to be awarded anything from the defendant. If the judge agrees, the case is over and the defendant wins. If the judge disagrees, the trial proceeds.

It is then the defendant's turn to call witnesses. They are questioned in the same manner as the plaintiff's witnesses. After the defendant rests its case, the plaintiff may ask the court to decide in its favor. If the court does rule in the plaintiff's favor, the case is over and the plaintiff has won. Otherwise, the case continues and the plaintiff is allowed to *rebut* the defendant's presentation of evidence.

To rebut a defendant's evidence is to counteract or neutralize the effect of that evidence. For example, a witness might testify that brick construction was improper because expansion joints were not provided in a specific length of wall. To rebut such a statement, a witness might testify that no joints were required in that particular case because the wall was only 19 feet long. This testimony explains and neutralizes the effect of the prior testimony. The defendant is then given the opportunity for rebuttal.

At the end of the trial, each party makes its final argument or *closing statement*. If there is a jury, the judge *charges* the jury (gives them instructions about how to reach a decision on the case). Finally, the jury or judge reaches a *verdict*, or decision, on the issues. In civil trials (those dealing with offenses that are not punishable by jail sentences), many states allow a less than unanimous verdict by a jury. This means that a verdict will be acceptable if as few as four out of six jurors agree to it. The standard that a plaintiff must meet to win its case is also less demanding in a civil case than in criminal cases (in which the defendant may be sentenced to prison). A civil plaintiff is only required to show that more likely than not, his or her version of the facts was correct. (This is called the *preponderance of the evidence* standard, and is not as stringent as the standard applicable to criminal trials, which is *beyond a reasonable doubt*.

Post-Trial

The losing party may ask for a new trial because of jury misconduct, newly discovered evidence, or errors of law by the judge. The losing party may also appeal the case to a higher court if he or she believes the decision is incorrect. The court that receives the matter on appeal will either affirm, reverse, or modify the lower court's decision. It could also *remand* the case, which means to send it back to the lower (original) court for further rulings on specific factual questions.

Collecting the Judgment

Although a party may win a money judgment after a trial, there may be difficulty in actually collecting the money. The court system has several means available to parties who are trying to collect their awards. It is up to the collecting party, however, to initiate the collection process. The court does not take the initiative in this regard.

One device available is a *judgment lien*, which may be placed on the losing party's real or personal property. Such a lien cannot be removed unless the judgment is paid. Additionally, the losing party may be forced to attend a court hearing so that a judge can assess the party's ability to pay and then establish a payment schedule. If the party refuses to attend such a hearing, he or she can be held *in contempt of court*, fined, or jailed.

Conclusion

Disputes are a fact of life in the construction industry. This chapter has assumed that parties who disagree have made every reasonable attempt to settle such a disagreement through negotiation before resorting to arbitration or litigation. Win or lose, most people would prefer not to go through either process.

Regardless, it is incumbent on those in the design and construction industry to have a basic understanding of dispute resolution procedures. Arbitration has been widely used and accepted in the design and construction industry for years. It is often favored by those who have experience using it. The alternative to arbitration is litigation. Litigation is the primary method for settling disputes in this country, and the applicable rules of law and procedure are well developed. They are conceived to produce a fair hearing for each party involved, and they normally do so.

It is important to understand the advantages and disadvantages of each process, and how each process works in a given situation. Such knowledge will enable prospective users to choose the best method for resolving disputes that arise in the course of doing business.

APPENDIX

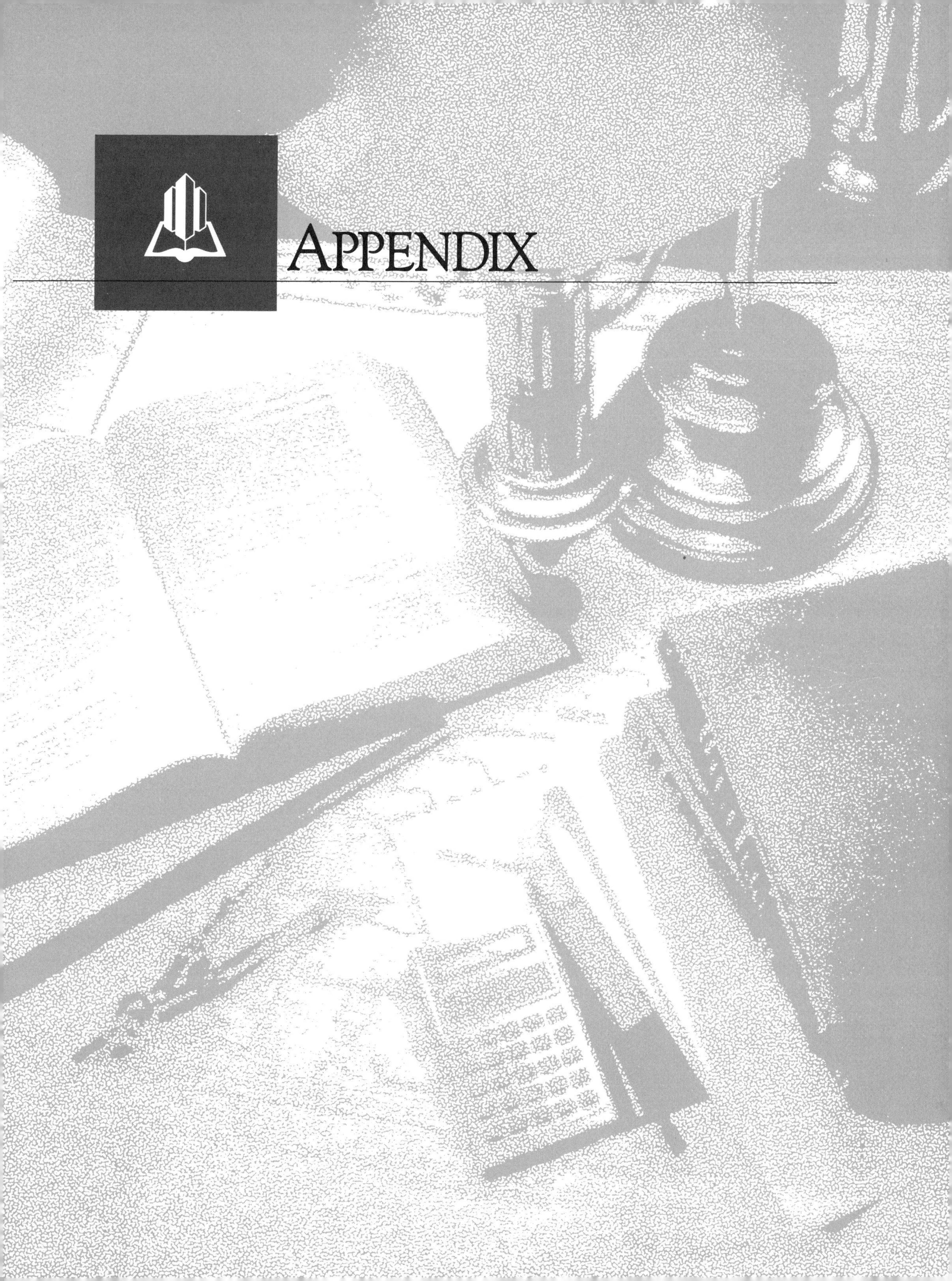

TABLE OF CONTENTS

APPENDIX

APPENDIX A CONSTRUCTION-RELATED ORGANIZATIONS

American Association of Cost Engineers
P.O. Box 1557
Morgantown, West Virginia 26507-1557
(304) 296-8444

American Building Contractors Association (Construction) (ABCA)
11100 Valley Blvd., Suite 120
Elmonte, California 91731
(818) 401-0071

American Consulting Engineers Council
1015 15th Street, N.W.
Washington, D.C. 20005
(202) 347-7474

American Institute of Architects
1735 New York Avenue, N.W.
Washington, D.C. 20006
(202) 626-7300

American Institute of Industrial Engineers
25 Technology Park
Atlanta/Norcross, Georgia 30092
(404) 449-0460

American Institute of Plant Engineers
3975 Erie Avenue
Cincinnati, Ohio 45208
(513) 561-6000

American Society of Civil Engineers
345 East 47th Street
New York, New York 10017
(212) 705-7496

American Society of Heating, Refrigerating, and Air Conditioning Engineers, Inc.
1791 Tullie Circle, N.E.
Atlanta, Georgia 30329
(404) 636-8400

American Society of Interior Designers
1430 Broadway, Floor 22
New York, New York 10018
(212) 944-9220

American Society of Landscape Architects
4401 Connecticut Avenue, N.W.
Washington, D.C. 20008
(202) 466-7730

American Society of Safety Engineers
850 Busse Highway
Park Ridge, Illinois 60068
(312) 692-4121

American Society for Testing Materials
1916 Race Street
Philadelphia, Pennsylvania 19103
(215) 299-5400

Associated Builders and Contractors, Inc.
729 15th Street, N.W.
Washington, D.C. 20005
(202) 637-8800

Associated General Contractors of America
1957 E Street, N.W.
Washington, D.C. 20006
(202) 393-2040

Basic Building Code (BBC)
Building Officials and Code Administrators, International (BOCA)
4051 Flossmoor Road
Country Club Hills, Illinois 60478
(312) 799-2300

Institute of Business Designers
1155 Merchandise Mart
Chicago, Illinois 60654
(312) 467-1950

Institute of Transportation Engineers
525 School Street, S.W., Suite 410
Washington, D.C. 20024-2729
(202) 554-8050

National Association of Home Builders of the United States
15th and M Streets, N.W.
Washington, D.C. 20005
(202) 822-0200

National Building Code (NBC)
National Conference of States on Building Codes and Standards, Inc.
481 Carlisle Drive
Herndon, Virginia 22070
(703) 437-0100

National Electrical Contractors Association
7315 Wisconsin Avenue, N.W.
Suite 1300
Bethesda, Maryland 20814
(301) 657-3110

National Institute of Building Science
1015 15th Street, N.W.
Suite 700
Washington, D.C. 20005
(202) 347-5710

National Society of Professional Engineers
1420 King Street
Alexandria, Virginia 22314
(703) 684-2800

Society of American Military Engineers
P.O. Box 21289
Alexandria, Virginia 22320-2289
(800) 336-3097

The Southern Standard Building Code (SSBC)
Southern Building Code Congress (SBCC)
900 Montclair Road
Birmingham, Alabama 35213
(205) 591-1853

Uniform Building Code (UBC)
International Conference of Building Officials (ICBO)
5360 South Workman Mill Road
Whittier, California 90601
(213) 699-0541

APPENDIX B

STANDARD FORM AGREEMENTS

THE AMERICAN INSTITUTE OF ARCHITECTS

AIA Document A101

Standard Form of Agreement Between Owner and Contractor

where the basis of payment is a

STIPULATED SUM

1987 EDITION

THIS DOCUMENT HAS IMPORTANT LEGAL CONSEQUENCES; CONSULTATION WITH AN ATTORNEY IS ENCOURAGED WITH RESPECT TO ITS COMPLETION OR MODIFICATION.

The 1987 Edition of AIA Document A201, General Conditions of the Contract for Construction, is adopted in this document by reference. Do not use with other general conditions unless this document is modified.

This document has been approved and endorsed by The Associated General Contractors of America.

AGREEMENT

made as of the day of in the year of Nineteen Hundred and

BETWEEN the Owner:
(Name and address)

and the Contractor:
(Name and address)

The Project is:
(Name and location)

The Architect is:
(Name and address)

The Owner and Contractor agree as set forth below.

ARTICLE 1
THE CONTRACT DOCUMENTS

The Contract Documents consist of this Agreement, Conditions of the Contract (General, Supplementary and other Conditions), Drawings, Specifications, Addenda issued prior to execution of this Agreement, other documents listed in this Agreement and Modifications issued after execution of this Agreement; these form the Contract, and are as fully a part of the Contract as if attached to this Agreement or repeated herein. The Contract represents the entire and integrated agreement between the parties hereto and supersedes prior negotiations, representations or agreements, either written or oral. An enumeration of the Contract Documents, other than Modifications, appears in Article 9.

ARTICLE 2
THE WORK OF THIS CONTRACT

The Contractor shall execute the entire Work described in the Contract Documents, except to the extent specifically indicated in the Contract Documents to be the responsibility of others, or as follows:

ARTICLE 3
DATE OF COMMENCEMENT AND SUBSTANTIAL COMPLETION

3.1 The date of commencement is the date from which the Contract Time of Paragraph 3.2 is measured, and shall be the date of this Agreement, as first written above, unless a different date is stated below or provision is made for the date to be fixed in a notice to proceed issued by the Owner.

(Insert the date of commencement, if it differs from the date of this Agreement or, if applicable, state that the date will be fixed in a notice to proceed.)

Unless the date of commencement is established by a notice to proceed issued by the Owner, the Contractor shall notify the Owner in writing not less than five days before commencing the Work to permit the timely filing of mortgages, mechanic's liens and other security interests.

3.2 The Contractor shall achieve Substantial Completion of the entire Work not later than

(Insert the calendar date or number of calendar days after the date of commencement. Also insert any requirements for earlier Substantial Completion of certain portions of the Work, if not stated elsewhere in the Contract Documents.)

, subject to adjustments of this Contract Time as provided in the Contract Documents.

(Insert provisions, if any, for liquidated damages relating to failure to complete on time.)

ARTICLE 4
CONTRACT SUM

4.1 The Owner shall pay the Contractor in current funds for the Contractor's performance of the Contract the Contract Sum of Dollars ($), subject to additions and deductions as provided in the Contract Documents.

4.2 The Contract Sum is based upon the following alternates, if any, which are described in the Contract Documents and are hereby accepted by the Owner:

(State the numbers or other identification of accepted alternates. If decisions on other alternates are to be made by the Owner subsequent to the execution of this Agreement, attach a schedule of such other alternates showing the amount for each and the date until which that amount is valid.)

4.3 Unit prices, if any, are as follows:

ARTICLE 5
PROGRESS PAYMENTS

5.1 Based upon Applications for Payment submitted to the Architect by the Contractor and Certificates for Payment issued by the Architect, the Owner shall make progress payments on account of the Contract Sum to the Contractor as provided below and elsewhere in the Contract Documents.

5.2 The period covered by each Application for Payment shall be one calendar month ending on the last day of the month, or as follows:

5.3 Provided an Application for Payment is received by the Architect not later than the day of a month, the Owner shall make payment to the Contractor not later than the day of the month. If an Application for Payment is received by the Architect after the application date fixed above, payment shall be made by the Owner not later than days after the Architect receives the Application for Payment.

5.4 Each Application for Payment shall be based upon the Schedule of Values submitted by the Contractor in accordance with the Contract Documents. The Schedule of Values shall allocate the entire Contract Sum among the various portions of the Work and be prepared in such form and supported by such data to substantiate its accuracy as the Architect may require. This Schedule, unless objected to by the Architect, shall be used as a basis for reviewing the Contractor's Applications for Payment.

5.5 Applications for Payment shall indicate the percentage of completion of each portion of the Work as of the end of the period covered by the Application for Payment.

5.6 Subject to the provisions of the Contract Documents, the amount of each progress payment shall be computed as follows:

5.6.1 Take that portion of the Contract Sum properly allocable to completed Work as determined by multiplying the percentage completion of each portion of the Work by the share of the total Contract Sum allocated to that portion of the Work in the Schedule of Values, less retainage of percent (%). Pending final determination of cost to the Owner of changes in the Work, amounts not in dispute may be included as provided in Subparagraph 7.3.7 of the General Conditions even though the Contract Sum has not yet been adjusted by Change Order;

5.6.2 Add that portion of the Contract Sum properly allocable to materials and equipment delivered and suitably stored at the site for subsequent incorporation in the completed construction (or, if approved in advance by the Owner, suitably stored off the site at a location agreed upon in writing), less retainage of percent (%);

5.6.3 Subtract the aggregate of previous payments made by the Owner; and

5.6.4 Subtract amounts, if any, for which the Architect has withheld or nullified a Certificate for Payment as provided in Paragraph 9.5 of the General Conditions.

5.7 The progress payment amount determined in accordance with Paragraph 5.6 shall be further modified under the following circumstances:

5.7.1 Add, upon Substantial Completion of the Work, a sum sufficient to increase the total payments to percent (%) of the Contract Sum, less such amounts as the Architect shall determine for incomplete Work and unsettled claims; and

5.7.2 Add, if final completion of the Work is thereafter materially delayed through no fault of the Contractor, any additional amounts payable in accordance with Subparagraph 9.10.3 of the General Conditions.

5.8 Reduction or limitation of retainage, if any, shall be as follows:

(If it is intended, prior to Substantial Completion of the entire Work, to reduce or limit the retainage resulting from the percentages inserted in Subparagraphs 5.6.1 and 5.6.2 above, and this is not explained elsewhere in the Contract Documents, insert here provisions for such reduction or limitation.)

ARTICLE 6
FINAL PAYMENT

Final payment, constituting the entire unpaid balance of the Contract Sum, shall be made by the Owner to the Contractor when (1) the Contract has been fully performed by the Contractor except for the Contractor's responsibility to correct nonconforming Work as provided in Subparagraph 12.2.2 of the General Conditions and to satisfy other requirements, if any, which necessarily survive final payment; and (2) a final Certificate for Payment has been issued by the Architect; such final payment shall be made by the Owner not more than 30 days after the issuance of the Architect's final Certificate for Payment, or as follows:

ARTICLE 7
MISCELLANEOUS PROVISIONS

7.1 Where reference is made in this Agreement to a provision of the General Conditions or another Contract Document, the reference refers to that provision as amended or supplemented by other provisions of the Contract Documents.

7.2 Payments due and unpaid under the Contract shall bear interest from the date payment is due at the rate stated below, or in the absence thereof, at the legal rate prevailing from time to time at the place where the Project is located.

(Insert rate of interest agreed upon, if any.)

(Usury laws and requirements under the Federal Truth in Lending Act, similar state and local consumer credit laws and other regulations at the Owner's and Contractor's principal places of business, the location of the Project and elsewhere may affect the validity of this provision. Legal advice should be obtained with respect to deletions or modifications, and also regarding requirements such as written disclosures or waivers.)

7.3 Other provisions:

ARTICLE 8
TERMINATION OR SUSPENSION

8.1 The Contract may be terminated by the Owner or the Contractor as provided in Article 14 of the General Conditions.

8.2 The Work may be suspended by the Owner as provided in Article 14 of the General Conditions.

ARTICLE 9
ENUMERATION OF CONTRACT DOCUMENTS

9.1 The Contract Documents, except for Modifications issued after execution of this Agreement, are enumerated as follows:

9.1.1 The Agreement is this executed Standard Form of Agreement Between Owner and Contractor, AIA Document A101, 1987 Edition.

9.1.2 The General Conditions are the General Conditions of the Contract for Construction, AIA Document A201, 1987 Edition.

9.1.3 The Supplementary and other Conditions of the Contract are those contained in the Project Manual dated , and are as follows:

Document	Title	Pages

9.1.4 The Specifications are those contained in the Project Manual dated as in Subparagraph 9.1.3, and are as follows:
(Either list the Specifications here or refer to an exhibit attached to this Agreement.)

Section	Title	Pages

9.1.5 The Drawings are as follows, and are dated unless a different date is shown below:
(Either list the Drawings here or refer to an exhibit attached to this Agreement.)

Number	**Title**	**Date**

9.1.6 The Addenda, if any, are as follows:

Number	**Date**	**Pages**

Portions of Addenda relating to bidding requirements are not part of the Contract Documents unless the bidding requirements are also enumerated in this Article 9.

9.1.7 Other documents, if any, forming part of the Contract Documents are as follows:

(List here any additional documents which are intended to form part of the Contract Documents. The General Conditions provide that bidding requirements such as advertisement or invitation to bid, Instructions to Bidders, sample forms and the Contractor's bid are not part of the Contract Documents unless enumerated in this Agreement. They should be listed here only if intended to be part of the Contract Documents.)

This Agreement is entered into as of the day and year first written above and is executed in at least three original copies of which one is to be delivered to the Contractor, one to the Architect for use in the administration of the Contract, and the remainder to the Owner.

OWNER	CONTRACTOR
(Signature)	*(Signature)*
(Printed name and title)	*(Printed name and title)*

THE AMERICAN INSTITUTE OF ARCHITECTS

AIA Document A111

Standard Form of Agreement Between Owner and Contractor

where the basis of payment is the

COST OF THE WORK PLUS A FEE

with or without a Guaranteed Maximum Price

1987 EDITION

THIS DOCUMENT HAS IMPORTANT LEGAL CONSEQUENCES; CONSULTATION WITH AN ATTORNEY IS ENCOURAGED WITH RESPECT TO ITS COMPLETION OR MODIFICATION.

The 1987 Edition of AIA Document A201, General Conditions of the Contract for Construction, is adopted in this document by reference. Do not use with other general conditions unless this document is modified.

This document has been approved and endorsed by The Associated General Contractors of America.

AGREEMENT

made as of the day of in the year of Nineteen Hundred and

BETWEEN the Owner:
(Name and address)

and the Contractor:
(Name and address)

the Project is:
(Name and address)

the Architect is:
(Name and address)

The Owner and Contractor agree as set forth below.

ARTICLE 1
THE CONTRACT DOCUMENTS

1.1 The Contract Documents consist of this Agreement, Conditions of the Contract (General, Supplementary and other Conditions), Drawings, Specifications, Addenda issued prior to execution of this Agreement, other documents listed in this Agreement and Modifications issued after execution of this Agreement; these form the Contract, and are as fully a part of the Contract as if attached to this Agreement or repeated herein. The Contract represents the entire and integrated agreement between the parties hereto and supersedes prior negotiations, representations or agreements, either written or oral. An enumeration of the Contract Documents, other than Modifications, appears in Article 16. If anything in the other Contract Documents is inconsistent with this Agreement, this Agreement shall govern.

ARTICLE 2
THE WORK OF THIS CONTRACT

2.1 The Contractor shall execute the entire Work described in the Contract Documents, except to the extent specifically indicated in the Contract Documents to be the responsibility of others, or as follows:

ARTICLE 3
RELATIONSHIP OF THE PARTIES

3.1 The Contractor accepts the relationship of trust and confidence established by this Agreement and covenants with the Owner to cooperate with the Architect and utilize the Contractor's best skill, efforts and judgment in furthering the interests of the Owner; to furnish efficient business administration and supervision; to make best efforts to furnish at all times an adequate supply of workers and materials; and to perform the Work in the best way and most expeditious and economical manner consistent with the interests of the Owner. The Owner agrees to exercise best efforts to enable the Contractor to perform the Work in the best way and most expeditious manner by furnishing and approving in a timely way information required by the Contractor and making payments to the Contractor in accordance with requirements of the Contract Documents.

ARTICLE 4
DATE OF COMMENCEMENT AND SUBSTANTIAL COMPLETION

4.1 The date of commencement is the date from which the Contract Time of Subparagraph 4.2 is measured; it shall be the date of this Agreement, as first written above, unless a different date is stated below or provision is made for the date to be fixed in a notice to proceed issued by the Owner.

(Insert the date of commencement, if it differs from the date of this Agreement or, if applicable, state that the date will be fixed in a notice to proceed.)

Unless the date of commencement is established by a notice to proceed issued by the Owner, the Contractor shall notify the Owner in writing not less than five days before commencing the Work to permit the timely filing of mortgages, mechanic's liens and other security interests.

4.2 The Contractor shall achieve Substantial Completion of the entire Work not later than

(Insert the calendar date or number of calendar days after the date of commencement. Also insert any requirements for earlier Substantial Completion of certain portions of the Work, if not stated elsewhere in the Contract Documents.)

, subject to adjustments of this Contract Time as provided in the Contract Documents.

(Insert provisions, if any, for liquidated damages relating to failure to complete on time.)

ARTICLE 5
CONTRACT SUM

5.1 The Owner shall pay the Contractor in current funds for the Contractor's performance of the Contract the Contract Sum consisting of the Cost of the Work as defined in Article 7 and the Contractor's Fee determined as follows:

(State a lump sum, percentage of Cost of the Work or other provision for determining the Contractor's Fee, and explain how the Contractor's Fee is to be adjusted for changes in the Work.)

5.2 GUARANTEED MAXIMUM PRICE (IF APPLICABLE)

5.2.1 The sum of the Cost of the Work and the Contractor's Fee is guaranteed by the Contractor not to exceed Dollars (**$**), subject to additions and deductions by Change Order as provided in the Contract Documents. Such maximum sum is referred to in the Contract Documents as the Guaranteed Maximum Price. Costs which would cause the Guaranteed Maximum Price to be exceeded shall be paid by the Contractor without reimbursement by the Owner.

(Insert specific provisions if the Contractor is to participate in any savings.)

5.2.2 The Guaranteed Maximum Price is based upon the following alternates, if any, which are described in the Contract Documents and are hereby accepted by the Owner:

(State the numbers or other identification of accepted alternates, but only if a Guaranteed Maximum Price is inserted in Subparagraph 5.2.1. If decisions on other alternates are to be made by the Owner subsequent to the execution of this Agreement, attach a schedule of such other alternates showing the amount for each and the date until which that amount is valid.)

5.2.3 The amounts agreed to for unit prices, if any, are as follows:

(State unit prices only if a Guaranteed Maximum Price is inserted in Subparagraph 5.2.1.)

ARTICLE 6
CHANGES IN THE WORK

6.1 CONTRACTS WITH A GUARANTEED MAXIMUM PRICE

6.1.1 Adjustments to the Guaranteed Maximum Price on account of changes in the Work may be determined by any of the methods listed in Subparagraph 7.3.3 of the General Conditions.

6.1.2 In calculating adjustments to subcontracts (except those awarded with the Owner's prior consent on the basis of cost plus a fee), the terms "cost" and "fee" as used in Clause 7.3.3.3 of the General Conditions and the terms "costs" and "a reasonable allowance for overhead and profit" as used in Subparagraph 7.3.6 of the General Conditions shall have the meanings assigned to them in the General Conditions and shall not be modified by Articles 5, 7 and 8 of this Agreement. Adjustments to subcontracts awarded with the Owner's prior consent on the basis of cost plus a fee shall be calculated in accordance with the terms of those subcontracts.

6.1.3 In calculating adjustments to this Contract, the terms "cost" and "costs" as used in the above-referenced provisions of the General Conditions shall mean the Cost of the Work as defined in Article 7 of this Agreement and the terms "fee" and "a reasonable allowance for overhead and profit" shall mean the Contractor's Fee as defined in Paragraph 5.1 of this Agreement.

6.2 CONTRACTS WITHOUT A GUARANTEED MAXIMUM PRICE

6.2.1 Increased costs for the items set forth in Article 7 which result from changes in the Work shall become part of the Cost of the Work, and the Contractor's Fee shall be adjusted as provided in Paragraph 5.1.

6.3 ALL CONTRACTS

6.3.1 If no specific provision is made in Paragraph 5.1 for adjustment of the Contractor's Fee in the case of changes in the Work, or if the extent of such changes is such, in the aggregate, that application of the adjustment provisions of Paragraph 5.1 will cause substantial inequity to the Owner or Contractor, the Contractor's Fee shall be equitably adjusted on the basis of the Fee established for the original Work.

ARTICLE 7
COSTS TO BE REIMBURSED

7.1 The term Cost of the Work shall mean costs necessarily incurred by the Contractor in the proper performance of the Work. Such costs shall be at rates not higher than the standard paid at the place of the Project except with prior consent of the Owner. The Cost of the Work shall include only the items set forth in this Article 7.

7.1.1 LABOR COSTS

7.1.1.1 Wages of construction workers directly employed by the Contractor to perform the construction of the Work at the site or, with the Owner's agreement, at off-site workshops.

7.1.1.2 Wages or salaries of the Contractor's supervisory and administrative personnel when stationed at the site with the Owner's agreement.

(If it is intended that the wages or salaries of certain personnel stationed at the Contractor's principal or other offices shall be included in the Cost of the Work, identify in Article 14 the personnel to be included and whether for all or only part of their time.)

7.1.1.3 Wages and salaries of the Contractor's supervisory or administrative personnel engaged, at factories, workshops or on the road, in expediting the production or transportation of materials or equipment required for the Work, but only for that portion of their time required for the Work.

7.1.1.4 Costs paid or incurred by the Contractor for taxes, insurance, contributions, assessments and benefits required by law or collective bargaining agreements and, for personnel not covered by such agreements, customary benefits such as sick leave, medical and health benefits, holidays, vacations and pensions, provided such costs are based on wages and salaries included in the Cost of the Work under Clauses 7.1.1.1 through 7.1.1.3.

7.1.2 SUBCONTRACT COSTS

Payments made by the Contractor to Subcontractors in accordance with the requirements of the subcontracts.

7.1.3 COSTS OF MATERIALS AND EQUIPMENT INCORPORATED IN THE COMPLETED CONSTRUCTION

7.1.3.1 Costs, including transportation, of materials and equipment incorporated or to be incorporated in the completed construction.

7.1.3.2 Costs of materials described in the preceding Clause 7.1.3.1 in excess of those actually installed but required to provide reasonable allowance for waste and for spoilage. Unused excess materials, if any, shall be handed over to the Owner at the completion of the Work or, at the Owner's option, shall be sold by the Contractor; amounts realized, if any, from such sales shall be credited to the Owner as a deduction from the Cost of the Work.

7.1.4 COSTS OF OTHER MATERIALS AND EQUIPMENT, TEMPORARY FACILITIES AND RELATED ITEMS

7.1.4.1 Costs, including transportation, installation, maintenance, dismantling and removal of materials, supplies, temporary facilities, machinery, equipment, and hand tools not customarily owned by the construction workers, which are provided by the Contractor at the site and fully consumed in the performance of the Work; and cost less salvage value on such items if not fully consumed, whether sold to others or retained by the Contractor. Cost for items previously used by the Contractor shall mean fair market value.

7.1.4.2 Rental charges for temporary facilities, machinery, equipment, and hand tools not customarily owned by the construction workers, which are provided by the Contractor at the site, whether rented from the Contractor or others, and costs of transportation, installation, minor repairs and replacements, dismantling and removal thereof. Rates and quantities of equipment rented shall be subject to the Owner's prior approval.

7.1.4.3 Costs of removal of debris from the site.

7.1.4.4 Costs of telegrams and long-distance telephone calls, postage and parcel delivery charges, telephone service at the site and reasonable petty cash expenses of the site office.

7.1.4.5 That portion of the reasonable travel and subsistence expenses of the Contractor's personnel incurred while traveling in discharge of duties connected with the Work.

7.1.5 MISCELLANEOUS COSTS

7.1.5.1 That portion directly attributable to this Contract of premiums for insurance and bonds.

7.1.5.2 Sales, use or similar taxes imposed by a governmental authority which are related to the Work and for which the Contractor is liable.

7.1.5.3 Fees and assessments for the building permit and for other permits, licenses and inspections for which the Contractor is required by the Contract Documents to pay.

7.1.5.4 Fees of testing laboratories for tests required by the Contract Documents, except those related to defective or nonconforming Work for which reimbursement is excluded by Subparagraph 13.5.3 of the General Conditions or other provisions of the Contract Documents and which do not fall within the scope of Subparagraphs 7.2.2 through 7.2.4 below.

7.1.5.5 Royalties and license fees paid for the use of a particular design, process or product required by the Contract Documents; the cost of defending suits or claims for infringement of patent rights arising from such requirement by the Contract Documents; payments made in accordance with legal judgments against the Contractor resulting from such suits or claims and payments of settlements made with the Owner's consent; provided, however, that such costs of legal defenses, judgment and settlements shall not be included in the calculation of the Contractor's Fee or of a Guaranteed Maximum Price, if any, and provided that such royalties, fees and costs are not excluded by the last sentence of Subparagraph 3.17.1 of the General Conditions or other provisions of the Contract Documents.

7.1.5.6 Deposits lost for causes other than the Contractor's fault or negligence.

7.1.6 OTHER COSTS

7.1.6.1 Other costs incurred in the performance of the Work if and to the extent approved in advance in writing by the Owner.

7.2 EMERGENCIES: REPAIRS TO DAMAGED, DEFECTIVE OR NONCONFORMING WORK

The Cost of the Work shall also include costs described in Paragraph 7.1 which are incurred by the Contractor:

7.2.1 In taking action to prevent threatened damage, injury or loss in case of an emergency affecting the safety of persons and property, as provided in Paragraph 10.3 of the General Conditions.

7.2.2 In repairing or correcting Work damaged or improperly executed by construction workers in the employ of the Contractor, provided such damage or improper execution did not result from the fault or negligence of the Contractor or the Contractor's foremen, engineers or superintendents, or other supervisory, administrative or managerial personnel of the Contractor.

7.2.3 In repairing damaged Work other than that described in Subparagraph 7.2.2, provided such damage did not result from the fault or negligence of the Contractor or the Contractor's personnel, and only to the extent that the cost of such repairs is not recoverable by the Contractor from others and the Contractor is not compensated therefor by insurance or otherwise.

7.2.4 In correcting defective or nonconforming Work performed or supplied by a Subcontractor or material supplier and not corrected by them, provided such defective or nonconforming Work did not result from the fault or neglect of the Contractor or the Contractor's personnel adequately to supervise and direct the Work of the Subcontractor or material supplier, and only to the extent that the cost of correcting the defective or nonconforming Work is not recoverable by the Contractor from the Subcontractor or material supplier.

ARTICLE 8

COSTS NOT TO BE REIMBURSED

8.1 The Cost of the Work shall not include:

8.1.1 Salaries and other compensation of the Contractor's personnel stationed at the Contractor's principal office or offices other than the site office, except as specifically provided in Clauses 7.1.1.2 and 7.1.1.3 or as may be provided in Article 14.

8.1.2 Expenses of the Contractor's principal office and offices other than the site office.

8.1.3 Overhead and general expenses, except as may be expressly included in Article 7.

8.1.4 The Contractor's capital expenses, including interest on the Contractor's capital employed for the Work.

8.1.5 Rental costs of machinery and equipment, except as specifically provided in Clause 7.1.4.2.

8.1.6 Except as provided in Subparagraphs 7.2.2 through 7.2.4 and Paragraph 13.5 of this Agreement, costs due to the fault or negligence of the Contractor, Subcontractors, anyone directly or indirectly employed by any of them, or for whose acts any of them may be liable, including but not limited to costs for the correction of damaged, defective or nonconforming Work, disposal and replacement of materials and equipment incorrectly ordered or supplied, and making good damage to property not forming part of the Work.

8.1.7 Any cost not specifically and expressly described in Article 7.

8.1.8 Costs which would cause the Guaranteed Maximum Price, if any, to be exceeded.

ARTICLE 9
DISCOUNTS, REBATES AND REFUNDS

9.1 Cash discounts obtained on payments made by the Contractor shall accrue to the Owner if (1) before making the payment, the Contractor included them in an Application for Payment and received payment therefor from the Owner, or (2) the Owner has deposited funds with the Contractor with which to make payments; otherwise, cash discounts shall accrue to the Contractor. Trade discounts, rebates, refunds and amounts received from sales of surplus materials and equipment shall accrue to the Owner, and the Contractor shall make provisions so that they can be secured.

9.2 Amounts which accrue to the Owner in accordance with the provisions of Paragraph 9.1 shall be credited to the Owner as a deduction from the Cost of the Work.

ARTICLE 10
SUBCONTRACTS AND OTHER AGREEMENTS

10.1 Those portions of the Work that the Contractor does not customarily perform with the Contractor's own personnel shall be performed under subcontracts or by other appropriate agreements with the Contractor. The Contractor shall obtain bids from Subcontractors and from suppliers of materials or equipment fabricated especially for the Work and shall deliver such bids to the Architect. The Owner will then determine, with the advice of the Contractor and subject to the reasonable objection of the Architect, which bids will be accepted. The Owner may designate specific persons or entities from whom the Contractor shall obtain bids; however, if a Guaranteed Maximum Price has been established, the Owner may not prohibit the Contractor from obtaining bids from others. The Contractor shall not be required to contract with anyone to whom the Contractor has reasonable objection.

10.2 If a Guaranteed Maximum Price has been established and a specific bidder among those whose bids are delivered by the Contractor to the Architect (1) is recommended to the Owner by the Contractor; (2) is qualified to perform that portion of the Work; and (3) has submitted a bid which conforms to the requirements of the Contract Documents without reservations or exceptions, but the Owner requires that another bid be accepted; then the Contractor may require that a Change Order be issued to adjust the Guaranteed Maximum Price by the difference between the bid of the person or entity recommended to the Owner by the Contractor and the amount of the subcontract or other agreement actually signed with the person or entity designated by the Owner.

10.3 Subcontracts or other agreements shall conform to the payment provisions of Paragraphs 12.7 and 12.8, and shall not be awarded on the basis of cost plus a fee without the prior consent of the Owner.

ARTICLE 11
ACCOUNTING RECORDS

11.1 The Contractor shall keep full and detailed accounts and exercise such controls as may be necessary for proper financial management under this Contract; the accounting and control systems shall be satisfactory to the Owner. The Owner and the Owner's accountants shall be afforded access to the Contractor's records, books, correspondence, instructions, drawings, receipts, subcontracts, purchase orders, vouchers, memoranda and other data relating to this Contract, and the Contractor shall preserve these for a period of three years after final payment, or for such longer period as may be required by law.

ARTICLE 12
PROGRESS PAYMENTS

12.1 Based upon Applications for Payment submitted to the Architect by the Contractor and Certificates for Payment issued by the Architect, the Owner shall make progress payments on account of the Contract Sum to the Contractor as provided below and elsewhere in the Contract Documents.

12.2 The period covered by each Application for Payment shall be one calendar month ending on the last day of the month, or as follows:

12.3 Provided an Application for Payment is received by the Architect not later than the day of a month, the Owner shall make payment to the Contractor not later than the day of the month. If an Application for Payment is received by the Architect after the application date fixed above, payment shall be made by the Owner not later than days after the Architect receives the Application for Payment.

12.4 With each Application for Payment the Contractor shall submit payrolls, petty cash accounts, receipted invoices or invoices with check vouchers attached, and any other evidence required by the Owner or Architect to demonstrate that cash disbursements already made by the Contractor on account of the Cost of the Work equal or exceed (1) progress payments already received by the Contractor; less (2) that portion of those payments attributable to the Contractor's Fee; plus (3) payrolls for the period covered by the present Application for Payment; plus (4) retainage provided in Subparagraph 12.5.4, if any, applicable to prior progress payments.

12.5 CONTRACTS WITH A GUARANTEED MAXIMUM PRICE

12.5.1 Each Application for Payment shall be based upon the most recent schedule of values submitted by the Contractor in accordance with the Contract Documents. The schedule of values shall allocate the entire Guaranteed Maximum Price among the various portions of the Work, except that the Contractor's Fee shall be shown as a single separate item. The schedule of values shall be prepared in such form and supported by such data to substantiate its accuracy as the Architect may require. This schedule, unless objected to by the Architect, shall be used as a basis for reviewing the Contractor's Applications for Payment.

12.5.2 Applications for Payment shall show the percentage completion of each portion of the Work as of the end of the period covered by the Application for Payment. The percentage completion shall be the lesser of (1) the percentage of that portion of the Work which has actually been completed or (2) the percentage obtained by dividing (a) the expense which has actually been incurred by the Contractor on account of that portion of the Work for which the Contractor has made or intends to make actual payment prior to the next Application for Payment by (b) the share of the Guaranteed Maximum Price allocated to that portion of the Work in the schedule of values.

12.5.3 Subject to other provisions of the Contract Documents, the amount of each progress payment shall be computed as follows:

12.5.3.1 Take that portion of the Guaranteed Maximum Price properly allocable to completed Work as determined by multiplying the percentage completion of each portion of the Work by the share of the Guaranteed Maximum Price allocated to that portion of the Work in the schedule of values. Pending final determination of cost to the Owner of changes in the Work, amounts not in dispute may be included as provided in Subparagraph 7.3.7 of the General Conditions, even though the Guaranteed Maximum Price has not yet been adjusted by Change Order.

12.5.3.2 Add that portion of the Guaranteed Maximum Price properly allocable to materials and equipment delivered and suitably stored at the site for subsequent incorporation in the Work or, if approved in advance by the Owner, suitably stored off the site at a location agreed upon in writing.

12.5.3.3 Add the Contractor's Fee, less retainage of percent (%). The Contractor's Fee shall be computed upon the Cost of the Work described in the two preceding Clauses at the rate stated in Paragraph 5.1 or, if the Contractor's Fee is stated as a fixed sum in that Paragraph, shall be an amount which bears the same ratio to that fixed-sum Fee as the Cost of the Work in the two preceding Clauses bears to a reasonable estimate of the probable Cost of the Work upon its completion.

12.5.3.4 Subtract the aggregate of previous payments made by the Owner.

12.5.3.5 Subtract the shortfall, if any, indicated by the Contractor in the documentation required by Paragraph 12.4 to substantiate prior Applications for Payment, or resulting from errors subsequently discovered by the Owner's accountants in such documentation.

12.5.3.6 Subtract amounts, if any, for which the Architect has withheld or nullified a Certificate for Payment as provided in Paragraph 9.5 of the General Conditions.

12.5.4 Additional retainage, if any, shall be as follows:
(If it is intended to retain additional amounts from progress payments to the Contractor beyond (1) the retainage from the Contractor's Fee provided in Clause 12.5.3.3, (2) the retainage from Subcontractors provided in Paragraph 12.7 below, and (3) the retainage, if any, provided by other provisions of the Contract, insert provision for such additional retainage here. Such provision, if made, should also describe any arrangement for limiting or reducing the amount retained after the Work reaches a certain state of completion.)

12.6 CONTRACTS WITHOUT A GUARANTEED MAXIMUM PRICE

12.6.1 Applications for Payment shall show the Cost of the Work actually incurred by the Contractor through the end of the period covered by the Application for Payment and for which the Contractor has made or intends to make actual payment prior to the next Application for Payment.

12.6.2 Subject to other provisions of the Contract Documents, the amount of each progress payment shall be computed as follows:

12.6.2.1 Take the Cost of the Work as described in Subparagraph 12.6.1.

12.6.2.2 Add the Contractor's Fee, less retainage of percent (%). The Contractor's Fee shall be computed upon the Cost of the Work described in the preceding Clause 12.6.2.1 at the rate stated in Paragraph 5.1 or, if the Contractor's Fee is stated as a fixed sum in that Paragraph, an amount which bears the same ratio to that fixed-sum Fee as the Cost of the Work in the preceding Clause bears to a reasonable estimate of the probable Cost of the Work upon its completion.

12.6.2.3 Subtract the aggregate of previous payments made by the Owner.

12.6.2.4 Subtract the shortfall, if any, indicated by the Contractor in the documentation required by Paragraph 12.4 or to substantiate prior Applications for Payment or resulting from errors subsequently discovered by the Owner's accountants in such documentation.

12.6.2.5 Subtract amounts, if any, for which the Architect has withheld or withdrawn a Certificate for Payment as provided in the Contract Documents.

12.6.3 Additional retainage, if any, shall be as follows:

12.7 Except with the Owner's prior approval, payments to Subcontractors included in the Contractor's Applications for Payment shall not exceed an amount for each Subcontractor calculated as follows:

12.7.1 Take that portion of the Subcontract Sum properly allocable to completed Work as determined by multiplying the percentage completion of each portion of the Subcontractor's Work by the share of the total Subcontract Sum allocated to that portion in the Subcontractor's schedule of values, less retainage of percent (%). Pending final determination of amounts to be paid to the Subcontractor for changes in the Work, amounts not in dispute may be included as provided in Subparagraph 7.3.7 of the General Conditions even though the Subcontract Sum has not yet been adjusted by Change Order.

12.7.2 Add that portion of the Subcontract Sum properly allocable to materials and equipment delivered and suitably stored at the site for subsequent incorporation in the Work or, if approved in advance by the Owner, suitably stored off the site at a location agreed upon in writing, less retainage of percent (%).

12.7.3 Subtract the aggregate of previous payments made by the Contractor to the Subcontractor.

12.7.4 Subtract amounts, if any, for which the Architect has withheld or nullified a Certificate for Payment by the Owner to the Contractor for reasons which are the fault of the Subcontractor.

12.7.5 Add, upon Substantial Completion of the entire Work of the Contractor, a sum sufficient to increase the total payments to the Subcontractor to percent (%) of the Subcontract Sum, less amounts, if any, for incomplete Work and unsettled claims; and, if final completion of the entire Work is thereafter materially delayed through no fault of the Subcontractor, add any additional amounts payable on account of Work of the Subcontractor in accordance with Subparagraph 9.10.3 of the General Conditions.

(If it is intended, prior to Substantial Completion of the entire Work of the Contractor, to reduce or limit the retainage from Subcontractors resulting from the percentages inserted in Subparagraphs 12.7.1 and 12.7.2 above, and this is not explained elsewhere in the Contract Documents, insert here provisions for such reduction or limitation.)

The Subcontract Sum is the total amount stipulated in the subcontract to be paid by the Contractor to the Subcontractor for the Subcontractor's performance of the subcontract.

12.8 Except with the Owner's prior approval, the Contractor shall not make advance payments to suppliers for materials or equipment which have not been delivered and stored at the site.

12.9 In taking action on the Contractor's Applications for Payment, the Architect shall be entitled to rely on the accuracy and completeness of the information furnished by the Contractor and shall not be deemed to represent that the Architect has made a detailed examination, audit or arithmetic verification of the documentation submitted in accordance with Paragraph 12.4 or other supporting data; that the Architect has made exhaustive or continuous on-site inspections or that the Architect has made examinations to ascertain how or for what purposes the Contractor has used amounts previously paid on account of the Contract. Such examinations, audits and verifications, if required by the Owner, will be performed by the Owner's accountants acting in the sole interest of the Owner.

ARTICLE 13
FINAL PAYMENT

13.1 Final payment shall be made by the Owner to the Contractor when (1) the Contract has been fully performed by the Contractor except for the Contractor's responsibility to correct defective or nonconforming Work, as provided in Subparagraph 12.2.2 of the General Conditions, and to satisfy other requirements, if any, which necessarily survive final payment; (2) a final Application for Payment and a final accounting for the Cost of the Work have been submitted by the Contractor and reviewed by the Owner's accountants; and (3) a final Certificate for Payment has then been issued by the Architect; such final payment shall be made by the Owner not more than 30 days after the issuance of the Architect's final Certificate for Payment, or as follows:

13.2 The amount of the final payment shall be calculated as follows:

13.2.1 Take the sum of the Cost of the Work substantiated by the Contractor's final accounting and the Contractor's Fee; but not more than the Guaranteed Maximum Price, if any.

13.2.2 Subtract amounts, if any, for which the Architect withholds, in whole or in part, a final Certificate for Payment as provided in Subparagraph 9.5.1 of the General Conditions or other provisions of the Contract Documents.

13.2.3 Subtract the aggregate of previous payments made by the Owner.

If the aggregate of previous payments made by the Owner exceeds the amount due the Contractor, the Contractor shall reimburse the difference to the Owner.

13.3 The Owner's accountants will review and report in writing on the Contractor's final accounting within 30 days after delivery of the final accounting to the Architect by the Contractor. Based upon such Cost of the Work as the Owner's accountants report to be substantiated by the Contractor's final accounting, and provided the other conditions of Paragraph 13.1 have been met, the Architect will, within seven days after receipt of the written report of the Owner's accountants, either issue to the Owner a final Certificate for Payment with a copy to the Contractor, or notify the Contractor and Owner in writing of the Architect's reasons for withholding a certificate as provided in Subparagraph 9.5.1 of the General Conditions. The time periods stated in this Paragraph 13.3 supersede those stated in Subparagraph 9.4.1 of the General Conditions.

13.4 If the Owner's accountants report the Cost of the Work as substantiated by the Contractor's final accounting to be less than claimed by the Contractor, the Contractor shall be entitled to demand arbitration of the disputed amount without a further decision of the Architect. Such demand for arbitration shall be made by the Contractor within 30 days after the Contractor's receipt of a copy of the Architect's final Certificate for Payment; failure to demand arbitration within this 30-day period shall result in the substantiated amount reported by the Owner's accountants becoming binding on the Contractor. Pending a final resolution by arbitration, the Owner shall pay the Contractor the amount certified in the Architect's final Certificate for Payment.

13.5 If, subsequent to final payment and at the Owner's request, the Contractor incurs costs described in Article 7 and not excluded by Article 8 to correct defective or nonconforming Work, the Owner shall reimburse the Contractor such costs and the Contractor's Fee applicable thereto on the same basis as if such costs had been incurred prior to final payment, but not in excess of the Guaranteed Maximum Price, if any. If the Contractor has participated in savings as provided in Paragraph 5.2, the amount of such savings shall be recalculated and appropriate credit given to the Owner in determining the net amount to be paid by the Owner to the Contractor.

ARTICLE 14
MISCELLANEOUS PROVISIONS

14.1 Where reference is made in this Agreement to a provision of the General Conditions or another Contract Document, the reference refers to that provision as amended or supplemented by other provisions of the Contract Documents.

14.2 Payments due and unpaid under the Contract shall bear interest from the date payment is due at the rate stated below, or in the absence thereof, at the legal rate prevailing from time to time at the place where the Project is located.

(Insert rate of interest agreed upon, if any.)

(Usury laws and requirements under the Federal Truth in Lending Act, similar state and local consumer credit laws and other regulations at the Owner's and Contractor's principal places of business, the location of the Project and elsewhere may affect the validity of this provision. Legal advice should be obtained with respect to deletions or modifications, and also regarding requirements such as written disclosures or waivers.)

14.3 Other provisions:

ARTICLE 15
TERMINATION OR SUSPENSION

15.1 The Contract may be terminated by the Contractor as provided in Article 14 of the General Conditions; however, the amount to be paid to the Contractor under Subparagraph 14.1.2 of the General Conditions shall not exceed the amount the Contractor would be entitled to receive under Paragraph 15.3 below, except that the Contractor's Fee shall be calculated as if the Work had been fully completed by the Contractor, including a reasonable estimate of the Cost of the Work for Work not actually completed.

15.2 If a Guaranteed Maximum Price is established in Article 5, the Contract may be terminated by the Owner for cause as provided in Article 14 of the General Conditions; however, the amount, if any, to be paid to the Contractor under Subparagraph 14.2.4 of the General Conditions shall not cause the Guaranteed Maximum Price to be exceeded, nor shall it exceed the amount the Contractor would be entitled to receive under Paragraph 15.3 below.

15.3 If no Guaranteed Maximum Price is established in Article 5, the Contract may be terminated by the Owner for cause as provided in Article 14 of the General Conditions; however, the Owner shall then pay the Contractor an amount calculated as follows:

15.3.1 Take the Cost of the Work incurred by the Contractor to the date of termination.

15.3.2 Add the Contractor's Fee computed upon the Cost of the Work to the date of termination at the rate stated in Paragraph 5.1 or, if the Contractor's Fee is stated as a fixed sum in that Paragraph, an amount which bears the same ratio to that fixed-sum Fee as the Cost of the Work at the time of termination bears to a reasonable estimate of the probable Cost of the Work upon its completion.

15.3.3 Subtract the aggregate of previous payments made by the Owner.

The Owner shall also pay the Contractor fair compensation, either by purchase or rental at the election of the Owner, for any equipment owned by the Contractor which the Owner elects to retain and which is not otherwise included in the Cost of the Work under Subparagraph 15.3.1. To the extent that the Owner elects to take legal assignment of subcontracts and purchase orders (including rental agreements), the Contractor shall, as a condition of receiving the payments referred to in this Article 15, execute and deliver all such papers and take all such steps, including the legal assignment of such subcontracts and other contractual rights of the Contractor, as the Owner may require for the purpose of fully vesting in the Owner the rights and benefits of the Contractor under such subcontracts or purchase orders.

15.4 The Work may be suspended by the Owner as provided in Article 14 of the General Conditions; in such case, the Guaranteed Maximum Price, if any, shall be increased as provided in Subparagraph 14.3.2 of the General Conditions except that the term "cost of performance of the Contract" in that Subparagraph shall be understood to mean the Cost of the Work and the term "profit" shall be understood to mean the Contractor's Fee as described in Paragraphs 5.1 and 6.3 of this Agreement.

ARTICLE 16
ENUMERATION OF CONTRACT DOCUMENTS

16.1 The Contract Documents, except for Modifications issued after execution of this Agreement, are enumerated as follows:

16.1.1 The Agreement is this executed Standard Form of Agreement Between Owner and Contractor, AIA Document A111, 1987 Edition.

16.1.2 The General Conditions are the General Conditions of the Contract for Construction, AIA Document A201, 1987 Edition.

16.1.3 The Supplementary and other Conditions of the Contract are those contained in the Project Manual dated
, and are as follows:

Document	**Title**	**Pages**

16.1.4 The Specifications are those contained in the Project Manual dated as in Paragraph 16.1.3, and are as follows:
(Either list the Specifications here or refer to an exhibit attached to this Agreement.)

Section	**Title**	**Pages**

16.1.5 The Drawings are as follows, and are dated unless a different date is shown below:

(Either list the Drawings here or refer to an exhibit attached to this Agreement.)

Number	Title	Date

16.1.6 The Addenda, if any, are as follows:

Number	Date	Pages

Portions of Addenda relating to bidding requirements are not part of the Contract Documents unless the bidding requirements are also enumerated in this Article 16.

16.1.7 Other Documents, if any, forming part of the Contract Documents are as follows:
(List here any additional documents which are intended to form part of the Contract Documents. The General Conditions provide that bidding requirements such as advertisement or invitation to bid, Instructions to Bidders, sample forms and the Contractor's bid are not part of the Contract Documents unless enumerated in this Agreement. They should be listed here only if intended to be part of the Contract Documents.)

This Agreement is entered into as of the day and year first written above and is executed in at least three original copies of which one is to be delivered to the Contractor, one to the Architect for use in the administration of the Contract, and the remainder to the Owner.

OWNER

(Signature)

(Printed name and title)

CONTRACTOR

(Signature)

(Printed name and title)

THE AMERICAN INSTITUTE OF ARCHITECTS

AIA Document A201

General Conditions of the Contract for Construction

THIS DOCUMENT HAS IMPORTANT LEGAL CONSEQUENCES; CONSULTATION WITH AN ATTORNEY IS ENCOURAGED WITH RESPECT TO ITS MODIFICATION

1987 EDITION
TABLE OF ARTICLES

This document has been approved and endorsed by the Associated General Contractors of America.

INDEX

 AIA DOCUMENT A201 • GENERAL CONDITIONS OF THE CONTRACT FOR CONSTRUCTION • FOURTEENTH EDITION

GENERAL CONDITIONS OF THE CONTRACT FOR CONSTRUCTION

ARTICLE 1

GENERAL PROVISIONS

1.1 BASIC DEFINITIONS

1.1.1 THE CONTRACT DOCUMENTS

The Contract Documents consist of the Agreement between Owner and Contractor (hereinafter the Agreement), Conditions of the Contract (General, Supplementary and other Conditions), Drawings, Specifications, addenda issued prior to execution of the Contract, other documents listed in the Agreement and Modifications issued after execution of the Contract. A Modification is (1) a written amendment to the Contract signed by both parties, (2) a Change Order, (3) a Construction Change Directive or (4) a written order for a minor change in the Work issued by the Architect. Unless specifically enumerated in the Agreement, the Contract Documents do not include other documents such as bidding requirements (advertisement or invitation to bid, Instructions to Bidders, sample forms, the Contractor's bid or portions of addenda relating to bidding requirements).

1.1.2 THE CONTRACT

The Contract Documents form the Contract for Construction. The Contract represents the entire and integrated agreement between the parties hereto and supersedes prior negotiations, representations or agreements, either written or oral. The Contract may be amended or modified only by a Modification. The Contract Documents shall not be construed to create a contractual relationship of any kind (1) between the Architect and Contractor, (2) between the Owner and a Subcontractor or Sub-subcontractor or (3) between any persons or entities other than the Owner and Contractor. The Architect shall, however, be entitled to performance and enforcement of obligations under the Contract intended to facilitate performance of the Architect's duties.

1.1.3 THE WORK

The term "Work" means the construction and services required by the Contract Documents, whether completed or partially completed, and includes all other labor, materials, equipment and services provided or to be provided by the Contractor to fulfill the Contractor's obligations. The Work may constitute the whole or a part of the Project.

1.1.4 THE PROJECT

The Project is the total construction of which the Work performed under the Contract Documents may be the whole or a part and which may include construction by the Owner or by separate contractors.

1.1.5 THE DRAWINGS

The Drawings are the graphic and pictorial portions of the Contract Documents, wherever located and whenever issued, showing the design, location and dimensions of the Work, generally including plans, elevations, sections, details, schedules and diagrams.

1.1.6 THE SPECIFICATIONS

The Specifications are that portion of the Contract Documents consisting of the written requirements for materials, equipment, construction systems, standards and workmanship for the Work, and performance of related services.

1.1.7 THE PROJECT MANUAL

The Project Manual is the volume usually assembled for the Work which may include the bidding requirements, sample forms, Conditions of the Contract and Specifications.

1.2 EXECUTION, CORRELATION AND INTENT

1.2.1 The Contract Documents shall be signed by the Owner and Contractor as provided in the Agreement. If either the Owner or Contractor or both do not sign all the Contract Documents, the Architect shall identify such unsigned Documents upon request.

1.2.2 Execution of the Contract by the Contractor is a representation that the Contractor has visited the site, become familiar with local conditions under which the Work is to be performed and correlated personal observations with requirements of the Contract Documents.

1.2.3 The intent of the Contract Documents is to include all items necessary for the proper execution and completion of the Work by the Contractor. The Contract Documents are complementary, and what is required by one shall be as binding as if required by all; performance by the Contractor shall be required only to the extent consistent with the Contract Documents and reasonably inferable from them as being necessary to produce the intended results.

1.2.4 Organization of the Specifications into divisions, sections and articles, and arrangement of Drawings shall not control the Contractor in dividing the Work among Subcontractors or in establishing the extent of Work to be performed by any trade.

1.2.5 Unless otherwise stated in the Contract Documents, words which have well-known technical or construction industry meanings are used in the Contract Documents in accordance with such recognized meanings.

1.3 OWNERSHIP AND USE OF ARCHITECT'S DRAWINGS, SPECIFICATIONS AND OTHER DOCUMENTS

1.3.1 The Drawings, Specifications and other documents prepared by the Architect are instruments of the Architect's service through which the Work to be executed by the Contractor is described. The Contractor may retain one contract record set. Neither the Contractor nor any Subcontractor, Sub-subcontractor or material or equipment supplier shall own or claim a copyright in the Drawings, Specifications and other documents prepared by the Architect, and unless otherwise indicated the Architect shall be deemed the author of them and will retain all common law, statutory and other reserved rights, in addition to the copyright. All copies of them, except the Contractor's record set, shall be returned or suitably accounted for to the Architect, on request, upon completion of the Work. The Drawings, Specifications and other documents prepared by the Architect, and copies thereof furnished to the Contractor, are for use solely with respect to this Project. They are not to be used by the Contractor or any Subcontractor, Sub-subcontractor or material or equipment supplier on other projects or for additions to this Project outside the scope of the

Work without the specific written consent of the Owner and Architect. The Contractor, Subcontractors, Sub-subcontractors and material or equipment suppliers are granted a limited license to use and reproduce applicable portions of the Drawings, Specifications and other documents prepared by the Architect appropriate to and for use in the execution of their Work under the Contract Documents. All copies made under this license shall bear the statutory copyright notice, if any, shown on the Drawings, Specifications and other documents prepared by the Architect. Submittal or distribution to meet official regulatory requirements or for other purposes in connection with this Project is not to be construed as publication in derogation of the Architect's copyright or other reserved rights.

1.4 CAPITALIZATION

1.4.1 Terms capitalized in these General Conditions include those which are (1) specifically defined, (2) the titles of numbered articles and identified references to Paragraphs, Subparagraphs and Clauses in the document or (3) the titles of other documents published by the American Institute of Architects.

1.5 INTERPRETATION

1.5.1 In the interest of brevity the Contract Documents frequently omit modifying words such as "all" and "any" and articles such as "the" and "an," but the fact that a modifier or an article is absent from one statement and appears in another is not intended to affect the interpretation of either statement.

ARTICLE 2

OWNER

2.1 DEFINITION

2.1.1 The Owner is the person or entity identified as such in the Agreement and is referred to throughout the Contract Documents as if singular in number. The term "Owner" means the Owner or the Owner's authorized representative.

2.1.2 The Owner upon reasonable written request shall furnish to the Contractor in writing information which is necessary and relevant for the Contractor to evaluate, give notice of or enforce mechanic's lien rights. Such information shall include a correct statement of the record legal title to the property on which the Project is located, usually referred to as the site, and the Owner's interest therein at the time of execution of the Agreement and, within five days after any change, information of such change in title, recorded or unrecorded.

2.2 INFORMATION AND SERVICES REQUIRED OF THE OWNER

2.2.1 The Owner shall, at the request of the Contractor, prior to execution of the Agreement and promptly from time to time thereafter, furnish to the Contractor reasonable evidence that financial arrangements have been made to fulfill the Owner's obligations under the Contract. *[Note: Unless such reasonable evidence were furnished on request prior to the execution of the Agreement, the prospective contractor would not be required to execute the Agreement or to commence the Work.]*

2.2.2 The Owner shall furnish surveys describing physical characteristics, legal limitations and utility locations for the site of the Project, and a legal description of the site.

2.2.3 Except for permits and fees which are the responsibility of the Contractor under the Contract Documents, the Owner shall secure and pay for necessary approvals, easements, assessments and charges required for construction, use or occupancy of permanent structures or for permanent changes in existing facilities.

2.2.4 Information or services under the Owner's control shall be furnished by the Owner with reasonable promptness to avoid delay in orderly progress of the Work.

2.2.5 Unless otherwise provided in the Contract Documents, the Contractor will be furnished, free of charge, such copies of Drawings and Project Manuals as are reasonably necessary for execution of the Work.

2.2.6 The foregoing are in addition to other duties and responsibilities of the Owner enumerated herein and especially those in respect to Article 6 (Construction by Owner or by Separate Contractors), Article 9 (Payments and Completion) and Article 11 (Insurance and Bonds).

2.3 OWNER'S RIGHT TO STOP THE WORK

2.3.1 If the Contractor fails to correct Work which is not in accordance with the requirements of the Contract Documents as required by Paragraph 12.2 or persistently fails to carry out Work in accordance with the Contract Documents, the Owner, by written order signed personally or by an agent specifically so empowered by the Owner in writing, may order the Contractor to stop the Work, or any portion thereof, until the cause for such order has been eliminated; however, the right of the Owner to stop the Work shall not give rise to a duty on the part of the Owner to exercise this right for the benefit of the Contractor or any other person or entity, except to the extent required by Subparagraph 6.1.3.

2.4 OWNER'S RIGHT TO CARRY OUT THE WORK

2.4.1 If the Contractor defaults or neglects to carry out the Work in accordance with the Contract Documents and fails within a seven-day period after receipt of written notice from the Owner to commence and continue correction of such default or neglect with diligence and promptness, the Owner may after such seven-day period give the Contractor a second written notice to correct such deficiencies within a second seven-day period. If the Contractor within such second seven-day period after receipt of such second notice fails to commence and continue to correct any deficiencies, the Owner may, without prejudice to other remedies the Owner may have, correct such deficiencies. In such case an appropriate Change Order shall be issued deducting from payments then or thereafter due the Contractor the cost of correcting such deficiencies, including compensation for the Architect's additional services and expenses made necessary by such default, neglect or failure. Such action by the Owner and amounts charged to the Contractor are both subject to prior approval of the Architect. If payments then or thereafter due the Contractor are not sufficient to cover such amounts, the Contractor shall pay the difference to the Owner.

ARTICLE 3

CONTRACTOR

3.1 DEFINITION

3.1.1 The Contractor is the person or entity identified as such in the Agreement and is referred to throughout the Contract Documents as if singular in number. The term "Contractor" means the Contractor or the Contractor's authorized representative.

3.2 REVIEW OF CONTRACT DOCUMENTS AND FIELD CONDITIONS BY CONTRACTOR

3.2.1 The Contractor shall carefully study and compare the Contract Documents with each other and with information furnished by the Owner pursuant to Subparagraph 2.2.2 and shall at once report to the Architect errors, inconsistencies or omissions discovered. The Contractor shall not be liable to the Owner or Architect for damage resulting from errors, inconsistencies or omissions in the Contract Documents unless the Contractor recognized such error, inconsistency or omission and knowingly failed to report it to the Architect. If the Contractor performs any construction activity knowing it involves a recognized error, inconsistency or omission in the Contract Documents without such notice to the Architect, the Contractor shall assume appropriate responsibility for such performance and shall bear an appropriate amount of the attributable costs for correction.

3.2.2 The Contractor shall take field measurements and verify field conditions and shall carefully compare such field measurements and conditions and other information known to the Contractor with the Contract Documents before commencing activities. Errors, inconsistencies or omissions discovered shall be reported to the Architect at once.

3.2.3 The Contractor shall perform the Work in accordance with the Contract Documents and submittals approved pursuant to Paragraph 3.12.

3.3 SUPERVISION AND CONSTRUCTION PROCEDURES

3.3.1 The Contractor shall supervise and direct the Work, using the Contractor's best skill and attention. The Contractor shall be solely responsible for and have control over construction means, methods, techniques, sequences and procedures and for coordinating all portions of the Work under the Contract, unless Contract Documents give other specific instructions concerning these matters.

3.3.2 The Contractor shall be responsible to the Owner for acts and omissions of the Contractor's employees, Subcontractors and their agents and employees, and other persons performing portions of the Work under a contract with the Contractor.

3.3.3 The Contractor shall not be relieved of obligations to perform the Work in accordance with the Contract Documents either by activities or duties of the Architect in the Architect's administration of the Contract, or by tests, inspections or approvals required or performed by persons other than the Contractor.

3.3.4 The Contractor shall be responsible for inspection of portions of Work already performed under this Contract to determine that such portions are in proper condition to receive subsequent Work.

3.4 LABOR AND MATERIALS

3.4.1 Unless otherwise provided in the Contract Documents, the Contractor shall provide and pay for labor, materials, equipment, tools, construction equipment and machinery, water, heat, utilities, transportation, and other facilities and services necessary for proper execution and completion of the Work, whether temporary or permanent and whether or not incorporated or to be incorporated in the Work.

3.4.2 The Contractor shall enforce strict discipline and good order among the Contractor's employees and other persons carrying out the Contract. The Contractor shall not permit employment of unfit persons or persons not skilled in tasks assigned to them.

3.5 WARRANTY

3.5.1 The Contractor warrants to the Owner and Architect that materials and equipment furnished under the Contract will be of good quality and new unless otherwise required or permitted by the Contract Documents, that the Work will be free from defects not inherent in the quality required or permitted, and that the Work will conform with the requirements of the Contract Documents. Work not conforming to these requirements, including substitutions not properly approved and authorized, may be considered defective. The Contractor's warranty excludes remedy for damage or defect caused by abuse, modifications not executed by the Contractor, improper or insufficient maintenance, improper operation, or normal wear and tear under normal usage. If required by the Architect, the Contractor shall furnish satisfactory evidence as to the kind and quality of materials and equipment.

3.6 TAXES

3.6.1 The Contractor shall pay sales, consumer, use and similar taxes for the Work or portions thereof provided by the Contractor which are legally enacted when bids are received or negotiations concluded, whether or not yet effective or merely scheduled to go into effect.

3.7 PERMITS, FEES AND NOTICES

3.7.1 Unless otherwise provided in the Contract Documents, the Contractor shall secure and pay for the building permit and other permits and governmental fees, licenses and inspections necessary for proper execution and completion of the Work which are customarily secured after execution of the Contract and which are legally required when bids are received or negotiations concluded.

3.7.2 The Contractor shall comply with and give notices required by laws, ordinances, rules, regulations and lawful orders of public authorities bearing on performance of the Work.

3.7.3 It is not the Contractor's responsibility to ascertain that the Contract Documents are in accordance with applicable laws, statutes, ordinances, building codes, and rules and regulations. However, if the Contractor observes that portions of the Contract Documents are at variance therewith, the Contractor shall promptly notify the Architect and Owner in writing, and necessary changes shall be accomplished by appropriate Modification.

3.7.4 If the Contractor performs Work knowing it to be contrary to laws, statutes, ordinances, building codes, and rules and regulations without such notice to the Architect and Owner, the Contractor shall assume full responsibility for such Work and shall bear the attributable costs.

3.8 ALLOWANCES

3.8.1 The Contractor shall include in the Contract Sum all allowances stated in the Contract Documents. Items covered by allowances shall be supplied for such amounts and by such persons or entities as the Owner may direct, but the Contractor shall not be required to employ persons or entities against which the Contractor makes reasonable objection.

3.8.2 Unless otherwise provided in the Contract Documents:

- **.1** materials and equipment under an allowance shall be selected promptly by the Owner to avoid delay in the Work;
- **.2** allowances shall cover the cost to the Contractor of materials and equipment delivered at the site and all required taxes, less applicable trade discounts;

.3 Contractor's costs for unloading and handling at the site, labor, installation costs, overhead, profit and other expenses contemplated for stated allowance amounts shall be included in the Contract Sum and not in the allowances;

.4 whenever costs are more than or less than allowances, the Contract Sum shall be adjusted accordingly by Change Order. The amount of the Change Order shall reflect (1) the difference between actual costs and the allowances under Clause 3.8.2.2 and (2) changes in Contractor's costs under Clause 3.8.2.3.

3.9 SUPERINTENDENT

3.9.1 The Contractor shall employ a competent superintendent and necessary assistants who shall be in attendance at the Project site during performance of the Work. The superintendent shall represent the Contractor, and communications given to the superintendent shall be as binding as if given to the Contractor. Important communications shall be confirmed in writing. Other communications shall be similarly confirmed on written request in each case.

3.10 CONTRACTOR'S CONSTRUCTION SCHEDULES

3.10.1 The Contractor, promptly after being awarded the Contract, shall prepare and submit for the Owner's and Architect's information a Contractor's construction schedule for the Work. The schedule shall not exceed time limits current under the Contract Documents, shall be revised at appropriate intervals as required by the conditions of the Work and Project, shall be related to the entire Project to the extent required by the Contract Documents, and shall provide for expeditious and practicable execution of the Work.

3.10.2 The Contractor shall prepare and keep current, for the Architect's approval, a schedule of submittals which is coordinated with the Contractor's construction schedule and allows the Architect reasonable time to review submittals.

3.10.3 The Contractor shall conform to the most recent schedules.

3.11 DOCUMENTS AND SAMPLES AT THE SITE

3.11.1 The Contractor shall maintain at the site for the Owner one record copy of the Drawings, Specifications, addenda, Change Orders and other Modifications, in good order and marked currently to record changes and selections made during construction, and in addition approved Shop Drawings, Product Data, Samples and similar required submittals. These shall be available to the Architect and shall be delivered to the Architect for submittal to the Owner upon completion of the Work.

3.12 SHOP DRAWINGS, PRODUCT DATA AND SAMPLES

3.12.1 Shop Drawings are drawings, diagrams, schedules and other data specially prepared for the Work by the Contractor or a Subcontractor, Sub-subcontractor, manufacturer, supplier or distributor to illustrate some portion of the Work.

3.12.2 Product Data are illustrations, standard schedules, performance charts, instructions, brochures, diagrams and other information furnished by the Contractor to illustrate materials or equipment for some portion of the Work.

3.12.3 Samples are physical examples which illustrate materials, equipment or workmanship and establish standards by which the Work will be judged.

3.12.4 Shop Drawings, Product Data, Samples and similar submittals are not Contract Documents. The purpose of their submittal is to demonstrate for those portions of the Work for which submittals are required the way the Contractor proposes to conform to the information given and the design concept expressed in the Contract Documents. Review by the Architect is subject to the limitations of Subparagraph 4.2.7.

3.12.5 The Contractor shall review, approve and submit to the Architect Shop Drawings, Product Data, Samples and similar submittals required by the Contract Documents with reasonable promptness and in such sequence as to cause no delay in the Work or in the activities of the Owner or of separate contractors. Submittals made by the Contractor which are not required by the Contract Documents may be returned without action.

3.12.6 The Contractor shall perform no portion of the Work requiring submittal and review of Shop Drawings, Product Data, Samples or similar submittals until the respective submittal has been approved by the Architect. Such Work shall be in accordance with approved submittals.

3.12.7 By approving and submitting Shop Drawings, Product Data, Samples and similar submittals, the Contractor represents that the Contractor has determined and verified materials, field measurements and field construction criteria related thereto, or will do so, and has checked and coordinated the information contained within such submittals with the requirements of the Work and of the Contract Documents.

3.12.8 The Contractor shall not be relieved of responsibility for deviations from requirements of the Contract Documents by the Architect's approval of Shop Drawings, Product Data, Samples or similar submittals unless the Contractor has specifically informed the Architect in writing of such deviation at the time of submittal and the Architect has given written approval to the specific deviation. The Contractor shall not be relieved of responsibility for errors or omissions in Shop Drawings, Product Data, Samples or similar submittals by the Architect's approval thereof.

3.12.9 The Contractor shall direct specific attention, in writing or on resubmitted Shop Drawings, Product Data, Samples or similar submittals, to revisions other than those requested by the Architect on previous submittals.

3.12.10 Informational submittals upon which the Architect is not expected to take responsive action may be so identified in the Contract Documents.

3.12.11 When professional certification of performance criteria of materials, systems or equipment is required by the Contract Documents, the Architect shall be entitled to rely upon the accuracy and completeness of such calculations and certifications.

3.13 USE OF SITE

3.13.1 The Contractor shall confine operations at the site to areas permitted by law, ordinances, permits and the Contract Documents and shall not unreasonably encumber the site with materials or equipment.

3.14 CUTTING AND PATCHING

3.14.1 The Contractor shall be responsible for cutting, fitting or patching required to complete the Work or to make its parts fit together properly.

3.14.2 The Contractor shall not damage or endanger a portion of the Work or fully or partially completed construction of the Owner or separate contractors by cutting, patching or otherwise altering such construction, or by excavation. The Contractor shall not cut or otherwise alter such construction by the

Owner or a separate contractor except with written consent of the Owner and of such separate contractor; such consent shall not be unreasonably withheld. The Contractor shall not unreasonably withhold from the Owner or a separate contractor the Contractor's consent to cutting or otherwise altering the Work.

3.15 CLEANING UP

3.15.1 The Contractor shall keep the premises and surrounding area free from accumulation of waste materials or rubbish caused by operations under the Contract. At completion of the Work the Contractor shall remove from and about the Project waste materials, rubbish, the Contractor's tools, construction equipment, machinery and surplus materials.

3.15.2 If the Contractor fails to clean up as provided in the Contract Documents, the Owner may do so and the cost thereof shall be charged to the Contractor.

3.16 ACCESS TO WORK

3.16.1 The Contractor shall provide the Owner and Architect access to the Work in preparation and progress wherever located.

3.17 ROYALTIES AND PATENTS

3.17.1 The Contractor shall pay all royalties and license fees. The Contractor shall defend suits or claims for infringement of patent rights and shall hold the Owner and Architect harmless from loss on account thereof, but shall not be responsible for such defense or loss when a particular design, process or product of a particular manufacturer or manufacturers is required by the Contract Documents. However, if the Contractor has reason to believe that the required design, process or product is an infringement of a patent, the Contractor shall be responsible for such loss unless such information is promptly furnished to the Architect.

3.18 INDEMNIFICATION

3.18.1 To the fullest extent permitted by law, the Contractor shall indemnify and hold harmless the Owner, Architect, Architect's consultants, and agents and employees of any of them from and against claims, damages, losses and expenses, including but not limited to attorneys' fees, arising out of or resulting from performance of the Work, provided that such claim, damage, loss or expense is attributable to bodily injury, sickness, disease or death, or to injury to or destruction of tangible property (other than the Work itself) including loss of use resulting therefrom, but only to the extent caused in whole or in part by negligent acts or omissions of the Contractor, a Subcontractor, anyone directly or indirectly employed by them or anyone for whose acts they may be liable, regardless of whether or not such claim, damage, loss or expense is caused in part by a party indemnified hereunder. Such obligation shall not be construed to negate, abridge, or reduce other rights or obligations of indemnity which would otherwise exist as to a party or person described in this Paragraph 3.18.

3.18.2 In claims against any person or entity indemnified under this Paragraph 3.18 by an employee of the Contractor, a Subcontractor, anyone directly or indirectly employed by them or anyone for whose acts they may be liable, the indemnification obligation under this Paragraph 3.18 shall not be limited by a limitation on amount or type of damages, compensation or benefits payable by or for the Contractor or a Subcontractor under workers' or workmen's compensation acts, disability benefit acts or other employee benefit acts.

3.18.3 The obligations of the Contractor under this Paragraph 3.18 shall not extend to the liability of the Architect, the Architect's consultants, and agents and employees of any of them arising out of (1) the preparation or approval of maps, drawings, opinions, reports, surveys, Change Orders, designs or specifications, or (2) the giving of or the failure to give directions or instructions by the Architect, the Architect's consultants, and agents and employees of any of them provided such giving or failure to give is the primary cause of the injury or damage.

ARTICLE 4

ADMINISTRATION OF THE CONTRACT

4.1 ARCHITECT

4.1.1 The Architect is the person lawfully licensed to practice architecture or an entity lawfully practicing architecture identified as such in the Agreement and is referred to throughout the Contract Documents as if singular in number. The term "Architect" means the Architect or the Architect's authorized representative.

4.1.2 Duties, responsibilities and limitations of authority of the Architect as set forth in the Contract Documents shall not be restricted, modified or extended without written consent of the Owner, Contractor and Architect. Consent shall not be unreasonably withheld.

4.1.3 In case of termination of employment of the Architect, the Owner shall appoint an architect against whom the Contractor makes no reasonable objection and whose status under the Contract Documents shall be that of the former architect.

4.1.4 Disputes arising under Subparagraphs 4.1.2 and 4.1.3 shall be subject to arbitration.

4.2 ARCHITECT'S ADMINISTRATION OF THE CONTRACT

4.2.1 The Architect will provide administration of the Contract as described in the Contract Documents, and will be the Owner's representative (1) during construction, (2) until final payment is due and (3) with the Owner's concurrence, from time to time during the correction period described in Paragraph 12.2. The Architect will advise and consult with the Owner. The Architect will have authority to act on behalf of the Owner only to the extent provided in the Contract Documents, unless otherwise modified by written instrument in accordance with other provisions of the Contract.

4.2.2 The Architect will visit the site at intervals appropriate to the stage of construction to become generally familiar with the progress and quality of the completed Work and to determine in general if the Work is being performed in a manner indicating that the Work, when completed, will be in accordance with the Contract Documents. However, the Architect will not be required to make exhaustive or continuous on-site inspections to check quality or quantity of the Work. On the basis of on-site observations as an architect, the Architect will keep the Owner informed of progress of the Work, and will endeavor to guard the Owner against defects and deficiencies in the Work.

4.2.3 The Architect will not have control over or charge of and will not be responsible for construction means, methods, techniques, sequences or procedures, or for safety precautions and programs in connection with the Work, since these are solely the Contractor's responsibility as provided in Paragraph 3.3. The Architect will not be responsible for the Contractor's failure to carry out the Work in accordance with the Contract Documents. The Architect will not have control over or charge of and will not be responsible for acts or omissions of the Contractor, Subcontractors, or their agents or employees, or of any other persons performing portions of the Work.

4.2.4 Communications Facilitating Contract Administration. Except as otherwise provided in the Contract Documents or when direct communications have been specially authorized, the Owner and Contractor shall endeavor to communicate through the Architect. Communications by and with the Architect's consultants shall be through the Architect. Communications by and with Subcontractors and material suppliers shall be through the Contractor. Communications by and with separate contractors shall be through the Owner.

4.2.5 Based on the Architect's observations and evaluations of the Contractor's Applications for Payment, the Architect will review and certify the amounts due the Contractor and will issue Certificates for Payment in such amounts.

4.2.6 The Architect will have authority to reject Work which does not conform to the Contract Documents. Whenever the Architect considers it necessary or advisable for implementation of the intent of the Contract Documents, the Architect will have authority to require additional inspection or testing of the Work in accordance with Subparagraphs 13.5.2 and 13.5.3, whether or not such Work is fabricated, installed or completed. However, neither this authority of the Architect nor a decision made in good faith either to exercise or not to exercise such authority shall give rise to a duty or responsibility of the Architect to the Contractor, Subcontractors, material and equipment suppliers, their agents or employees, or other persons performing portions of the Work.

4.2.7 The Architect will review and approve or take other appropriate action upon the Contractor's submittals such as Shop Drawings, Product Data and Samples, but only for the limited purpose of checking for conformance with information given and the design concept expressed in the Contract Documents. The Architect's action will be taken with such reasonable promptness as to cause no delay in the Work or in the activities of the Owner, Contractor or separate contractors, while allowing sufficient time in the Architect's professional judgment to permit adequate review. Review of such submittals is not conducted for the purpose of determining the accuracy and completeness of other details such as dimensions and quantities, or for substantiating instructions for installation or performance of equipment or systems, all of which remain the responsibility of the Contractor as required by the Contract Documents. The Architect's review of the Contractor's submittals shall not relieve the Contractor of the obligations under Paragraphs 3.3, 3.5 and 3.12. The Architect's review shall not constitute approval of safety precautions or, unless otherwise specifically stated by the Architect, of any construction means, methods, techniques, sequences or procedures. The Architect's approval of a specific item shall not indicate approval of an assembly of which the item is a component.

4.2.8 The Architect will prepare Change Orders and Construction Change Directives, and may authorize minor changes in the Work as provided in Paragraph 7.4.

4.2.9 The Architect will conduct inspections to determine the date or dates of Substantial Completion and the date of final completion, will receive and forward to the Owner for the Owner's review and records written warranties and related documents required by the Contract and assembled by the Contractor, and will issue a final Certificate for Payment upon compliance with the requirements of the Contract Documents.

4.2.10 If the Owner and Architect agree, the Architect will provide one or more project representatives to assist in carrying out the Architect's responsibilities at the site. The duties, responsibilities and limitations of authority of such project representatives shall be as set forth in an exhibit to be incorporated in the Contract Documents.

4.2.11 The Architect will interpret and decide matters concerning performance under and requirements of the Contract Documents on written request of either the Owner or Contractor. The Architect's response to such requests will be made with reasonable promptness and within any time limits agreed upon. If no agreement is made concerning the time within which interpretations required of the Architect shall be furnished in compliance with this Paragraph 4.2, then delay shall not be recognized on account of failure by the Architect to furnish such interpretations until 15 days after written request is made for them.

4.2.12 Interpretations and decisions of the Architect will be consistent with the intent of and reasonably inferable from the Contract Documents and will be in writing or in the form of drawings. When making such interpretations and decisions, the Architect will endeavor to secure faithful performance by both Owner and Contractor, will not show partiality to either and will not be liable for results of interpretations or decisions so rendered in good faith.

4.2.13 The Architect's decisions on matters relating to aesthetic effect will be final if consistent with the intent expressed in the Contract Documents.

4.3 CLAIMS AND DISPUTES

4.3.1 Definition. A Claim is a demand or assertion by one of the parties seeking, as a matter of right, adjustment or interpretation of Contract terms, payment of money, extension of time or other relief with respect to the terms of the Contract. The term "Claim" also includes other disputes and matters in question between the Owner and Contractor arising out of or relating to the Contract. Claims must be made by written notice. The responsibility to substantiate Claims shall rest with the party making the Claim.

4.3.2 Decision of Architect. Claims, including those alleging an error or omission by the Architect, shall be referred initially to the Architect for action as provided in Paragraph 4.4. A decision by the Architect, as provided in Subparagraph 4.4.4, shall be required as a condition precedent to arbitration or litigation of a Claim between the Contractor and Owner as to all such matters arising prior to the date final payment is due, regardless of (1) whether such matters relate to execution and progress of the Work or (2) the extent to which the Work has been completed. The decision by the Architect in response to a Claim shall not be a condition precedent to arbitration or litigation in the event (1) the position of Architect is vacant, (2) the Architect has not received evidence or has failed to render a decision within agreed time limits, (3) the Architect has failed to take action required under Subparagraph 4.4.4 within 30 days after the Claim is made, (4) 45 days have passed after the Claim has been referred to the Architect or (5) the Claim relates to a mechanic's lien.

4.3.3 Time Limits on Claims. Claims by either party must be made within 21 days after occurrence of the event giving rise to such Claim or within 21 days after the claimant first recognizes the condition giving rise to the Claim, whichever is later. Claims must be made by written notice. An additional Claim made after the initial Claim has been implemented by Change Order will not be considered unless submitted in a timely manner.

4.3.4 Continuing Contract Performance. Pending final resolution of a Claim including arbitration, unless otherwise agreed in writing the Contractor shall proceed diligently with performance of the Contract and the Owner shall continue to make payments in accordance with the Contract Documents.

4.3.5 Waiver of Claims: Final Payment. The making of final payment shall constitute a waiver of Claims by the Owner except those arising from:

.1 liens, Claims, security interests or encumbrances arising out of the Contract and unsettled;

.2 failure of the Work to comply with the requirements of the Contract Documents; or

.3 terms of special warranties required by the Contract Documents.

4.3.6 Claims for Concealed or Unknown Conditions. If conditions are encountered at the site which are (1) subsurface or otherwise concealed physical conditions which differ materially from those indicated in the Contract Documents or (2) unknown physical conditions of an unusual nature, which differ materially from those ordinarily found to exist and generally recognized as inherent in construction activities of the character provided for in the Contract Documents, then notice by the observing party shall be given to the other party promptly before conditions are disturbed and in no event later than 21 days after first observance of the conditions. The Architect will promptly investigate such conditions and, if they differ materially and cause an increase or decrease in the Contractor's cost of, or time required for, performance of any part of the Work, will recommend an equitable adjustment in the Contract Sum or Contract Time, or both. If the Architect determines that the conditions at the site are not materially different from those indicated in the Contract Documents and that no change in the terms of the Contract is justified, the Architect shall so notify the Owner and Contractor in writing, stating the reasons. Claims by either party in opposition to such determination must be made within 21 days after the Architect has given notice of the decision. If the Owner and Contractor cannot agree on an adjustment in the Contract Sum or Contract Time, the adjustment shall be referred to the Architect for initial determination, subject to further proceedings pursuant to Paragraph 4.4.

4.3.7 Claims for Additional Cost. If the Contractor wishes to make Claim for an increase in the Contract Sum, written notice as provided herein shall be given before proceeding to execute the Work. Prior notice is not required for Claims relating to an emergency endangering life or property arising under Paragraph 10.3. If the Contractor believes additional cost is involved for reasons including but not limited to (1) a written interpretation from the Architect, (2) an order by the Owner to stop the Work where the Contractor was not at fault, (3) a written order for a minor change in the Work issued by the Architect, (4) failure of payment by the Owner, (5) termination of the Contract by the Owner, (6) Owner's suspension or (7) other reasonable grounds, Claim shall be filed in accordance with the procedure established herein.

4.3.8 Claims for Additional Time

4.3.8.1 If the Contractor wishes to make Claim for an increase in the Contract Time, written notice as provided herein shall be given. The Contractor's Claim shall include an estimate of cost and of probable effect of delay on progress of the Work. In the case of a continuing delay only one Claim is necessary.

4.3.8.2 If adverse weather conditions are the basis for a Claim for additional time, such Claim shall be documented by data substantiating that weather conditions were abnormal for the period of time and could not have been reasonably anticipated, and that weather conditions had an adverse effect on the scheduled construction.

4.3.9 Injury or Damage to Person or Property. If either party to the Contract suffers injury or damage to person or property because of an act or omission of the other party, of any of the other party's employees or agents, or of others for whose acts such party is legally liable, written notice of such injury or damage, whether or not insured, shall be given to the other party within a reasonable time not exceeding 21 days after first observance. The notice shall provide sufficient detail to enable the other party to investigate the matter. If a Claim for additional cost or time related to this Claim is to be asserted, it shall be filed as provided in Subparagraphs 4.3.7 or 4.3.8.

4.4 RESOLUTION OF CLAIMS AND DISPUTES

4.4.1 The Architect will review Claims and take one or more of the following preliminary actions within ten days of receipt of a Claim: (1) request additional supporting data from the claimant, (2) submit a schedule to the parties indicating when the Architect expects to take action, (3) reject the Claim in whole or in part, stating reasons for rejection, (4) recommend approval of the Claim by the other party or (5) suggest a compromise. The Architect may also, but is not obligated to, notify the surety, if any, of the nature and amount of the Claim.

4.4.2 If a Claim has been resolved, the Architect will prepare or obtain appropriate documentation.

4.4.3 If a Claim has not been resolved, the party making the Claim shall, within ten days after the Architect's preliminary response, take one or more of the following actions: (1) submit additional supporting data requested by the Architect, (2) modify the initial Claim or (3) notify the Architect that the initial Claim stands.

4.4.4 If a Claim has not been resolved after consideration of the foregoing and of further evidence presented by the parties or requested by the Architect, the Architect will notify the parties in writing that the Architect's decision will be made within seven days, which decision shall be final and binding on the parties but subject to arbitration. Upon expiration of such time period, the Architect will render to the parties the Architect's written decision relative to the Claim, including any change in the Contract Sum or Contract Time or both. If there is a surety and there appears to be a possibility of a Contractor's default, the Architect may, but is not obligated to, notify the surety and request the surety's assistance in resolving the controversy.

4.5 ARBITRATION

4.5.1 Controversies and Claims Subject to Arbitration. Any controversy or Claim arising out of or related to the Contract, or the breach thereof, shall be settled by arbitration in accordance with the Construction Industry Arbitration Rules of the American Arbitration Association, and judgment upon the award rendered by the arbitrator or arbitrators may be entered in any court having jurisdiction thereof, except controversies or Claims relating to aesthetic effect and except those waived as provided for in Subparagraph 4.3.5. Such controversies or Claims upon which the Architect has given notice and rendered a decision as provided in Subparagraph 4.4.4 shall be subject to arbitration upon written demand of either party. Arbitration may be commenced when 45 days have passed after a Claim has been referred to the Architect as provided in Paragraph 4.3 and no decision has been rendered.

4.5.2 Rules and Notices for Arbitration. Claims between the Owner and Contractor not resolved under Paragraph 4.4 shall, if subject to arbitration under Subparagraph 4.5.1, be decided by arbitration in accordance with the Construction Industry Arbitration Rules of the American Arbitration Association currently in effect, unless the parties mutually agree otherwise. Notice of demand for arbitration shall be filed in writing with the other party to the Agreement between the Owner and Contractor and with the American Arbitration Association, and a copy shall be filed with the Architect.

4.5.3 Contract Performance During Arbitration. During arbitration proceedings, the Owner and Contractor shall comply with Subparagraph 4.3.4.

4.5.4 When Arbitration May Be Demanded. Demand for arbitration of any Claim may not be made until the earlier of (1) the date on which the Architect has rendered a final written decision on the Claim, (2) the tenth day after the parties have presented evidence to the Architect or have been given reasonable opportunity to do so, if the Architect has not rendered a final written decision by that date, or (3) any of the five events described in Subparagraph 4.3.2.

4.5.4.1 When a written decision of the Architect states that (1) the decision is final but subject to arbitration and (2) a demand for arbitration of a Claim covered by such decision must be made within 30 days after the date on which the party making the demand receives the final written decision, then failure to demand arbitration within said 30 days' period shall result in the Architect's decision becoming final and binding upon the Owner and Contractor. If the Architect renders a decision after arbitration proceedings have been initiated, such decision may be entered as evidence, but shall not supersede arbitration proceedings unless the decision is acceptable to all parties concerned.

4.5.4.2 A demand for arbitration shall be made within the time limits specified in Subparagraphs 4.5.1 and 4.5.4 and Clause 4.5.4.1 as applicable, and in other cases within a reasonable time after the Claim has arisen, and in no event shall it be made after the date when institution of legal or equitable proceedings based on such Claim would be barred by the applicable statute of limitations as determined pursuant to Paragraph 13.7.

4.5.5 Limitation on Consolidation or Joinder. No arbitration arising out of or relating to the Contract Documents shall include, by consolidation or joinder or in any other manner, the Architect, the Architect's employees or consultants, except by written consent containing specific reference to the Agreement and signed by the Architect, Owner, Contractor and any other person or entity sought to be joined. No arbitration shall include, by consolidation or joinder or in any other manner, parties other than the Owner, Contractor, a separate contractor as described in Article 6 and other persons substantially involved in a common question of fact or law whose presence is required if complete relief is to be accorded in arbitration. No person or entity other than the Owner, Contractor or a separate contractor as described in Article 6 shall be included as an original third party or additional third party to an arbitration whose interest or responsibility is insubstantial. Consent to arbitration involving an additional person or entity shall not constitute consent to arbitration of a dispute not described therein or with a person or entity not named or described therein. The foregoing agreement to arbitrate and other agreements to arbitrate with an additional person or entity duly consented to by parties to the Agreement shall be specifically enforceable under applicable law in any court having jurisdiction thereof.

4.5.6 Claims and Timely Assertion of Claims. A party who files a notice of demand for arbitration must assert in the demand all Claims then known to that party on which arbitration is permitted to be demanded. When a party fails to include a Claim through oversight, inadvertence or excusable neglect, or when a Claim has matured or been acquired subsequently, the arbitrator or arbitrators may permit amendment.

4.5.7 Judgment on Final Award. The award rendered by the arbitrator or arbitrators shall be final, and judgment may be entered upon it in accordance with applicable law in any court having jurisdiction thereof.

ARTICLE 5

SUBCONTRACTORS

5.1 DEFINITIONS

5.1.1 A Subcontractor is a person or entity who has a direct contract with the Contractor to perform a portion of the Work at the site. The term "Subcontractor" is referred to throughout the Contract Documents as if singular in number and means a Subcontractor or an authorized representative of the Subcontractor. The term "Subcontractor" does not include a separate contractor or subcontractors of a separate contractor.

5.1.2 A Sub-subcontractor is a person or entity who has a direct or indirect contract with a Subcontractor to perform a portion of the Work at the site. The term "Sub-subcontractor" is referred to throughout the Contract Documents as if singular in number and means a Sub-subcontractor or an authorized representative of the Sub-subcontractor.

5.2 AWARD OF SUBCONTRACTS AND OTHER CONTRACTS FOR PORTIONS OF THE WORK

5.2.1 Unless otherwise stated in the Contract Documents or the bidding requirements, the Contractor, as soon as practicable after award of the Contract, shall furnish in writing to the Owner through the Architect the names of persons or entities (including those who are to furnish materials or equipment fabricated to a special design) proposed for each principal portion of the Work. The Architect will promptly reply to the Contractor in writing stating whether or not the Owner or the Architect, after due investigation, has reasonable objection to any such proposed person or entity. Failure of the Owner or Architect to reply promptly shall constitute notice of no reasonable objection.

5.2.2 The Contractor shall not contract with a proposed person or entity to whom the Owner or Architect has made reasonable and timely objection. The Contractor shall not be required to contract with anyone to whom the Contractor has made reasonable objection.

5.2.3 If the Owner or Architect has reasonable objection to a person or entity proposed by the Contractor, the Contractor shall propose another to whom the Owner or Architect has no reasonable objection. The Contract Sum shall be increased or decreased by the difference in cost occasioned by such change and an appropriate Change Order shall be issued. However, no increase in the Contract Sum shall be allowed for such change unless the Contractor has acted promptly and responsively in submitting names as required.

5.2.4 The Contractor shall not change a Subcontractor, person or entity previously selected if the Owner or Architect makes reasonable objection to such change.

5.3 SUBCONTRACTUAL RELATIONS

5.3.1 By appropriate agreement, written where legally required for validity, the Contractor shall require each Subcontractor, to the extent of the Work to be performed by the Subcontractor, to be bound to the Contractor by terms of the Contract Documents, and to assume toward the Contractor all the obligations and responsibilities which the Contractor, by these Documents, assumes toward the Owner and Architect. Each subcontract agreement shall preserve and protect the rights of the Owner and Architect under the Contract Documents with respect to the Work to be performed by the Subcontractor so that subcontracting thereof will not prejudice such rights, and shall allow to the Subcontractor, unless specifically provided otherwise in the subcontract agreement, the benefit of all rights, remedies and redress against the Contractor that the Contractor, by the Contract Documents, has against the Owner. Where appropriate, the Contractor shall require each Subcontractor to enter into similar agreements with Sub-subcontractors. The Contractor shall make available to each proposed Subcontractor, prior to the execution of the subcontract agreement, copies of the Contract Documents to which the Subcontractor will be bound, and, upon written request of the Subcontractor, identify to the Subcontractor terms and conditions of the proposed subcontract agreement which may be at variance with the Contract Documents. Subcontractors shall similarly make copies of applicable portions of such documents available to their respective proposed Sub-subcontractors.

5.4 CONTINGENT ASSIGNMENT OF SUBCONTRACTS

5.4.1 Each subcontract agreement for a portion of the Work is assigned by the Contractor to the Owner provided that:

.1 assignment is effective only after termination of the Contract by the Owner for cause pursuant to Paragraph 14.2 and only for those subcontract agreements which the Owner accepts by notifying the Subcontractor in writing; and

.2 assignment is subject to the prior rights of the surety, if any, obligated under bond relating to the Contract.

5.4.2 If the Work has been suspended for more than 30 days, the Subcontractor's compensation shall be equitably adjusted.

ARTICLE 6

CONSTRUCTION BY OWNER OR BY SEPARATE CONTRACTORS

6.1 OWNER'S RIGHT TO PERFORM CONSTRUCTION AND TO AWARD SEPARATE CONTRACTS

6.1.1 The Owner reserves the right to perform construction or operations related to the Project with the Owner's own forces, and to award separate contracts in connection with other portions of the Project or other construction or operations on the site under Conditions of the Contract identical or substantially similar to these including those portions related to insurance and waiver of subrogation. If the Contractor claims that delay or additional cost is involved because of such action by the Owner, the Contractor shall make such Claim as provided elsewhere in the Contract Documents.

6.1.2 When separate contracts are awarded for different portions of the Project or other construction or operations on the site, the term "Contractor" in the Contract Documents in each case shall mean the Contractor who executes each separate Owner-Contractor Agreement.

6.1.3 The Owner shall provide for coordination of the activities of the Owner's own forces and of each separate contractor with the Work of the Contractor, who shall cooperate with them. The Contractor shall participate with other separate contractors and the Owner in reviewing their construction schedules when directed to do so. The Contractor shall make any revisions to the construction schedule and Contract Sum deemed necessary after a joint review and mutual agreement. The construction schedules shall then constitute the schedules to be used by the Contractor, separate contractors and the Owner until subsequently revised.

6.1.4 Unless otherwise provided in the Contract Documents, when the Owner performs construction or operations related to the Project with the Owner's own forces, the Owner shall be deemed to be subject to the same obligations and to have the same rights which apply to the Contractor under the Conditions of the Contract, including, without excluding others, those stated in Article 3, this Article 6 and Articles 10, 11 and 12.

6.2 MUTUAL RESPONSIBILITY

6.2.1 The Contractor shall afford the Owner and separate contractors reasonable opportunity for introduction and storage of their materials and equipment and performance of their activities and shall connect and coordinate the Contractor's construction and operations with theirs as required by the Contract Documents.

6.2.2 If part of the Contractor's Work depends for proper execution or results upon construction or operations by the Owner or a separate contractor, the Contractor shall, prior to proceeding with that portion of the Work, promptly report to the Architect apparent discrepancies or defects in such other construction that would render it unsuitable for such proper execution and results. Failure of the Contractor so to report shall constitute an acknowledgment that the Owner's or separate contractors' completed or partially completed construction is fit and proper to receive the Contractor's Work, except as to defects not then reasonably discoverable.

6.2.3 Costs caused by delays or by improperly timed activities or defective construction shall be borne by the party responsible therefor.

6.2.4 The Contractor shall promptly remedy damage wrongfully caused by the Contractor to completed or partially completed construction or to property of the Owner or separate contractors as provided in Subparagraph 10.2.5.

6.2.5 Claims and other disputes and matters in question between the Contractor and a separate contractor shall be subject to the provisions of Paragraph 4.3 provided the separate contractor has reciprocal obligations.

6.2.6 The Owner and each separate contractor shall have the same responsibilities for cutting and patching as are described for the Contractor in Paragraph 3.14.

6.3 OWNER'S RIGHT TO CLEAN UP

6.3.1 If a dispute arises among the Contractor, separate contractors and the Owner as to the responsibility under their respective contracts for maintaining the premises and surrounding area free from waste materials and rubbish as described in Paragraph 3.15, the Owner may clean up and allocate the cost among those responsible as the Architect determines to be just.

ARTICLE 7

CHANGES IN THE WORK

7.1 CHANGES

7.1.1 Changes in the Work may be accomplished after execution of the Contract, and without invalidating the Contract, by Change Order, Construction Change Directive or order for a minor change in the Work, subject to the limitations stated in this Article 7 and elsewhere in the Contract Documents.

7.1.2 A Change Order shall be based upon agreement among the Owner, Contractor and Architect; a Construction Change Directive requires agreement by the Owner and Architect and may or may not be agreed to by the Contractor; an order for a minor change in the Work may be issued by the Architect alone.

7.1.3 Changes in the Work shall be performed under applicable provisions of the Contract Documents, and the Contractor shall proceed promptly, unless otherwise provided in the Change Order, Construction Change Directive or order for a minor change in the Work.

7.1.4 If unit prices are stated in the Contract Documents or subsequently agreed upon, and if quantities originally contemplated are so changed in a proposed Change Order or Construction Change Directive that application of such unit prices to quantities of Work proposed will cause substantial inequity to the Owner or Contractor, the applicable unit prices shall be equitably adjusted.

7.2 CHANGE ORDERS

7.2.1 A Change Order is a written instrument prepared by the Architect and signed by the Owner, Contractor and Architect, stating their agreement upon all of the following:

.1 a change in the Work;

.2 the amount of the adjustment in the Contract Sum, if any; and

.3 the extent of the adjustment in the Contract Time, if any.

7.2.2 Methods used in determining adjustments to the Contract Sum may include those listed in Subparagraph 7.3.3.

7.3 CONSTRUCTION CHANGE DIRECTIVES

7.3.1 A Construction Change Directive is a written order prepared by the Architect and signed by the Owner and Architect, directing a change in the Work and stating a proposed basis for adjustment, if any, in the Contract Sum or Contract Time, or both. The Owner may by Construction Change Directive, without invalidating the Contract, order changes in the Work within the general scope of the Contract consisting of additions, deletions or other revisions, the Contract Sum and Contract Time being adjusted accordingly.

7.3.2 A Construction Change Directive shall be used in the absence of total agreement on the terms of a Change Order.

7.3.3 If the Construction Change Directive provides for an adjustment to the Contract Sum, the adjustment shall be based on one of the following methods:

.1 mutual acceptance of a lump sum properly itemized and supported by sufficient substantiating data to permit evaluation;

.2 unit prices stated in the Contract Documents or subsequently agreed upon;

.3 cost to be determined in a manner agreed upon by the parties and a mutually acceptable fixed or percentage fee; or

.4 as provided in Subparagraph 7.3.6.

7.3.4 Upon receipt of a Construction Change Directive, the Contractor shall promptly proceed with the change in the Work involved and advise the Architect of the Contractor's agreement or disagreement with the method, if any, provided in the Construction Change Directive for determining the proposed adjustment in the Contract Sum or Contract Time.

7.3.5 A Construction Change Directive signed by the Contractor indicates the agreement of the Contractor therewith, including adjustment in Contract Sum and Contract Time or the method for determining them. Such agreement shall be effective immediately and shall be recorded as a Change Order.

7.3.6 If the Contractor does not respond promptly or disagrees with the method for adjustment in the Contract Sum, the method and the adjustment shall be determined by the Architect on the basis of reasonable expenditures and savings of those performing the Work attributable to the change, including, in case of an increase in the Contract Sum, a reasonable allowance for overhead and profit. In such case, and also under Clause 7.3.3.3, the Contractor shall keep and present, in such form as the Architect may prescribe, an itemized accounting together with appropriate supporting data. Unless otherwise provided in the Contract Documents, costs for the purposes of this Subparagraph 7.3.6 shall be limited to the following:

.1 costs of labor, including social security, old age and unemployment insurance, fringe benefits required by agreement or custom, and workers' or workmen's compensation insurance;

.2 costs of materials, supplies and equipment, including cost of transportation, whether incorporated or consumed;

.3 rental costs of machinery and equipment, exclusive of hand tools, whether rented from the Contractor or others;

.4 costs of premiums for all bonds and insurance, permit fees, and sales, use or similar taxes related to the Work; and

.5 additional costs of supervision and field office personnel directly attributable to the change.

7.3.7 Pending final determination of cost to the Owner, amounts not in dispute may be included in Applications for Payment. The amount of credit to be allowed by the Contractor to the Owner for a deletion or change which results in a net decrease in the Contract Sum shall be actual net cost as confirmed by the Architect. When both additions and credits covering related Work or substitutions are involved in a change, the allowance for overhead and profit shall be figured on the basis of net increase, if any, with respect to that change.

7.3.8 If the Owner and Contractor do not agree with the adjustment in Contract Time or the method for determining it, the adjustment or the method shall be referred to the Architect for determination.

7.3.9 When the Owner and Contractor agree with the determination made by the Architect concerning the adjustments in the Contract Sum and Contract Time, or otherwise reach agreement upon the adjustments, such agreement shall be effective immediately and shall be recorded by preparation and execution of an appropriate Change Order.

7.4 MINOR CHANGES IN THE WORK

7.4.1 The Architect will have authority to order minor changes in the Work not involving adjustment in the Contract Sum or extension of the Contract Time and not inconsistent with the intent of the Contract Documents. Such changes shall be effected by written order and shall be binding on the Owner and Contractor. The Contractor shall carry out such written orders promptly.

ARTICLE 8

TIME

8.1 DEFINITIONS

8.1.1 Unless otherwise provided, Contract Time is the period of time, including authorized adjustments, allotted in the Contract Documents for Substantial Completion of the Work.

8.1.2 The date of commencement of the Work is the date established in the Agreement. The date shall not be postponed by the failure to act of the Contractor or of persons or entities for whom the Contractor is responsible.

8.1.3 The date of Substantial Completion is the date certified by the Architect in accordance with Paragraph 9.8.

8.1.4 The term "day" as used in the Contract Documents shall mean calendar day unless otherwise specifically defined.

8.2 PROGRESS AND COMPLETION

8.2.1 Time limits stated in the Contract Documents are of the essence of the Contract. By executing the Agreement the Contractor confirms that the Contract Time is a reasonable period for performing the Work.

8.2.2 The Contractor shall not knowingly, except by agreement or instruction of the Owner in writing, prematurely commence operations on the site or elsewhere prior to the effective date of insurance required by Article 11 to be furnished by the Contractor. The date of commencement of the Work shall not be changed by the effective date of such insurance. Unless the date of commencement is established by a notice to proceed given by the Owner, the Contractor shall notify the Owner in writing not less than five days or other agreed period before commencing the Work to permit the timely filing of mortgages, mechanic's liens and other security interests.

8.2.3 The Contractor shall proceed expeditiously with adequate forces and shall achieve Substantial Completion within the Contract Time.

8.3 DELAYS AND EXTENSIONS OF TIME

8.3.1 If the Contractor is delayed at any time in progress of the Work by an act or neglect of the Owner or Architect, or of an employee of either, or of a separate contractor employed by the Owner, or by changes ordered in the Work, or by labor disputes, fire, unusual delay in deliveries, unavoidable casualties or other causes beyond the Contractor's control, or by delay authorized by the Owner pending arbitration, or by other causes which the Architect determines may justify delay, then the Contract Time shall be extended by Change Order for such reasonable time as the Architect may determine.

8.3.2 Claims relating to time shall be made in accordance with applicable provisions of Paragraph 4.3.

8.3.3 This Paragraph 8.3 does not preclude recovery of damages for delay by either party under other provisions of the Contract Documents.

ARTICLE 9

PAYMENTS AND COMPLETION

9.1 CONTRACT SUM

9.1.1 The Contract Sum is stated in the Agreement and, including authorized adjustments, is the total amount payable by the Owner to the Contractor for performance of the Work under the Contract Documents.

9.2 SCHEDULE OF VALUES

9.2.1 Before the first Application for Payment, the Contractor shall submit to the Architect a schedule of values allocated to various portions of the Work, prepared in such form and supported by such data to substantiate its accuracy as the Architect may require. This schedule, unless objected to by the Architect, shall be used as a basis for reviewing the Contractor's Applications for Payment.

9.3 APPLICATIONS FOR PAYMENT

9.3.1 At least ten days before the date established for each progress payment, the Contractor shall submit to the Architect an itemized Application for Payment for operations completed in accordance with the schedule of values. Such application shall be notarized, if required, and supported by such data substantiating the Contractor's right to payment as the Owner or Architect may require, such as copies of requisitions from Subcontractors and material suppliers, and reflecting retainage if provided for elsewhere in the Contract Documents.

9.3.1.1 Such applications may include requests for payment on account of changes in the Work which have been properly authorized by Construction Change Directives but not yet included in Change Orders.

9.3.1.2 Such applications may not include requests for payment of amounts the Contractor does not intend to pay to a Subcontractor or material supplier because of a dispute or other reason.

9.3.2 Unless otherwise provided in the Contract Documents, payments shall be made on account of materials and equipment delivered and suitably stored at the site for subsequent incorporation in the Work. If approved in advance by the Owner, payment may similarly be made for materials and equipment suitably stored off the site at a location agreed upon in writing. Payment for materials and equipment stored on or off the site shall be conditioned upon compliance by the Contractor with procedures satisfactory to the Owner to establish the Owner's title to such materials and equipment or otherwise protect the Owner's interest, and shall include applicable insurance, storage and transportation to the site for such materials and equipment stored off the site.

9.3.3 The Contractor warrants that title to all Work covered by an Application for Payment will pass to the Owner no later than the time of payment. The Contractor further warrants that upon submittal of an Application for Payment all Work for which Certificates for Payment have been previously issued and payments received from the Owner shall, to the best of the Contractor's knowledge, information and belief, be free and clear of liens, claims, security interests or encumbrances in favor of the Contractor, Subcontractors, material suppliers, or other persons or entities making a claim by reason of having provided labor, materials and equipment relating to the Work.

9.4 CERTIFICATES FOR PAYMENT

9.4.1 The Architect will, within seven days after receipt of the Contractor's Application for Payment, either issue to the

Owner a Certificate for Payment, with a copy to the Contractor, for such amount as the Architect determines is properly due, or notify the Contractor and Owner in writing of the Architect's reasons for withholding certification in whole or in part as provided in Subparagraph 9.5.1.

9.4.2 The issuance of a Certificate for Payment will constitute a representation by the Architect to the Owner, based on the Architect's observations at the site and the data comprising the Application for Payment, that the Work has progressed to the point indicated and that, to the best of the Architect's knowledge, information and belief, quality of the Work is in accordance with the Contract Documents. The foregoing representations are subject to an evaluation of the Work for conformance with the Contract Documents upon Substantial Completion, to results of subsequent tests and inspections, to minor deviations from the Contract Documents correctable prior to completion and to specific qualifications expressed by the Architect. The issuance of a Certificate for Payment will further constitute a representation that the Contractor is entitled to payment in the amount certified. However, the issuance of a Certificate for Payment will not be a representation that the Architect has (1) made exhaustive or continuous on-site inspections to check the quality or quantity of the Work, (2) reviewed construction means, methods, techniques, sequences or procedures, (3) reviewed copies of requisitions received from Subcontractors and material suppliers and other data requested by the Owner to substantiate the Contractor's right to payment or (4) made examination to ascertain how or for what purpose the Contractor has used money previously paid on account of the Contract Sum.

9.5 DECISIONS TO WITHHOLD CERTIFICATION

9.5.1 The Architect may decide not to certify payment and may withhold a Certificate for Payment in whole or in part, to the extent reasonably necessary to protect the Owner, if in the Architect's opinion the representations to the Owner required by Subparagraph 9.4.2 cannot be made. If the Architect is unable to certify payment in the amount of the Application, the Architect will notify the Contractor and Owner as provided in Subparagraph 9.4.1. If the Contractor and Architect cannot agree on a revised amount, the Architect will promptly issue a Certificate for Payment for the amount for which the Architect is able to make such representations to the Owner. The Architect may also decide not to certify payment or, because of subsequently discovered evidence or subsequent observations, may nullify the whole or a part of a Certificate for Payment previously issued, to such extent as may be necessary in the Architect's opinion to protect the Owner from loss because of:

- **.1** defective Work not remedied;
- **.2** third party claims filed or reasonable evidence indicating probable filing of such claims;
- **.3** failure of the Contractor to make payments properly to Subcontractors or for labor, materials or equipment;
- **.4** reasonable evidence that the Work cannot be completed for the unpaid balance of the Contract Sum;
- **.5** damage to the Owner or another contractor;
- **.6** reasonable evidence that the Work will not be completed within the Contract Time, and that the unpaid balance would not be adequate to cover actual or liquidated damages for the anticipated delay; or
- **.7** persistent failure to carry out the Work in accordance with the Contract Documents.

9.5.2 When the above reasons for withholding certification are removed, certification will be made for amounts previously withheld.

9.6 PROGRESS PAYMENTS

9.6.1 After the Architect has issued a Certificate for Payment, the Owner shall make payment in the manner and within the time provided in the Contract Documents, and shall so notify the Architect.

9.6.2 The Contractor shall promptly pay each Subcontractor, upon receipt of payment from the Owner, out of the amount paid to the Contractor on account of such Subcontractor's portion of the Work, the amount to which said Subcontractor is entitled, reflecting percentages actually retained from payments to the Contractor on account of such Subcontractor's portion of the Work. The Contractor shall, by appropriate agreement with each Subcontractor, require each Subcontractor to make payments to Sub-subcontractors in similar manner.

9.6.3 The Architect will, on request, furnish to a Subcontractor, if practicable, information regarding percentages of completion or amounts applied for by the Contractor and action taken thereon by the Architect and Owner on account of portions of the Work done by such Subcontractor.

9.6.4 Neither the Owner nor Architect shall have an obligation to pay or to see to the payment of money to a Subcontractor except as may otherwise be required by law.

9.6.5 Payment to material suppliers shall be treated in a manner similar to that provided in Subparagraphs 9.6.2, 9.6.3 and 9.6.4.

9.6.6 A Certificate for Payment, a progress payment, or partial or entire use or occupancy of the Project by the Owner shall not constitute acceptance of Work not in accordance with the Contract Documents.

9.7 FAILURE OF PAYMENT

9.7.1 If the Architect does not issue a Certificate for Payment, through no fault of the Contractor, within seven days after receipt of the Contractor's Application for Payment, or if the Owner does not pay the Contractor within seven days after the date established in the Contract Documents the amount certified by the Architect or awarded by arbitration, then the Contractor may, upon seven additional days' written notice to the Owner and Architect, stop the Work until payment of the amount owing has been received. The Contract Time shall be extended appropriately and the Contract Sum shall be increased by the amount of the Contractor's reasonable costs of shut-down, delay and start-up, which shall be accomplished as provided in Article 7.

9.8 SUBSTANTIAL COMPLETION

9.8.1 Substantial Completion is the stage in the progress of the Work when the Work or designated portion thereof is sufficiently complete in accordance with the Contract Documents so the Owner can occupy or utilize the Work for its intended use.

9.8.2 When the Contractor considers that the Work, or a portion thereof which the Owner agrees to accept separately, is substantially complete, the Contractor shall prepare and submit to the Architect a comprehensive list of items to be completed or corrected. The Contractor shall proceed promptly to complete and correct items on the list. Failure to include an item on such list does not alter the responsibility of the Contractor to complete all Work in accordance with the Contract Documents. Upon receipt of the Contractor's list, the Architect will make an inspection to determine whether the Work or designated portion thereof is substantially complete. If the Architect's inspection discloses any item, whether or not included on the Contractor's list, which is not in accordance with the requirements of the Contract Documents, the Contractor shall, before issuance of the Certificate of Substantial Completion, complete or correct such item upon notification by the Architect. The Contractor shall then submit a request for another inspection by the Architect to determine Substantial Completion. When the Work or designated portion thereof is substantially complete, the Architect will prepare a Certificate of Substantial Completion which shall establish the date of Substantial Completion, shall establish responsibilities of the Owner and Contractor for security, maintenance, heat, utilities, damage to the Work and insurance, and shall fix the time within which the Contractor shall finish all items on the list accompanying the Certificate. Warranties required by the Contract Documents shall commence on the date of Substantial Completion of the Work or designated portion thereof unless otherwise provided in the Certificate of Substantial Completion. The Certificate of Substantial Completion shall be submitted to the Owner and Contractor for their written acceptance of responsibilities assigned to them in such Certificate.

9.8.3 Upon Substantial Completion of the Work or designated portion thereof and upon application by the Contractor and certification by the Architect, the Owner shall make payment, reflecting adjustment in retainage, if any, for such Work or portion thereof as provided in the Contract Documents.

9.9 PARTIAL OCCUPANCY OR USE

9.9.1 The Owner may occupy or use any completed or partially completed portion of the Work at any stage when such portion is designated by separate agreement with the Contractor, provided such occupancy or use is consented to by the insurer as required under Subparagraph 11.3.11 and authorized by public authorities having jurisdiction over the Work. Such partial occupancy or use may commence whether or not the portion is substantially complete, provided the Owner and Contractor have accepted in writing the responsibilities assigned to each of them for payments, retainage if any, security, maintenance, heat, utilities, damage to the Work and insurance, and have agreed in writing concerning the period for correction of the Work and commencement of warranties required by the Contract Documents. When the Contractor considers a portion substantially complete, the Contractor shall prepare and submit a list to the Architect as provided under Subparagraph 9.8.2. Consent of the Contractor to partial occupancy or use shall not be unreasonably withheld. The stage of the progress of the Work shall be determined by written agreement between the Owner and Contractor or, if no agreement is reached, by decision of the Architect.

9.9.2 Immediately prior to such partial occupancy or use, the Owner, Contractor and Architect shall jointly inspect the area to be occupied or portion of the Work to be used in order to determine and record the condition of the Work.

9.9.3 Unless otherwise agreed upon, partial occupancy or use of a portion or portions of the Work shall not constitute acceptance of Work not complying with the requirements of the Contract Documents.

9.10 FINAL COMPLETION AND FINAL PAYMENT

9.10.1 Upon receipt of written notice that the Work is ready for final inspection and acceptance and upon receipt of a final Application for Payment, the Architect will promptly make such inspection and, when the Architect finds the Work acceptable under the Contract Documents and the Contract fully performed, the Architect will promptly issue a final Certificate for Payment stating that to the best of the Architect's knowledge, information and belief, and on the basis of the Architect's observations and inspections, the Work has been completed in accordance with terms and conditions of the Contract Documents and that the entire balance found to be due the Contractor and noted in said final Certificate is due and payable. The Architect's final Certificate for Payment will constitute a further representation that conditions listed in Subparagraph 9.10.2 as precedent to the Contractor's being entitled to final payment have been fulfilled.

9.10.2 Neither final payment nor any remaining retained percentage shall become due until the Contractor submits to the Architect (1) an affidavit that payrolls, bills for materials and equipment, and other indebtedness connected with the Work for which the Owner or the Owner's property might be responsible or encumbered (less amounts withheld by Owner) have been paid or otherwise satisfied, (2) a certificate evidencing that insurance required by the Contract Documents to remain in force after final payment is currently in effect and will not be cancelled or allowed to expire until at least 30 days' prior written notice has been given to the Owner, (3) a written statement that the Contractor knows of no substantial reason that the insurance will not be renewable to cover the period required by the Contract Documents, (4) consent of surety, if any, to final payment and (5), if required by the Owner, other data establishing payment or satisfaction of obligations, such as receipts, releases and waivers of liens, claims, security interests or encumbrances arising out of the Contract, to the extent and in such form as may be designated by the Owner. If a Subcontractor refuses to furnish a release or waiver required by the Owner, the Contractor may furnish a bond satisfactory to the Owner to indemnify the Owner against such lien. If such lien remains unsatisfied after payments are made, the Contractor shall refund to the Owner all money that the Owner may be compelled to pay in discharging such lien, including all costs and reasonable attorneys' fees.

9.10.3 If, after Substantial Completion of the Work, final completion thereof is materially delayed through no fault of the Contractor or by issuance of Change Orders affecting final completion, and the Architect so confirms, the Owner shall, upon application by the Contractor and certification by the Architect, and without terminating the Contract, make payment of the balance due for that portion of the Work fully completed and accepted. If the remaining balance for Work not fully completed or corrected is less than retainage stipulated in the Contract Documents, and if bonds have been furnished, the written consent of surety to payment of the balance due for that portion of the Work fully completed and accepted shall be submitted by the Contractor to the Architect prior to certification of such payment. Such payment shall be made under terms and conditions governing final payment, except that it shall not constitute a waiver of claims. The making of final payment shall constitute a waiver of claims by the Owner as provided in Subparagraph 4.3.5.

9.10.4 Acceptance of final payment by the Contractor, a Subcontractor or material supplier shall constitute a waiver of claims by that payee except those previously made in writing and identified by that payee as unsettled at the time of final Application for Payment. Such waivers shall be in addition to the waiver described in Subparagraph 4.3.5.

ARTICLE 10

PROTECTION OF PERSONS AND PROPERTY

10.1 SAFETY PRECAUTIONS AND PROGRAMS

10.1.1 The Contractor shall be responsible for initiating, maintaining and supervising all safety precautions and programs in connection with the performance of the Contract.

10.1.2 In the event the Contractor encounters on the site material reasonably believed to be asbestos or polychlorinated biphenyl (PCB) which has not been rendered harmless, the Contractor shall immediately stop Work in the area affected and report the condition to the Owner and Architect in writing. The Work in the affected area shall not thereafter be resumed except by written agreement of the Owner and Contractor if in fact the material is asbestos or polychlorinated biphenyl (PCB) and has not been rendered harmless. The Work in the affected area shall be resumed in the absence of asbestos or polychlorinated biphenyl (PCB), or when it has been rendered harmless, by written agreement of the Owner and Contractor, or in accordance with final determination by the Architect on which arbitration has not been demanded, or by arbitration under Article 4.

10.1.3 The Contractor shall not be required pursuant to Article 7 to perform without consent any Work relating to asbestos or polychlorinated biphenyl (PCB).

10.1.4 To the fullest extent permitted by law, the Owner shall indemnify and hold harmless the Contractor, Architect, Architect's consultants and agents and employees of any of them from and against claims, damages, losses and expenses, including but not limited to attorneys' fees, arising out of or resulting from performance of the Work in the affected area if in fact the material is asbestos or polychlorinated biphenyl (PCB) and has not been rendered harmless, provided that such claim, damage, loss or expense is attributable to bodily injury, sickness, disease or death, or to injury to or destruction of tangible property (other than the Work itself) including loss of use resulting therefrom, but only to the extent caused in whole or in part by negligent acts or omissions of the Owner, anyone directly or indirectly employed by the Owner or anyone for whose acts the Owner may be liable, regardless of whether or not such claim, damage, loss or expense is caused in part by a party indemnified hereunder. Such obligation shall not be construed to negate, abridge, or reduce other rights or obligations of indemnity which would otherwise exist as to a party or person described in this Subparagraph 10.1.4.

10.2 SAFETY OF PERSONS AND PROPERTY

10.2.1 The Contractor shall take reasonable precautions for safety of, and shall provide reasonable protection to prevent damage, injury or loss to:

.1 employees on the Work and other persons who may be affected thereby;

.2 the Work and materials and equipment to be incorporated therein, whether in storage on or off the site, under care, custody or control of the Contractor or the Contractor's Subcontractors or Sub-subcontractors; and

.3 other property at the site or adjacent thereto, such as trees, shrubs, lawns, walks, pavements, roadways, structures and utilities not designated for removal, relocation or replacement in the course of construction.

10.2.2 The Contractor shall give notices and comply with applicable laws, ordinances, rules, regulations and lawful orders of public authorities bearing on safety of persons or property or their protection from damage, injury or loss.

10.2.3 The Contractor shall erect and maintain, as required by existing conditions and performance of the Contract, reasonable safeguards for safety and protection, including posting danger signs and other warnings against hazards, promulgating safety regulations and notifying owners and users of adjacent sites and utilities.

10.2.4 When use or storage of explosives or other hazardous materials or equipment or unusual methods are necessary for execution of the Work, the Contractor shall exercise utmost care and carry on such activities under supervision of properly qualified personnel.

10.2.5 The Contractor shall promptly remedy damage and loss (other than damage or loss insured under property insurance required by the Contract Documents) to property referred to in Clauses 10.2.1.2 and 10.2.1.3 caused in whole or in part by the Contractor, a Subcontractor, a Sub-subcontractor, or anyone directly or indirectly employed by any of them, or by anyone for whose acts they may be liable and for which the Contractor is responsible under Clauses 10.2.1.2 and 10.2.1.3, except damage or loss attributable to acts or omissions of the Owner or Architect or anyone directly or indirectly employed by either of them, or by anyone for whose acts either of them may be liable, and not attributable to the fault or negligence of the Contractor. The foregoing obligations of the Contractor are in addition to the Contractor's obligations under Paragraph 3.18.

10.2.6 The Contractor shall designate a responsible member of the Contractor's organization at the site whose duty shall be the prevention of accidents. This person shall be the Contractor's superintendent unless otherwise designated by the Contractor in writing to the Owner and Architect.

10.2.7 The Contractor shall not load or permit any part of the construction or site to be loaded so as to endanger its safety.

10.3 EMERGENCIES

10.3.1 In an emergency affecting safety of persons or property, the Contractor shall act, at the Contractor's discretion, to prevent threatened damage, injury or loss. Additional compensation or extension of time claimed by the Contractor on account of an emergency shall be determined as provided in Paragraph 4.3 and Article 7.

ARTICLE 11

INSURANCE AND BONDS

11.1 CONTRACTOR'S LIABILITY INSURANCE

11.1.1 The Contractor shall purchase from and maintain in a company or companies lawfully authorized to do business in the jurisdiction in which the Project is located such insurance as will protect the Contractor from claims set forth below which may arise out of or result from the Contractor's operations under the Contract and for which the Contractor may be legally liable, whether such operations be by the Contractor or by a Subcontractor or by anyone directly or indirectly employed by any of them, or by anyone for whose acts any of them may be liable:

.1 claims under workers' or workmen's compensation, disability benefit and other similar employee benefit acts which are applicable to the Work to be performed;

.2 claims for damages because of bodily injury, occupational sickness or disease, or death of the Contractor's employees;

.3 claims for damages because of bodily injury, sickness or disease, or death of any person other than the Contractor's employees;

.4 claims for damages insured by usual personal injury liability coverage which are sustained (1) by a person as a result of an offense directly or indirectly related to employment of such person by the Contractor, or (2) by another person;

.5 claims for damages, other than to the Work itself, because of injury to or destruction of tangible property, including loss of use resulting therefrom;

.6 claims for damages because of bodily injury, death of a person or property damage arising out of ownership, maintenance or use of a motor vehicle; and

.7 claims involving contractual liability insurance applicable to the Contractor's obligations under Paragraph 3.18.

11.1.2 The insurance required by Subparagraph 11.1.1 shall be written for not less than limits of liability specified in the Contract Documents or required by law, whichever coverage is greater. Coverages, whether written on an occurrence or claims-made basis, shall be maintained without interruption from date of commencement of the Work until date of final payment and termination of any coverage required to be maintained after final payment.

11.1.3 Certificates of Insurance acceptable to the Owner shall be filed with the Owner prior to commencement of the Work. These Certificates and the insurance policies required by this Paragraph 11.1 shall contain a provision that coverages afforded under the policies will not be cancelled or allowed to expire until at least 30 days' prior written notice has been given to the Owner. If any of the foregoing insurance coverages are required to remain in force after final payment and are reasonably available, an additional certificate evidencing continuation of such coverage shall be submitted with the final Application for Payment as required by Subparagraph 9.10.2. Information concerning reduction of coverage shall be furnished by the Contractor with reasonable promptness in accordance with the Contractor's information and belief.

11.2 OWNER'S LIABILITY INSURANCE

11.2.1 The Owner shall be responsible for purchasing and maintaining the Owner's usual liability insurance. Optionally, the Owner may purchase and maintain other insurance for self-protection against claims which may arise from operations under the Contract. The Contractor shall not be responsible for purchasing and maintaining this optional Owner's liability insurance unless specifically required by the Contract Documents.

11.3 PROPERTY INSURANCE

11.3.1 Unless otherwise provided, the Owner shall purchase and maintain, in a company or companies lawfully authorized to do business in the jurisdiction in which the Project is located, property insurance in the amount of the initial Contract Sum as well as subsequent modifications thereto for the entire Work at the site on a replacement cost basis without voluntary deductibles. Such property insurance shall be maintained, unless otherwise provided in the Contract Documents or otherwise agreed in writing by all persons and entities who are beneficiaries of such insurance, until final payment has been made as provided in Paragraph 9.10 or until no person or entity other than the Owner has an insurable interest in the property required by this Paragraph 11.3 to be covered, whichever is earlier. This insurance shall include interests of the Owner, the Contractor, Subcontractors and Sub-subcontractors in the Work.

11.3.1.1 Property insurance shall be on an all-risk policy form and shall insure against the perils of fire and extended coverage and physical loss or damage including, without duplication of coverage, theft, vandalism, malicious mischief, collapse, falsework, temporary buildings and debris removal including demolition occasioned by enforcement of any applicable legal requirements, and shall cover reasonable compensation for Architect's services and expenses required as a result of such insured loss. Coverage for other perils shall not be required unless otherwise provided in the Contract Documents.

11.3.1.2 If the Owner does not intend to purchase such property insurance required by the Contract and with all of the coverages in the amount described above, the Owner shall so inform the Contractor in writing prior to commencement of the Work. The Contractor may then effect insurance which will protect the interests of the Contractor, Subcontractors and Sub-subcontractors in the Work, and by appropriate Change Order the cost thereof shall be charged to the Owner. If the Contractor is damaged by the failure or neglect of the Owner to purchase or maintain insurance as described above, without so notifying the Contractor, then the Owner shall bear all reasonable costs properly attributable thereto.

11.3.1.3 If the property insurance requires minimum deductibles and such deductibles are identified in the Contract Documents, the Contractor shall pay costs not covered because of such deductibles. If the Owner or insurer increases the required minimum deductibles above the amounts so identified or if the Owner elects to purchase this insurance with voluntary deductible amounts, the Owner shall be responsible for payment of the additional costs not covered because of such increased or voluntary deductibles. If deductibles are not identified in the Contract Documents, the Owner shall pay costs not covered because of deductibles.

11.3.1.4 Unless otherwise provided in the Contract Documents, this property insurance shall cover portions of the Work stored off the site after written approval of the Owner at the value established in the approval, and also portions of the Work in transit.

11.3.2 Boiler and Machinery Insurance. The Owner shall purchase and maintain boiler and machinery insurance required by the Contract Documents or by law, which shall specifically cover such insured objects during installation and until final acceptance by the Owner; this insurance shall include interests of the Owner, Contractor, Subcontractors and Sub-subcontractors in the Work, and the Owner and Contractor shall be named insureds.

11.3.3 Loss of Use Insurance. The Owner, at the Owner's option, may purchase and maintain such insurance as will insure the Owner against loss of use of the Owner's property due to fire or other hazards, however caused. The Owner waives all rights of action against the Contractor for loss of use of the Owner's property, including consequential losses due to fire or other hazards however caused.

11.3.4 If the Contractor requests in writing that insurance for risks other than those described herein or for other special hazards be included in the property insurance policy, the Owner shall, if possible, include such insurance, and the cost thereof shall be charged to the Contractor by appropriate Change Order.

11.3.5 If during the Project construction period the Owner insures properties, real or personal or both, adjoining or adjacent to the site by property insurance under policies separate from those insuring the Project, or if after final payment property insurance is to be provided on the completed Project through a policy or policies other than those insuring the Project during the construction period, the Owner shall waive all rights in accordance with the terms of Subparagraph 11.3.7 for damages caused by fire or other perils covered by this separate property insurance. All separate policies shall provide this waiver of subrogation by endorsement or otherwise.

11.3.6 Before an exposure to loss may occur, the Owner shall file with the Contractor a copy of each policy that includes insurance coverages required by this Paragraph 11.3. Each policy shall contain all generally applicable conditions, definitions, exclusions and endorsements related to this Project. Each policy shall contain a provision that the policy will not be cancelled or allowed to expire until at least 30 days' prior written notice has been given to the Contractor.

11.3.7 Waivers of Subrogation. The Owner and Contractor waive all rights against (1) each other and any of their subcontractors, sub-subcontractors, agents and employees, each of the other, and (2) the Architect, Architect's consultants, separate contractors described in Article 6, if any, and any of their subcontractors, sub-subcontractors, agents and employees, for damages caused by fire or other perils to the extent covered by property insurance obtained pursuant to this Paragraph 11.3 or other property insurance applicable to the Work, except such rights as they have to proceeds of such insurance held by the Owner as fiduciary. The Owner or Contractor, as appropriate, shall require of the Architect, Architect's consultants, separate contractors described in Article 6, if any, and the subcontractors, sub-subcontractors, agents and employees of any of them, by appropriate agreements, written where legally required for validity, similar waivers each in favor of other parties enumerated herein. The policies shall provide such waivers of subrogation by endorsement or otherwise. A waiver of subrogation shall be effective as to a person or entity even though that person or entity would otherwise have a duty of indemnification, contractual or otherwise, did not pay the insurance premium directly or indirectly, and whether or not the person or entity had an insurable interest in the property damaged.

11.3.8 A loss insured under Owner's property insurance shall be adjusted by the Owner as fiduciary and made payable to the Owner as fiduciary for the insureds, as their interests may appear, subject to requirements of any applicable mortgagee clause and of Subparagraph 11.3.10. The Contractor shall pay Subcontractors their just shares of insurance proceeds received by the Contractor, and by appropriate agreements, written where legally required for validity, shall require Subcontractors to make payments to their Sub-subcontractors in similar manner.

11.3.9 If required in writing by a party in interest, the Owner as fiduciary shall, upon occurrence of an insured loss, give bond for proper performance of the Owner's duties. The cost of required bonds shall be charged against proceeds received as fiduciary. The Owner shall deposit in a separate account proceeds so received, which the Owner shall distribute in accordance with such agreement as the parties in interest may reach, or in accordance with an arbitration award in which case the procedure shall be as provided in Paragraph 4.5. If after such loss no other special agreement is made, replacement of damaged property shall be covered by appropriate Change Order.

11.3.10 The Owner as fiduciary shall have power to adjust and settle a loss with insurers unless one of the parties in interest shall object in writing within five days after occurrence of loss to the Owner's exercise of this power; if such objection be made, arbitrators shall be chosen as provided in Paragraph 4.5. The Owner as fiduciary shall, in that case, make settlement with insurers in accordance with directions of such arbitrators. If distribution of insurance proceeds by arbitration is required, the arbitrators will direct such distribution.

11.3.11 Partial occupancy or use in accordance with Paragraph 9.9 shall not commence until the insurance company or companies providing property insurance have consented to such partial occupancy or use by endorsement or otherwise. The Owner and the Contractor shall take reasonable steps to obtain consent of the insurance company or companies and shall, without mutual written consent, take no action with respect to partial occupancy or use that would cause cancellation, lapse or reduction of insurance.

11.4 PERFORMANCE BOND AND PAYMENT BOND

11.4.1 The Owner shall have the right to require the Contractor to furnish bonds covering faithful performance of the Contract and payment of obligations arising thereunder as stipulated in bidding requirements or specifically required in the Contract Documents on the date of execution of the Contract.

11.4.2 Upon the request of any person or entity appearing to be a potential beneficiary of bonds covering payment of obligations arising under the Contract, the Contractor shall promptly furnish a copy of the bonds or shall permit a copy to be made.

ARTICLE 12

UNCOVERING AND CORRECTION OF WORK

12.1 UNCOVERING OF WORK

12.1.1 If a portion of the Work is covered contrary to the Architect's request or to requirements specifically expressed in the Contract Documents, it must, if required in writing by the Architect, be uncovered for the Architect's observation and be replaced at the Contractor's expense without change in the Contract Time.

12.1.2 If a portion of the Work has been covered which the Architect has not specifically requested to observe prior to its being covered, the Architect may request to see such Work and it shall be uncovered by the Contractor. If such Work is in accordance with the Contract Documents, costs of uncovering and replacement shall, by appropriate Change Order, be charged to the Owner. If such Work is not in accordance with the Contract Documents, the Contractor shall pay such costs unless the condition was caused by the Owner or a separate contractor in which event the Owner shall be responsible for payment of such costs.

12.2 CORRECTION OF WORK

12.2.1 The Contractor shall promptly correct Work rejected by the Architect or failing to conform to the requirements of the Contract Documents, whether observed before or after Substantial Completion and whether or not fabricated, installed or completed. The Contractor shall bear costs of correcting such rejected Work, including additional testing and inspections and compensation for the Architect's services and expenses made necessary thereby.

12.2.2 If, within one year after the date of Substantial Completion of the Work or designated portion thereof, or after the date for commencement of warranties established under Subparagraph 9.9.1, or by terms of an applicable special warranty required by the Contract Documents, any of the Work is found to be not in accordance with the requirements of the Contract Documents, the Contractor shall correct it promptly after receipt of written notice from the Owner to do so unless the Owner has previously given the Contractor a written acceptance of such condition. This period of one year shall be extended with respect to portions of Work first performed after Substantial Completion by the period of time between Substantial Completion and the actual performance of the Work. This obligation under this Subparagraph 12.2.2 shall survive acceptance of the Work under the Contract and termination of the Contract. The Owner shall give such notice promptly after discovery of the condition.

12.2.3 The Contractor shall remove from the site portions of the Work which are not in accordance with the requirements of the Contract Documents and are neither corrected by the Contractor nor accepted by the Owner.

12.2.4 If the Contractor fails to correct nonconforming Work within a reasonable time, the Owner may correct it in accordance with Paragraph 2.4. If the Contractor does not proceed with correction of such nonconforming Work within a reasonable time fixed by written notice from the Architect, the Owner may remove it and store the salvable materials or equipment at the Contractor's expense. If the Contractor does not pay costs of such removal and storage within ten days after written notice, the Owner may upon ten additional days' written notice sell such materials and equipment at auction or at private sale and shall account for the proceeds thereof, after deducting costs and damages that should have been borne by the Contractor, including compensation for the Architect's services and expenses made necessary thereby. If such proceeds of sale do not cover costs which the Contractor should have borne, the Contract Sum shall be reduced by the deficiency. If payments then or thereafter due the Contractor are not sufficient to cover such amount, the Contractor shall pay the difference to the Owner.

12.2.5 The Contractor shall bear the cost of correcting destroyed or damaged construction, whether completed or partially completed, of the Owner or separate contractors caused by the Contractor's correction or removal of Work which is not in accordance with the requirements of the Contract Documents.

12.2.6 Nothing contained in this Paragraph 12.2 shall be construed to establish a period of limitation with respect to other obligations which the Contractor might have under the Contract Documents. Establishment of the time period of one year as described in Subparagraph 12.2.2 relates only to the specific obligation of the Contractor to correct the Work, and has no relationship to the time within which the obligation to comply with the Contract Documents may be sought to be enforced, nor to the time within which proceedings may be commenced to establish the Contractor's liability with respect to the Contractor's obligations other than specifically to correct the Work.

12.3 ACCEPTANCE OF NONCONFORMING WORK

12.3.1 If the Owner prefers to accept Work which is not in accordance with the requirements of the Contract Documents, the Owner may do so instead of requiring its removal and correction, in which case the Contract Sum will be reduced as appropriate and equitable. Such adjustment shall be effected whether or not final payment has been made.

ARTICLE 13

MISCELLANEOUS PROVISIONS

13.1 GOVERNING LAW

13.1.1 The Contract shall be governed by the law of the place where the Project is located.

13.2 SUCCESSORS AND ASSIGNS

13.2.1 The Owner and Contractor respectively bind themselves, their partners, successors, assigns and legal representatives to the other party hereto and to partners, successors, assigns and legal representatives of such other party in respect to covenants, agreements and obligations contained in the Contract Documents. Neither party to the Contract shall assign the Contract as a whole without written consent of the other. If either party attempts to make such an assignment without such consent, that party shall nevertheless remain legally responsible for all obligations under the Contract.

13.3 WRITTEN NOTICE

13.3.1 Written notice shall be deemed to have been duly served if delivered in person to the individual or a member of the firm or entity or to an officer of the corporation for which it was intended, or if delivered at or sent by registered or certified mail to the last business address known to the party giving notice.

13.4 RIGHTS AND REMEDIES

13.4.1 Duties and obligations imposed by the Contract Documents and rights and remedies available thereunder shall be in addition to and not a limitation of duties, obligations, rights and remedies otherwise imposed or available by law.

13.4.2 No action or failure to act by the Owner, Architect or Contractor shall constitute a waiver of a right or duty afforded them under the Contract, nor shall such action or failure to act constitute approval of or acquiescence in a breach thereunder, except as may be specifically agreed in writing.

13.5 TESTS AND INSPECTIONS

13.5.1 Tests, inspections and approvals of portions of the Work required by the Contract Documents or by laws, ordinances, rules, regulations or orders of public authorities having jurisdiction shall be made at an appropriate time. Unless otherwise provided, the Contractor shall make arrangements for such tests, inspections and approvals with an independent testing laboratory or entity acceptable to the Owner, or with the appropriate public authority, and shall bear all related costs of tests, inspections and approvals. The Contractor shall give the Architect timely notice of when and where tests and inspections are to be made so the Architect may observe such procedures. The Owner shall bear costs of tests, inspections or approvals which do not become requirements until after bids are received or negotiations concluded.

13.5.2 If the Architect, Owner or public authorities having jurisdiction determine that portions of the Work require additional testing, inspection or approval not included under Subparagraph 13.5.1, the Architect will, upon written authorization from the Owner, instruct the Contractor to make arrangements for such additional testing, inspection or approval by an entity acceptable to the Owner, and the Contractor shall give timely notice to the Architect of when and where tests and inspections are to be made so the Architect may observe such procedures.

The Owner shall bear such costs except as provided in Subparagraph 13.5.3.

13.5.3 If such procedures for testing, inspection or approval under Subparagraphs 13.5.1 and 13.5.2 reveal failure of the portions of the Work to comply with requirements established by the Contract Documents, the Contractor shall bear all costs made necessary by such failure including those of repeated procedures and compensation for the Architect's services and expenses.

13.5.4 Required certificates of testing, inspection or approval shall, unless otherwise required by the Contract Documents, be secured by the Contractor and promptly delivered to the Architect.

13.5.5 If the Architect is to observe tests, inspections or approvals required by the Contract Documents, the Architect will do so promptly and, where practicable, at the normal place of testing.

13.5.6 Tests or inspections conducted pursuant to the Contract Documents shall be made promptly to avoid unreasonable delay in the Work.

13.6 INTEREST

13.6.1 Payments due and unpaid under the Contract Documents shall bear interest from the date payment is due at such rate as the parties may agree upon in writing or, in the absence thereof, at the legal rate prevailing from time to time at the place where the Project is located.

13.7 COMMENCEMENT OF STATUTORY LIMITATION PERIOD

13.7.1 As between the Owner and Contractor:

.1 Before Substantial Completion. As to acts or failures to act occurring prior to the relevant date of Substantial Completion, any applicable statute of limitations shall commence to run and any alleged cause of action shall be deemed to have accrued in any and all events not later than such date of Substantial Completion;

.2 Between Substantial Completion and Final Certificate for Payment. As to acts or failures to act occurring subsequent to the relevant date of Substantial Completion and prior to issuance of the final Certificate for Payment, any applicable statute of limitations shall commence to run and any alleged cause of action shall be deemed to have accrued in any and all events not later than the date of issuance of the final Certificate for Payment; and

.3 After Final Certificate for Payment. As to acts or failures to act occurring after the relevant date of issuance of the final Certificate for Payment, any applicable statute of limitations shall commence to run and any alleged cause of action shall be deemed to have accrued in any and all events not later than the date of any act or failure to act by the Contractor pursuant to any warranty provided under Paragraph 3.5, the date of any correction of the Work or failure to correct the Work by the Contractor under Paragraph 12.2, or the date of actual commission of any other act or failure to perform any duty or obligation by the Contractor or Owner, whichever occurs last.

ARTICLE 14

TERMINATION OR SUSPENSION OF THE CONTRACT

14.1 TERMINATION BY THE CONTRACTOR

14.1.1 The Contractor may terminate the Contract if the Work is stopped for a period of 30 days through no act or fault of the Contractor or a Subcontractor, Sub-subcontractor or their agents or employees or any other persons performing portions of the Work under contract with the Contractor, for any of the following reasons:

.1 issuance of an order of a court or other public authority having jurisdiction;

.2 an act of government, such as a declaration of national emergency, making material unavailable;

.3 because the Architect has not issued a Certificate for Payment and has not notified the Contractor of the reason for withholding certification as provided in Subparagraph 9.4.1, or because the Owner has not made payment on a Certificate for Payment within the time stated in the Contract Documents;

.4 if repeated suspensions, delays or interruptions by the Owner as described in Paragraph 14.3 constitute in the aggregate more than 100 percent of the total number of days scheduled for completion, or 120 days in any 365-day period, whichever is less; or

.5 the Owner has failed to furnish to the Contractor promptly, upon the Contractor's request, reasonable evidence as required by Subparagraph 2.2.1.

14.1.2 If one of the above reasons exists, the Contractor may, upon seven additional days' written notice to the Owner and Architect, terminate the Contract and recover from the Owner payment for Work executed and for proven loss with respect to materials, equipment, tools, and construction equipment and machinery, including reasonable overhead, profit and damages.

14.1.3 If the Work is stopped for a period of 60 days through no act or fault of the Contractor or a Subcontractor or their agents or employees or any other persons performing portions of the Work under contract with the Contractor because the Owner has persistently failed to fulfill the Owner's obligations under the Contract Documents with respect to matters important to the progress of the Work, the Contractor may, upon seven additional days' written notice to the Owner and the Architect, terminate the Contract and recover from the Owner as provided in Subparagraph 14.1.2.

14.2 TERMINATION BY THE OWNER FOR CAUSE

14.2.1 The Owner may terminate the Contract if the Contractor:

.1 persistently or repeatedly refuses or fails to supply enough properly skilled workers or proper materials;

.2 fails to make payment to Subcontractors for materials or labor in accordance with the respective agreements between the Contractor and the Subcontractors;

.3 persistently disregards laws, ordinances, or rules, regulations or orders of a public authority having jurisdiction; or

.4 otherwise is guilty of substantial breach of a provision of the Contract Documents.

14.2.2 When any of the above reasons exist, the Owner, upon certification by the Architect that sufficient cause exists to justify such action, may without prejudice to any other rights or remedies of the Owner and after giving the Contractor and the Contractor's surety, if any, seven days' written notice, terminate employment of the Contractor and may, subject to any prior rights of the surety:

.1 take possession of the site and of all materials, equipment, tools, and construction equipment and machinery thereon owned by the Contractor;

.2 accept assignment of subcontracts pursuant to Paragraph 5.4; and

.3 finish the Work by whatever reasonable method the Owner may deem expedient.

14.2.3 When the Owner terminates the Contract for one of the reasons stated in Subparagraph 14.2.1, the Contractor shall not be entitled to receive further payment until the Work is finished.

14.2.4 If the unpaid balance of the Contract Sum exceeds costs of finishing the Work, including compensation for the Architect's services and expenses made necessary thereby, such excess shall be paid to the Contractor. If such costs exceed the unpaid balance, the Contractor shall pay the difference to the Owner. The amount to be paid to the Contractor or Owner, as the case may be, shall be certified by the Architect, upon application, and this obligation for payment shall survive termination of the Contract.

14.3 SUSPENSION BY THE OWNER FOR CONVENIENCE

14.3.1 The Owner may, without cause, order the Contractor in writing to suspend, delay or interrupt the Work in whole or in part for such period of time as the Owner may determine.

14.3.2 An adjustment shall be made for increases in the cost of performance of the Contract, including profit on the increased cost of performance, caused by suspension, delay or interruption. No adjustment shall be made to the extent:

.1 that performance is, was or would have been so suspended, delayed or interrupted by another cause for which the Contractor is responsible; or

.2 that an equitable adjustment is made or denied under another provision of this Contract.

14.3.3 Adjustments made in the cost of performance may have a mutually agreed fixed or percentage fee.

THE AMERICAN INSTITUTE OF ARCHITECTS

AIA Document B141

Standard Form of Agreement Between Owner and Architect

1987 EDITION

THIS DOCUMENT HAS IMPORTANT LEGAL CONSEQUENCES; CONSULTATION WITH AN ATTORNEY IS ENCOURAGED WITH RESPECT TO ITS COMPLETION OR MODIFICATION.

AGREEMENT

made as of the day of in the year of Nineteen Hundred and

BETWEEN the Owner:
(Name and address)

and the Architect:
(Name and address)

For the following Project:
(Include detailed description of Project, location, address and scope.)

The Owner and Architect agree as set forth below.

TERMS AND CONDITIONS OF AGREEMENT BETWEEN OWNER AND ARCHITECT

ARTICLE 1
ARCHITECT'S RESPONSIBILITIES

1.1 ARCHITECT'S SERVICES

1.1.1 The Architect's services consist of those services performed by the Architect, Architect's employees and Architect's consultants as enumerated in Articles 2 and 3 of this Agreement and any other services included in Article 12.

1.1.2 The Architect's services shall be performed as expeditiously as is consistent with professional skill and care and the orderly progress of the Work. Upon request of the Owner, the Architect shall submit for the Owner's approval a schedule for the performance of the Architect's services which may be adjusted as the Project proceeds, and shall include allowances for periods of time required for the Owner's review and for approval of submissions by authorities having jurisdiction over the Project. Time limits established by this schedule approved by the Owner shall not, except for reasonable cause, be exceeded by the Architect or Owner.

1.1.3 The services covered by this Agreement are subject to the time limitations contained in Subparagraph 11.5.1.

ARTICLE 2
SCOPE OF ARCHITECT'S BASIC SERVICES

2.1 DEFINITION

2.1.1 The Architect's Basic Services consist of those described in Paragraphs 2.2 through 2.6 and any other services identified in Article 12 as part of Basic Services, and include normal structural, mechanical and electrical engineering services.

2.2 SCHEMATIC DESIGN PHASE

2.2.1 The Architect shall review the program furnished by the Owner to ascertain the requirements of the Project and shall arrive at a mutual understanding of such requirements with the Owner.

2.2.2 The Architect shall provide a preliminary evaluation of the Owner's program, schedule and construction budget requirements, each in terms of the other, subject to the limitations set forth in Subparagraph 5.2.1.

2.2.3 The Architect shall review with the Owner alternative approaches to design and construction of the Project.

2.2.4 Based on the mutually agreed-upon program, schedule and construction budget requirements, the Architect shall prepare, for approval by the Owner, Schematic Design Documents consisting of drawings and other documents illustrating the scale and relationship of Project components.

2.2.5 The Architect shall submit to the Owner a preliminary estimate of Construction Cost based on current area, volume or other unit costs.

2.3 DESIGN DEVELOPMENT PHASE

2.3.1 Based on the approved Schematic Design Documents and any adjustments authorized by the Owner in the program, schedule or construction budget, the Architect shall prepare, for approval by the Owner, Design Development Documents consisting of drawings and other documents to fix and describe the size and character of the Project as to architectural, structural, mechanical and electrical systems, materials and such other elements as may be appropriate.

2.3.2 The Architect shall advise the Owner of any adjustments to the preliminary estimate of Construction Cost.

2.4 CONSTRUCTION DOCUMENTS PHASE

2.4.1 Based on the approved Design Development Documents and any further adjustments in the scope or quality of the Project or in the construction budget authorized by the Owner, the Architect shall prepare, for approval by the Owner, Construction Documents consisting of Drawings and Specifications setting forth in detail the requirements for the construction of the Project.

2.4.2 The Architect shall assist the Owner in the preparation of the necessary bidding information, bidding forms, the Conditions of the Contract, and the form of Agreement between the Owner and Contractor.

2.4.3 The Architect shall advise the Owner of any adjustments to previous preliminary estimates of Construction Cost indicated by changes in requirements or general market conditions.

2.4.4 The Architect shall assist the Owner in connection with the Owner's responsibility for filing documents required for the approval of governmental authorities having jurisdiction over the Project.

2.5 BIDDING OR NEGOTIATION PHASE

2.5.1 The Architect, following the Owner's approval of the Construction Documents and of the latest preliminary estimate of Construction Cost, shall assist the Owner in obtaining bids or negotiated proposals and assist in awarding and preparing contracts for construction.

2.6 CONSTRUCTION PHASE—ADMINISTRATION OF THE CONSTRUCTION CONTRACT

2.6.1 The Architect's responsibility to provide Basic Services for the Construction Phase under this Agreement commences with the award of the Contract for Construction and terminates at the earlier of the issuance to the Owner of the final Certificate for Payment or 60 days after the date of Substantial Completion of the Work.

2.6.2 The Architect shall provide administration of the Contract for Construction as set forth below and in the edition of AIA Document A201, General Conditions of the Contract for Construction, current as of the date of this Agreement, unless otherwise provided in this Agreement.

2.6.3 Duties, responsibilities and limitations of authority of the Architect shall not be restricted, modified or extended without written agreement of the Owner and Architect with consent of the Contractor, which consent shall not be unreasonably withheld.

2.6.4 The Architect shall be a representative of and shall advise and consult with the Owner (1) during construction until final payment to the Contractor is due, and (2) as an Additional Service at the Owner's direction from time to time during the correction period described in the Contract for Construction. The Architect shall have authority to act on behalf of the Owner only to the extent provided in this Agreement unless otherwise modified by written instrument.

2.6.5 The Architect shall visit the site at intervals appropriate to the stage of construction or as otherwise agreed by the Owner and Architect in writing to become generally familiar with the progress and quality of the Work completed and to determine in general if the Work is being performed in a manner indicating that the Work when completed will be in accordance with the Contract Documents. However, the Architect shall not be required to make exhaustive or continuous on-site inspections to check the quality or quantity of the Work. On the basis of on-site observations as an architect, the Architect shall keep the Owner informed of the progress and quality of the Work, and shall endeavor to guard the Owner against defects and deficiencies in the Work. *(More extensive site representation may be agreed to as an Additional Service, as described in Paragraph 3.2.)*

2.6.6 The Architect shall not have control over or charge of and shall not be responsible for construction means, methods, techniques, sequences or procedures, or for safety precautions and programs in connection with the Work, since these are solely the Contractor's responsibility under the Contract for Construction. The Architect shall not be responsible for the Contractor's schedules or failure to carry out the Work in accordance with the Contract Documents. The Architect shall not have control over or charge of acts or omissions of the Contractor, Subcontractors, or their agents or employees, or of any other persons performing portions of the Work.

2.6.7 The Architect shall at all times have access to the Work wherever it is in preparation or progress.

2.6.8 Except as may otherwise be provided in the Contract Documents or when direct communications have been specially authorized, the Owner and Contractor shall communicate through the Architect. Communications by and with the Architect's consultants shall be through the Architect.

2.6.9 Based on the Architect's observations and evaluations of the Contractor's Applications for Payment, the Architect shall review and certify the amounts due the Contractor.

2.6.10 The Architect's certification for payment shall constitute a representation to the Owner, based on the Architect's observations at the site as provided in Subparagraph 2.6.5 and on the data comprising the Contractor's Application for Payment, that the Work has progressed to the point indicated and that, to the best of the Architect's knowledge, information and belief, quality of the Work is in accordance with the Contract Documents. The foregoing representations are subject to an evaluation of the Work for conformance with the Contract Documents upon Substantial Completion, to results of subsequent tests and inspections, to minor deviations from the Contract Documents correctable prior to completion and to specific qualifications expressed by the Architect. The issuance of a Certificate for Payment shall further constitute a representation that the Contractor is entitled to payment in the amount certified. However, the issuance of a Certificate for Payment shall not be a representation that the Architect has (1) made exhaustive or continuous on-site inspections to check the quality or quantity of the Work, (2) reviewed construction means, methods, techniques, sequences or procedures, (3) reviewed copies of requisitions received from Subcontractors and material suppliers and other data requested by the Owner to substantiate the Contractor's right to payment or (4) ascertained how or for what purpose the Contractor has used money previously paid on account of the Contract Sum.

2.6.11 The Architect shall have authority to reject Work which does not conform to the Contract Documents. Whenever the Architect considers it necessary or advisable for implementation of the intent of the Contract Documents, the Architect will have authority to require additional inspection or testing of the Work in accordance with the provisions of the Contract Documents, whether or not such Work is fabricated, installed or completed. However, neither this authority of the Architect nor a decision made in good faith either to exercise or not to exercise such authority shall give rise to a duty or responsibility of the Architect to the Contractor, Subcontractors, material and equipment suppliers, their agents or employees or other persons performing portions of the Work.

2.6.12 The Architect shall review and approve or take other appropriate action upon Contractor's submittals such as Shop Drawings, Product Data and Samples, but only for the limited purpose of checking for conformance with information given and the design concept expressed in the Contract Documents. The Architect's action shall be taken with such reasonable promptness as to cause no delay in the Work or in the construction of the Owner or of separate contractors, while allowing sufficient time in the Architect's professional judgment to permit adequate review. Review of such submittals is not conducted for the purpose of determining the accuracy and completeness of other details such as dimensions and quantities or for substantiating instructions for installation or performance of equipment or systems designed by the Contractor, all of which remain the responsibility of the Contractor to the extent required by the Contract Documents. The Architect's review shall not constitute approval of safety precautions or, unless otherwise specifically stated by the Architect, of construction means, methods, techniques, sequences or procedures. The Architect's approval of a specific item shall not indicate approval of an assembly of which the item is a component. When professional certification of performance characteristics of materials, systems or equipment is required by the Contract Documents, the Architect shall be entitled to rely upon such certification to establish that the materials, systems or equipment will meet the performance criteria required by the Contract Documents.

2.6.13 The Architect shall prepare Change Orders and Construction Change Directives, with supporting documentation and data if deemed necessary by the Architect as provided in Subparagraphs 3.1.1 and 3.3.3, for the Owner's approval and execution in accordance with the Contract Documents, and may authorize minor changes in the Work not involving an adjustment in the Contract Sum or an extension of the Contract Time which are not inconsistent with the intent of the Contract Documents.

2.6.14 The Architect shall conduct inspections to determine the date or dates of Substantial Completion and the date of final completion, shall receive and forward to the Owner for the Owner's review and records written warranties and related documents required by the Contract Documents and assembled by the Contractor, and shall issue a final Certificate for Payment upon compliance with the requirements of the Contract Documents.

2.6.15 The Architect shall interpret and decide matters concerning performance of the Owner and Contractor under the requirements of the Contract Documents on written request of either the Owner or Contractor. The Architect's response to such requests shall be made with reasonable promptness and within any time limits agreed upon.

2.6.16 Interpretations and decisions of the Architect shall be consistent with the intent of and reasonably inferable from the Contract Documents and shall be in writing or in the form of drawings. When making such interpretations and initial decisions, the Architect shall endeavor to secure faithful performance by both Owner and Contractor, shall not show partiality to either, and shall not be liable for results of interpretations or decisions so rendered in good faith.

2.6.17 The Architect's decisions on matters relating to aesthetic effect shall be final if consistent with the intent expressed in the Contract Documents.

2.6.18 The Architect shall render written decisions within a reasonable time on all claims, disputes or other matters in question between the Owner and Contractor relating to the execution or progress of the Work as provided in the Contract Documents.

2.6.19 The Architect's decisions on claims, disputes or other matters, including those in question between the Owner and Contractor, except for those relating to aesthetic effect as provided in Subparagraph 2.6.17, shall be subject to arbitration as provided in this Agreement and in the Contract Documents.

ARTICLE 3
ADDITIONAL SERVICES

3.1 GENERAL

3.1.1 The services described in this Article 3 are not included in Basic Services unless so identified in Article 12, and they shall be paid for by the Owner as provided in this Agreement, in addition to the compensation for Basic Services. The services described under Paragraphs 3.2 and 3.4 shall only be provided if authorized or confirmed in writing by the Owner. If services described under Contingent Additional Services in Paragraph 3.3 are required due to circumstances beyond the Architect's control, the Architect shall notify the Owner prior to commencing such services. If the Owner deems that such services described under Paragraph 3.3 are not required, the Owner shall give prompt written notice to the Architect. If the Owner indicates in writing that all or part of such Contingent Additional Services are not required, the Architect shall have no obligation to provide those services.

3.2 PROJECT REPRESENTATION BEYOND BASIC SERVICES

3.2.1 If more extensive representation at the site than is described in Subparagraph 2.6.5 is required, the Architect shall provide one or more Project Representatives to assist in carrying out such additional on-site responsibilities.

3.2.2 Project Representatives shall be selected, employed and directed by the Architect, and the Architect shall be compensated therefor as agreed by the Owner and Architect. The duties, responsibilities and limitations of authority of Project Representatives shall be as described in the edition of AIA Document B352 current as of the date of this Agreement, unless otherwise agreed.

3.2.3 Through the observations by such Project Representatives, the Architect shall endeavor to provide further protection for the Owner against defects and deficiencies in the Work, but the furnishing of such project representation shall not modify the rights, responsibilities or obligations of the Architect as described elsewhere in this Agreement.

3.3 CONTINGENT ADDITIONAL SERVICES

3.3.1 Making revisions in Drawings, Specifications or other documents when such revisions are:

.1 inconsistent with approvals or instructions previously given by the Owner, including revisions made necessary by adjustments in the Owner's program or Project budget;

.2 required by the enactment or revision of codes, laws or regulations subsequent to the preparation of such documents; or

.3 due to changes required as a result of the Owner's failure to render decisions in a timely manner.

3.3.2 Providing services required because of significant changes in the Project including, but not limited to, size, quality, complexity, the Owner's schedule, or the method of bidding or negotiating and contracting for construction, except for services required under Subparagraph 5.2.5.

3.3.3 Preparing Drawings, Specifications and other documentation and supporting data, evaluating Contractor's proposals, and providing other services in connection with Change Orders and Construction Change Directives.

3.3.4 Providing services in connection with evaluating substitutions proposed by the Contractor and making subsequent revisions to Drawings, Specifications and other documentation resulting therefrom.

3.3.5 Providing consultation concerning replacement of Work damaged by fire or other cause during construction, and furnishing services required in connection with the replacement of such Work.

3.3.6 Providing services made necessary by the default of the Contractor, by major defects or deficiencies in the Work of the Contractor, or by failure of performance of either the Owner or Contractor under the Contract for Construction.

3.3.7 Providing services in evaluating an extensive number of claims submitted by the Contractor or others in connection with the Work.

3.3.8 Providing services in connection with a public hearing, arbitration proceeding or legal proceeding except where the Architect is party thereto.

3.3.9 Preparing documents for alternate, separate or sequential bids or providing services in connection with bidding, negotiation or construction prior to the completion of the Construction Documents Phase.

3.4 OPTIONAL ADDITIONAL SERVICES

3.4.1 Providing analyses of the Owner's needs and programming the requirements of the Project.

3.4.2 Providing financial feasibility or other special studies.

3.4.3 Providing planning surveys, site evaluations or comparative studies of prospective sites.

3.4.4 Providing special surveys, environmental studies and submissions required for approvals of governmental authorities or others having jurisdiction over the Project.

3.4.5 Providing services relative to future facilities, systems and equipment.

3.4.6 Providing services to investigate existing conditions or facilities or to make measured drawings thereof.

3.4.7 Providing services to verify the accuracy of drawings or other information furnished by the Owner.

3.4.8 Providing coordination of construction performed by separate contractors or by the Owner's own forces and coordination of services required in connection with construction performed and equipment supplied by the Owner.

3.4.9 Providing services in connection with the work of a construction manager or separate consultants retained by the Owner.

3.4.10 Providing detailed estimates of Construction Cost.

3.4.11 Providing detailed quantity surveys or inventories of material, equipment and labor.

3.4.12 Providing analyses of owning and operating costs.

3.4.13 Providing interior design and other similar services required for or in connection with the selection, procurement or installation of furniture, furnishings and related equipment.

3.4.14 Providing services for planning tenant or rental spaces.

3.4.15 Making investigations, inventories of materials or equipment, or valuations and detailed appraisals of existing facilities.

3.4.16 Preparing a set of reproducible record drawings showing significant changes in the Work made during construction based on marked-up prints, drawings and other data furnished by the Contractor to the Architect.

3.4.17 Providing assistance in the utilization of equipment or systems such as testing, adjusting and balancing, preparation of operation and maintenance manuals, training personnel for operation and maintenance, and consultation during operation.

3.4.18 Providing services after issuance to the Owner of the final Certificate for Payment, or in the absence of a final Certificate for Payment, more than 60 days after the date of Substantial Completion of the Work.

3.4.19 Providing services of consultants for other than architectural, structural, mechanical and electrical engineering portions of the Project provided as a part of Basic Services.

3.4.20 Providing any other services not otherwise included in this Agreement or not customarily furnished in accordance with generally accepted architectural practice.

ARTICLE 4

OWNER'S RESPONSIBILITIES

4.1 The Owner shall provide full information regarding requirements for the Project, including a program which shall set forth the Owner's objectives, schedule, constraints and criteria, including space requirements and relationships, flexibility, expandability, special equipment, systems and site requirements.

4.2 The Owner shall establish and update an overall budget for the Project, including the Construction Cost, the Owner's other costs and reasonable contingencies related to all of these costs.

4.3 If requested by the Architect, the Owner shall furnish evidence that financial arrangements have been made to fulfill the Owner's obligations under this Agreement.

4.4 The Owner shall designate a representative authorized to act on the Owner's behalf with respect to the Project. The Owner or such authorized representative shall render decisions in a timely manner pertaining to documents submitted by the Architect in order to avoid unreasonable delay in the orderly and sequential progress of the Architect's services.

4.5 The Owner shall furnish surveys describing physical characteristics, legal limitations and utility locations for the site of the Project, and a written legal description of the site. The surveys and legal information shall include, as applicable, grades and lines of streets, alleys, pavements and adjoining property and structures; adjacent drainage; rights-of-way, restrictions, easements, encroachments, zoning, deed restrictions, boundaries and contours of the site; locations, dimensions and necessary data pertaining to existing buildings, other improvements and trees; and information concerning available utility services and lines, both public and private, above and below grade, including inverts and depths. All the information on the survey shall be referenced to a project benchmark.

4.6 The Owner shall furnish the services of geotechnical engineers when such services are requested by the Architect. Such services may include but are not limited to test borings, test pits, determinations of soil bearing values, percolation tests, evaluations of hazardous materials, ground corrosion and resistivity tests, including necessary operations for anticipating subsoil conditions, with reports and appropriate professional recommendations.

4.6.1 The Owner shall furnish the services of other consultants when such services are reasonably required by the scope of the Project and are requested by the Architect.

4.7 The Owner shall furnish structural, mechanical, chemical, air and water pollution tests, tests for hazardous materials, and other laboratory and environmental tests, inspections and reports required by law or the Contract Documents.

4.8 The Owner shall furnish all legal, accounting and insurance counseling services as may be necessary at any time for the Project, including auditing services the Owner may require to verify the Contractor's Applications for Payment or to ascertain how or for what purposes the Contractor has used the money paid by or on behalf of the Owner.

4.9 The services, information, surveys and reports required by Paragraphs 4.5 through 4.8 shall be furnished at the Owner's expense, and the Architect shall be entitled to rely upon the accuracy and completeness thereof.

4.10 Prompt written notice shall be given by the Owner to the Architect if the Owner becomes aware of any fault or defect in the Project or nonconformance with the Contract Documents.

4.11 The proposed language of certificates or certifications requested of the Architect or Architect's consultants shall be submitted to the Architect for review and approval at least 14 days prior to execution. The Owner shall not request certifications that would require knowledge or services beyond the scope of this Agreement.

ARTICLE 5
CONSTRUCTION COST

5.1 DEFINITION

5.1.1 The Construction Cost shall be the total cost or estimated cost to the Owner of all elements of the Project designed or specified by the Architect.

5.1.2 The Construction Cost shall include the cost at current market rates of labor and materials furnished by the Owner and equipment designed, specified, selected or specially provided for by the Architect, plus a reasonable allowance for the Contractor's overhead and profit. In addition, a reasonable allowance for contingencies shall be included for market conditions at the time of bidding and for changes in the Work during construction.

5.1.3 Construction Cost does not include the compensation of the Architect and Architect's consultants, the costs of the land, rights-of-way, financing or other costs which are the responsibility of the Owner as provided in Article 4.

5.2 RESPONSIBILITY FOR CONSTRUCTION COST

5.2.1 Evaluations of the Owner's Project budget, preliminary estimates of Construction Cost and detailed estimates of Construction Cost, if any, prepared by the Architect, represent the Architect's best judgment as a design professional familiar with the construction industry. It is recognized, however, that neither the Architect nor the Owner has control over the cost of labor, materials or equipment, over the Contractor's methods of determining bid prices, or over competitive bidding, market or negotiating conditions. Accordingly, the Architect cannot and does not warrant or represent that bids or negotiated prices will not vary from the Owner's Project budget or from any estimate of Construction Cost or evaluation prepared or agreed to by the Architect.

5.2.2 No fixed limit of Construction Cost shall be established as a condition of this Agreement by the furnishing, proposal or establishment of a Project budget, unless such fixed limit has been agreed upon in writing and signed by the parties hereto. If such a fixed limit has been established, the Architect shall be permitted to include contingencies for design, bidding and price escalation, to determine what materials, equipment, component systems and types of construction are to be included in the Contract Documents, to make reasonable adjustments in the scope of the Project and to include in the Contract Documents alternate bids to adjust the Construction Cost to the fixed limit. Fixed limits, if any, shall be increased in the amount of an increase in the Contract Sum occurring after execution of the Contract for Construction.

5.2.3 If the Bidding or Negotiation Phase has not commenced within 90 days after the Architect submits the Construction Documents to the Owner, any Project budget or fixed limit of Construction Cost shall be adjusted to reflect changes in the general level of prices in the construction industry between the date of submission of the Construction Documents to the Owner and the date on which proposals are sought.

5.2.4 If a fixed limit of Construction Cost (adjusted as provided in Subparagraph 5.2.3) is exceeded by the lowest bona fide bid or negotiated proposal, the Owner shall:

.1 give written approval of an increase in such fixed limit;

.2 authorize rebidding or renegotiating of the Project within a reasonable time;

.3 if the Project is abandoned, terminate in accordance with Paragraph 8.3; or

.4 cooperate in revising the Project scope and quality as required to reduce the Construction Cost.

5.2.5 If the Owner chooses to proceed under Clause 5.2.4.4, the Architect, without additional charge, shall modify the Contract Documents as necessary to comply with the fixed limit, if established as a condition of this Agreement. The modification of Contract Documents shall be the limit of the Architect's responsibility arising out of the establishment of a fixed limit. The Architect shall be entitled to compensation in accordance with this Agreement for all services performed whether or not the Construction Phase is commenced.

ARTICLE 6
USE OF ARCHITECT'S DRAWINGS, SPECIFICATIONS AND OTHER DOCUMENTS

6.1 The Drawings, Specifications and other documents prepared by the Architect for this Project are instruments of the Architect's service for use solely with respect to this Project and, unless otherwise provided, the Architect shall be deemed the author of these documents and shall retain all common law, statutory and other reserved rights, including the copyright. The Owner shall be permitted to retain copies, including reproducible copies, of the Architect's Drawings, Specifications and other documents for information and reference in connection with the Owner's use and occupancy of the Project. The Architect's Drawings, Specifications or other documents shall not be used by the Owner or others on other projects, for additions to this Project or for completion of this Project by others, unless the Architect is adjudged to be in default under this Agreement, except by agreement in writing and with appropriate compensation to the Architect.

6.2 Submission or distribution of documents to meet official regulatory requirements or for similar purposes in connection with the Project is not to be construed as publication in derogation of the Architect's reserved rights.

ARTICLE 7
ARBITRATION

7.1 Claims, disputes or other matters in question between the parties to this Agreement arising out of or relating to this Agreement or breach thereof shall be subject to and decided by arbitration in accordance with the Construction Industry Arbitration Rules of the American Arbitration Association currently in effect unless the parties mutually agree otherwise.

7.2 Demand for arbitration shall be filed in writing with the other party to this Agreement and with the American Arbitration Association. A demand for arbitration shall be made within a reasonable time after the claim, dispute or other matter in question has arisen. In no event shall the demand for arbitration be made after the date when institution of legal or equitable proceedings based on such claim, dispute or other matter in question would be barred by the applicable statutes of limitations.

7.3 No arbitration arising out of or relating to this Agreement shall include, by consolidation, joinder or in any other manner, an additional person or entity not a party to this Agreement, except by written consent containing a specific reference to this Agreement signed by the Owner, Architect, and any other person or entity sought to be joined. Consent to arbitration involving an additional person or entity shall not constitute consent to arbitration of any claim, dispute or other matter in question not described in the written consent or with a person or entity not named or described therein. The foregoing agreement to arbitrate and other agreements to arbitrate with an additional person or entity duly consented to by the parties to this Agreement shall be specifically enforceable in accordance with applicable law in any court having jurisdiction thereof.

7.4 The award rendered by the arbitrator or arbitrators shall be final, and judgment may be entered upon it in accordance with applicable law in any court having jurisdiction thereof.

ARTICLE 8

TERMINATION, SUSPENSION OR ABANDONMENT

8.1 This Agreement may be terminated by either party upon not less than seven days' written notice should the other party fail substantially to perform in accordance with the terms of this Agreement through no fault of the party initiating the termination.

8.2 If the Project is suspended by the Owner for more than 30 consecutive days, the Architect shall be compensated for services performed prior to notice of such suspension. When the Project is resumed, the Architect's compensation shall be equitably adjusted to provide for expenses incurred in the interruption and resumption of the Architect's services.

8.3 This Agreement may be terminated by the Owner upon not less than seven days' written notice to the Architect in the event that the Project is permanently abandoned. If the Project is abandoned by the Owner for more than 90 consecutive days, the Architect may terminate this Agreement by giving written notice.

8.4 Failure of the Owner to make payments to the Architect in accordance with this Agreement shall be considered substantial nonperformance and cause for termination.

8.5 If the Owner fails to make payment when due the Architect for services and expenses, the Architect may, upon seven days' written notice to the Owner, suspend performance of services under this Agreement. Unless payment in full is received by the Architect within seven days of the date of the notice, the suspension shall take effect without further notice. In the event of a suspension of services, the Architect shall have no liability to the Owner for delay or damage caused the Owner because of such suspension of services.

8.6 In the event of termination not the fault of the Architect, the Architect shall be compensated for services performed prior to termination, together with Reimbursable Expenses then due and all Termination Expenses as defined in Paragraph 8.7.

8.7 Termination Expenses are in addition to compensation for Basic and Additional Services, and include expenses which are directly attributable to termination. Termination Expenses shall be computed as a percentage of the total compensation for Basic Services and Additional Services earned to the time of termination, as follows:

.1 Twenty percent of the total compensation for Basic and Additional Services earned to date if termination occurs before or during the predesign, site analysis, or Schematic Design Phases; or

.2 Ten percent of the total compensation for Basic and Additional Services earned to date if termination occurs during the Design Development Phase; or

.3 Five percent of the total compensation for Basic and Additional Services earned to date if termination occurs during any subsequent phase.

ARTICLE 9

MISCELLANEOUS PROVISIONS

9.1 Unless otherwise provided, this Agreement shall be governed by the law of the principal place of business of the Architect.

9.2 Terms in this Agreement shall have the same meaning as those in AIA Document A201, General Conditions of the Contract for Construction, current as of the date of this Agreement.

9.3 Causes of action between the parties to this Agreement pertaining to acts or failures to act shall be deemed to have accrued and the applicable statutes of limitations shall commence to run not later than either the date of Substantial Completion for acts or failures to act occurring prior to Substantial Completion, or the date of issuance of the final Certificate for Payment for acts or failures to act occurring after Substantial Completion.

9.4 The Owner and Architect waive all rights against each other and against the contractors, consultants, agents and employees of the other for damages, but only to the extent covered by property insurance during construction, except such rights as they may have to the proceeds of such insurance as set forth in the edition of AIA Document A201, General Conditions of the Contract for Construction, current as of the date of this Agreement. The Owner and Architect each shall require similar waivers from their contractors, consultants and agents.

9.5 The Owner and Architect, respectively, bind themselves, their partners, successors, assigns and legal representatives to the other party to this Agreement and to the partners, successors, assigns and legal representatives of such other party with respect to all covenants of this Agreement. Neither Owner nor Architect shall assign this Agreement without the written consent of the other.

9.6 This Agreement represents the entire and integrated agreement between the Owner and Architect and supersedes all prior negotiations, representations or agreements, either written or oral. This Agreement may be amended only by written instrument signed by both Owner and Architect.

9.7 Nothing contained in this Agreement shall create a contractual relationship with or a cause of action in favor of a third party against either the Owner or Architect.

9.8 Unless otherwise provided in this Agreement, the Architect and Architect's consultants shall have no responsibility for the discovery, presence, handling, removal or disposal of or exposure of persons to hazardous materials in any form at the Project site, including but not limited to asbestos, asbestos products, polychlorinated biphenyl (PCB) or other toxic substances.

9.9 The Architect shall have the right to include representations of the design of the Project, including photographs of the exterior and interior, among the Architect's promotional and professional materials. The Architect's materials shall not include the Owner's confidential or proprietary information if the Owner has previously advised the Architect in writing of the specific information considered by the Owner to be confidential or proprietary. The Owner shall provide professional credit for the Architect on the construction sign and in the promotional materials for the Project.

ARTICLE 10
PAYMENTS TO THE ARCHITECT

10.1 DIRECT PERSONNEL EXPENSE

10.1.1 Direct Personnel Expense is defined as the direct salaries of the Architect's personnel engaged on the Project and the portion of the cost of their mandatory and customary contributions and benefits related thereto, such as employment taxes and other statutory employee benefits, insurance, sick leave, holidays, vacations, pensions and similar contributions and benefits.

10.2 REIMBURSABLE EXPENSES

10.2.1 Reimbursable Expenses are in addition to compensation for Basic and Additional Services and include expenses incurred by the Architect and Architect's employees and consultants in the interest of the Project, as identified in the following Clauses.

10.2.1.1 Expense of transportation in connection with the Project; expenses in connection with authorized out-of-town travel; long-distance communications; and fees paid for securing approval of authorities having jurisdiction over the Project.

10.2.1.2 Expense of reproductions, postage and handling of Drawings, Specifications and other documents.

10.2.1.3 If authorized in advance by the Owner, expense of overtime work requiring higher than regular rates.

10.2.1.4 Expense of renderings, models and mock-ups requested by the Owner.

10.2.1.5 Expense of additional insurance coverage or limits, including professional liability insurance, requested by the Owner in excess of that normally carried by the Architect and Architect's consultants.

10.2.1.6 Expense of computer-aided design and drafting equipment time when used in connection with the Project.

10.3 PAYMENTS ON ACCOUNT OF BASIC SERVICES

10.3.1 An initial payment as set forth in Paragraph 11.1 is the minimum payment under this Agreement.

10.3.2 Subsequent payments for Basic Services shall be made monthly and, where applicable, shall be in proportion to services performed within each phase of service, on the basis set forth in Subparagraph 11.2.2.

10.3.3 If and to the extent that the time initially established in Subparagraph 11.5.1 of this Agreement is exceeded or extended through no fault of the Architect, compensation for any services rendered during the additional period of time shall be computed in the manner set forth in Subparagraph 11.3.2.

10.3.4 When compensation is based on a percentage of Construction Cost and any portions of the Project are deleted or otherwise not constructed, compensation for those portions of the Project shall be payable to the extent services are performed on those portions, in accordance with the schedule set forth in Subparagraph 11.2.2, based on (1) the lowest bona fide bid or negotiated proposal, or (2) if no such bid or proposal is received, the most recent preliminary estimate of Construction Cost or detailed estimate of Construction Cost for such portions of the Project.

10.4 PAYMENTS ON ACCOUNT OF ADDITIONAL SERVICES

10.4.1 Payments on account of the Architect's Additional Services and for Reimbursable Expenses shall be made monthly upon presentation of the Architect's statement of services rendered or expenses incurred.

10.5 PAYMENTS WITHHELD

10.5.1 No deductions shall be made from the Architect's compensation on account of penalty, liquidated damages or other sums withheld from payments to contractors, or on account of the cost of changes in the Work other than those for which the Architect has been found to be liable.

10.6 ARCHITECT'S ACCOUNTING RECORDS

10.6.1 Records of Reimbursable Expenses and expenses pertaining to Additional Services and services performed on the basis of a multiple of Direct Personnel Expense shall be available to the Owner or the Owner's authorized representative at mutually convenient times.

ARTICLE 11
BASIS OF COMPENSATION

The Owner shall compensate the Architect as follows:

11.1 AN INITIAL PAYMENT of Dollars ($
shall be made upon execution of this Agreement and credited to the Owner's account at final payment.

11.2 BASIC COMPENSATION

11.2.1 FOR BASIC SERVICES, as described in Article 2, and any other services included in Article 12 as part of Basic Services, Basic Compensation shall be computed as follows:

(Insert basis of compensation, including stipulated sums, multiples or percentages, and identify phases to which particular methods of compensation apply, if necessary.)

11.2.2 Where compensation is based on a stipulated sum or percentage of Construction Cost, progress payments for Basic Services in each phase shall total the following percentages of the total Basic Compensation payable:

(Insert additional phases as appropriate.)

Schematic Design Phase:	percent (%)
Design Development Phase:	percent (%)
Construction Documents Phase:	percent (%)
Bidding or Negotiation Phase:	percent (%)
Construction Phase:	percent (%)
Total Basic Compensation:	one hundred percent (100%)

11.3 COMPENSATION FOR ADDITIONAL SERVICES

11.3.1 FOR PROJECT REPRESENTATION BEYOND BASIC SERVICES, as described in Paragraph 3.2, compensation shall be computed as follows:

11.3.2 FOR ADDITIONAL SERVICES OF THE ARCHITECT, as described in Articles 3 and 12, other than (1) Additional Project Representation, as described in Paragraph 3.2, and (2) services included in Article 12 as part of Basic Services, but excluding services of consultants, compensation shall be computed as follows:

(Insert basis of compensation, including rates and/or multiples of Direct Personnel Expense for Principals and employees, and identify Principals and classify employees, if required. Identify specific services to which particular methods of compensation apply, if necessary.)

11.3.3 FOR ADDITIONAL SERVICES OF CONSULTANTS, including additional structural, mechanical and electrical engineering services and those provided under Subparagraph 3.4.19 or identified in Article 12 as part of Additional Services, a multiple of ( ) times the amounts billed to the Architect for such services.

(Identify specific types of consultants in Article 12, if required.)

11.4 REIMBURSABLE EXPENSES

11.4.1 FOR REIMBURSABLE EXPENSES, as described in Paragraph 10.2, and any other items included in Article 12 as Reimbursable Expenses, a multiple of ( ) times the expenses incurred by the Architect, the Architect's employees and consultants in the interest of the Project.

11.5 ADDITIONAL PROVISIONS

11.5.1 IF THE BASIC SERVICES covered by this Agreement have not been completed within ( ) months of the date hereof, through no fault of the Architect, extension of the Architect's services beyond that time shall be compensated as provided in Subparagraphs 10.3.3 and 11.3.2.

11.5.2 Payments are due and payable ( ) days from the date of the Architect's invoice. Amounts unpaid ( ) days after the invoice date shall bear interest at the rate entered below, or in the absence thereof at the legal rate prevailing from time to time at the principal place of business of the Architect.

(Insert rate of interest agreed upon.)

(Usury laws and requirements under the Federal Truth in Lending Act, similar state and local consumer credit laws and other regulations at the Owner's and Architect's principal places of business, the location of the Project and elsewhere may affect the validity of this provision. Specific legal advice should be obtained with respect to deletions or modifications, and also regarding requirements such as written disclosures or waivers.)

11.5.3 The rates and multiples set forth for Additional Services shall be annually adjusted in accordance with normal salary review practices of the Architect.

ARTICLE 12
OTHER CONDITIONS OR SERVICES

(Insert descriptions of other services, identify Additional Services included within Basic Compensation and modifications to the payment and compensation terms included in this Agreement.)

This Agreement entered into as of the day and year first written above.

OWNER	ARCHITECT
________________ *(Signature)*	________________ *(Signature)*
________________ *(Printed name and title)*	________________ *(Printed name and title)*

This document has important legal consequences; consultation with an attorney is encouraged with respect to its completion or modification.

STANDARD FORM OF AGREEMENT BETWEEN OWNER AND ENGINEER FOR PROFESSIONAL SERVICES

Prepared by

ENGINEERS JOINT CONTRACT DOCUMENTS COMMITTEE

and

Issued and Published Jointly By

PROFESSIONAL ENGINEERS IN PRIVATE PRACTICE
A practice division of the
NATIONAL SOCIETY OF PROFESSIONAL ENGINEERS

AMERICAN CONSULTING ENGINEERS COUNCIL

AMERICAN SOCIETY OF CIVIL ENGINEERS

EJCDC No. 1910-1 (1984 Edition)
Reprinted 11/88

Guide Sheet for Completing Standard Form of Agreement Between Owner and Engineer for Professional Services

1. **Page 1**—Proper name of partnership, corporation or governmental body which is Owner should be inserted on first page and signature page. Make certain person signing for Owner has authority to do so and that there is an indication of position or the capacity of the signer. (See Commentary, 1981 edit., paragraph III 2.)

2. **Page 1**—Insert as complete a description of the Project as possible. Include, to the extent known, a general description of the land where the Project is to be located, any special requirements as to performance, capacity or function and any special source of funds for which the Project must qualify. If a Construction Cost limitation is to be established as a condition to the Agreement as provided in ¶6.2.2, a specific statement to that effect setting forth the amount of the limitation initially agreed to by Owner and Engineer should be included as part of the Project description. If a limitation is agreed upon after the Agreement has been signed, a formal amendment to the Agreement should be signed by Owner and Engineer to document the understanding. Identify studies, reports or analyses previously prepared which are being furnished by Owner to Engineer for guidance, such as reports and studies referred to in **¶3.4**. Identify other special aspects or peculiarities of the Project. (See Commentary, 1981 edit., paragraph III 4.)

3. **Section 1—Basic Services**—Supplemental description in Exhibit A "Further Description of Basic Engineering Services and Related Matters". The Agreement contemplates that an exhibit will be prepared to describe in more detail the Basic Services to be rendered during Phases 1 through 6 (see language following **¶¶ 1.2.7, 1.3.5, 1.4.5, 1.5.5** and **1.7.6**; also **¶¶ 1.6.1, 1.7.1, 2.1** and **2.2**). A suggested format for such an exhibit appears on page vii of this Guide Sheet. The paragraphs of the exhibit follow the sequence of the description of Basic Services and should contain a further or expanded description of each phase of Basic Services as applicable to the particular Project undertaken (the printed language of the Agreement being generally applicable to most projects). Exhibit A should also include an agreed time for performance of Phase 1, 2 and 3 Basic Services as indicated in **¶¶ 4.2, 4.3** and **4.4**; and, if there are specific time requirements for performance of services during the remaining Phases, these should also be included. If Basic Services are expanded to include services normally treated as Additional Services, care should be taken to strike out or modify comparable language in the various paragraphs of Section 2—Additional Services of Engineer and to make appropriate changes in the provisions for payment for professional services (see Section 5). If Engineer rather than Owner is to furnish services of the types described in **¶¶3.4** and **3.5**, this information should also be included in Exhibit A. Reference is also made to Exhibit A in **¶5.1.1** of each alternate method of payment. In those places where supplemental or amending data will not appear in Exhibit A reference to the Exhibit should be crossed out and initialed by both parties.

Note that **¶1.6.1** refers to the EJCDC's Standard General Conditions and all of the provisions of this Agreement are based on the assumption that standard EJCDC documents for design and construction services will be used throughout (see **¶1.4.4**) and that the Specifications will follow the 16 Division format of the CSI (see **¶1.4.1**). If that is not the case, extensive changes in this Agreement, particularly with respect to Basic Services will probably be required.

4. **Paragraph 1.6.2.1—Visits to Site; Concerning Project Representative Services**—see discussion in Commentary, 1981 edit., Paragraphs II 1 (b) and III 7 and 20. In the 1984 edition of this Agreement, regular resident services at the site are treated as part of Basic Services, although earlier editions treated them as Additional Services. This is primarily because most practitioners are unwilling to assume the responsibilities associated with full Construction Phase services without providing regular resident services at the site. They feel that the risks arising from a division of responsibility are unacceptable and it is in the best interests of all concerned that these services be performed by the Engineer who has designed the Project. There is a separate subparagraph in **¶5.1.1** in which the particular method of compensation for resident services may be set out. If a decision regarding employment of a Resident Project Representative is not to be made until after the Agreement is signed, it may be wise to treat such services as Additional Services Requiring Authorization in Advance by adding an additional subparagraph to Paragraph 2.1. Helpful language for this purpose may be found in Paragraph **2.3** of the 1979 edition which is worded substantially as follows:

Resident Services During Construction.

1. If requested by OWNER or recommended by ENGINEER and agreed to in writing by the other, a Resident Project Representative will be furnished and will act as directed by ENGINEER in order to assist ENGINEER in observing performance of the work of Contractor(s). Such services will be paid for by OWNER as indicated in paragraph 5.1.

2. The duties and responsibilities and the limitations on the authority of the Resident Project Representative and assistants will be set forth in Exhibit B which is to be identified, attached to and made a part of this Agreement before such services begin.

3. Through more extensive on-site observations of the work in progress and field checks of materials and equipment by

the Resident Project Representative (if furnished) and assistants, ENGINEER shall endeavor to provide further protection for OWNER against defects and deficiencies in the work of Contractor(s); but the furnishing of such resident Project representation will not extend ENGINEER's responsibilities or authority beyond the limitations set forth in ¶**1.6.2.3** or elsewhere in ¶**1.6**, unless otherwise specifically noted.

Postponing the decision on resident services until after the Agreement is signed will necessitate changing ¶**1.6.2** and all other provisions of ¶**1.6** should be reviewed and modified as appropriate.

Regardless of when the decision is reached, the job site responsibilities of the Resident Project Representative should be clarified in writing and this should be attached to the Agreement. Appropriate language for this purpose is provided as Exhibit B to the Agreement which is introduced in ¶**1.6.2.2**. The language of Exhibit B has been taken directly from EJCDC document No. 1910-1A (1983 edit.) which is entitled "Suggested Listing of Duties, Responsibilities and Limitations of Authority of Resident Project Representative". It may well be that the circumstances of a particular Project will dictate changes in Exhibit B, but it may be used as a guide or check-off list in preparing a document specially tailored to the particular assignment.

5. **Section 3—Owner's Responsibilities.** If OWNER employs a construction manager or construction coordinator for the Project, the relationships between OWNER, ENGINEER and the construction manager or coordinator will require careful analysis and detailed discussions to determine the effect thereof on the duties, responsibilities and legal exposure of Engineer. Legal counsel should be consulted. Extensive revisions of the Standard Owner-Engineer Agreement will be required. The portions of the Standard General Conditions that deal with ENGINEER's functions must be changed. This most frequently occurs when there are multi-prime construction contracts. See ¶**3.11**.

If Owner employs a separate representative to provide on-site services during construction, ¶**3.10** calls for the attachment of an exhibit to the Agreement detailing the functions of such a person and the extent to which that person's employment will affect the duties and responsibilities of Engineer and the Resident Project Representative. This will probably necessitate changes in ¶**1.6** especially ¶¶**1.6.2** and **1.6.9**.

6. **Section 4—Period of Service.** This Section has been prepared in recognition of the thought expressed in ¶**4.1** that there will be a continuous period of service through completion of the Construction Phase with timely responses from Owner to Engineer's submittals and with prompt authority to proceed with each Phase of services after the preceding Phase has been completed. The stipulated time for performing Phase 1, 2 and 3 services referred to in ¶¶ **4.2, 4.3** and **4.4** should be determined in recognition of this understanding, and the agreed upon time periods should be set forth in Exhibit A. The blank space in ¶**4.10** should be filled in with whatever is considered a reasonable time in the particular Project for the taking of bids or receiving proposals, awarding the contracts and starting construction. If provision is to be included to the effect that all engineering services or the construction work is to be completed by a certain date, the necessary additional language to amend ¶¶**4.7, 4.8** or **4.10** should appear in Exhibit A. Any limitation as to the number of hours to be spent during the Operational Phase should also appear in Exhibit A. The three-month and one-year periods provided in ¶**4.11** may not be appropriate for all Projects and should be modified as circumstances dictate. Paragraph **4.12** recognizes the possibility of engineering services during Phases 3, 4 and 5 being separately sequenced because of fast-tracking under one prime construction contract or where the construction work is to be performed under several prime contracts with possibly different starting times. In such cases it is expected that the special timing required for services rendered during Phases 3, 4, 5 and 6 will be set forth in Exhibit A.

7. **Section 5—Methods of Payment.** Five different methods of payment are covered by the attached form of Agreement: these are Lump Sum, Percentage, Cost Plus a Fixed Fee, Direct Labor Costs times a factor and Salary Costs times a factor. Each appears on a separate set of pages so that the pages for the method selected may be retained and the others destroyed without affecting the page numbers of the printed form. If some other method or combination of methods of payment is to be used, separately-prepared pages may be inserted in lieu of the printed ones. Each printed form contains blank spaces that should be filled in, and it is expected that whatever method is selected the printed provisions will be supplemented or amended in Exhibit A where appropriate.

When payment is on the basis of a Lump Sum, a Percentage Fee or Cost Plus a Fixed Fee it is customary to provide for higher compensation for Engineer in a Project involving several prime contracts or in the case of construction contracts which contain cost-plus or incentive-savings provisions. Accordingly, blank spaces and alternate choices have been provided. The blank spaces for the number of prime contractors is intended to be filled in with the number normally anticipated for a Project of the type involved, such as a governmentally financed one where the law requires separate prime contracts. Paragraph **2.1.8** is intended to cover an additional number, if any, not anticipated at the time Engineer makes a fee commitment with Owner. In filling the blank spaces with respect to the anticipated extra services due to cost-plus or incentive-compensation arrangements with Contractor bear in mind the provisions of ¶**3.9** which require Owner to furnish auditing services if it is desired to ascertain how or for what purposes Contractor has used the moneys paid under the Construction Contract. Bear in mind also the provisions of ¶**3.9** which obligate Owner to provide such inspection services as Owner may require to ascertain if Contractor is performing the work in accordance with legal requirements, such as safety codes, etc.

Note that the terms Construction Cost, Direct Labor Costs, Salary Costs (formerly called Payroll Costs) and Reimbursable Expenses are specifically defined and that in ¶**5.1.1** of each alternate method of payment reference is made to Exhibit A "Further Description of Basic Engineering Services and Related Matters".

Engineers frequently employ outside or independent professional associates or consultants to assist in the performance of Basic Services which is permitted under ¶**7.5.2**. There is a separate subparagraph in ¶**5.1.1** in the alternative pages for the Cost Plus a Fixed Fee, Direct Labor and Salary Costs methods of payment in which the particular method of compensation for such services may be set out. In the earlier editions of this Agreement Engineer's expenses for such services was considered Reimbursable Expense. [EJCDC has prepared three standard forms of agreement which may be used when Engineer employs an independent professional associate or consultant; these are Engineer-Architect Agreement, No. 1910-10; Engineer-Associate Engineer Agreement, No. 1910-13; and, Engineer-Consultant Agreement, No. 1910-14.]

Where compensation for Basic Services is on the basis of Cost Plus a Fixed Fee, Direct Labor or Salary Costs, Owners frequently wish some indication of the amount that Engineers expect to charge for engineering services. Accordingly, a further paragraph has been added to ¶**5.1** to provide an estimate of that amount and an agreement to discuss compensation when it becomes apparent that such amount may be exceeded. This amount will obviously vary depending on the number of prime contracts, cost-plus or incentive compensation arrangements that actually are used. Blank spaces must be filled in as applicable and inapplicable language crossed out. Note that the estimate covers all amounts payable for Basic Services, including services of Engineer's independent professional associates and consultants and Resident Project Representative as well as services during the Operational Phase, but does not include services of professional associates, consultants and others which are considered Additional Services. Also, some Owners may require a schedule of salary and wage rates for each classification of personnel expected to be used on the Project.

8. **Paragraph 5.1.2—Payment for Additional Services** (see Commentary, 1981 edit., paragraph III 13). Space has been provided in ¶¶**5.1.2.1** through **5.1.2.3** for different rates of compensation for different types of Additional Services. Any inapplicable paragraph should be crossed out and initials inserted in the margin by both parties. If one method of compensation is to apply to all Additional Services, be certain that the cross references are correctly adjusted.

9. Note that ¶**5.1.3** provides for reimbursement of the "actual" cost of Reimbursable Expenses. For most items there is no markup although this sometimes customary. In the Lump Sum and Percentage methods of payment Reimbursable Expenses do not include overtime charges and expenses of computer time and other specialized equipment while in the other methods of payment such costs are included at a factored amount.

10. The blank space in ¶**5.3.2** should be filled in after consultation with Owner.

11. **Paragraph 5.4.1** contemplates the identification of the key personnel in Engineer's organization who are to be assigned to the Project (see Commentary, 1981 edit., paragraph III 13) as to the proper meaning of the term "principals"); this applies to both the Direct Labor Cost and Salary Cost definitions. In developing the factor for the Salary Cost method of compensation, bear in mind that ¶**5.4.1** is set up so that the factor is applied to the salary, wages and fringe benefits of all personnel, and Salary Costs of principals (whether they be corporate officers at fixed salaries or partners with drawing accounts who share in the profits) are to be itemized. When using Salary Costs as a basis for compensation, instead of the detailed accounting required to substantiate the amount paid for customary and statutory benefits of personnel, a percentage of salaries and wages may be agreed to in advance and inserted in the blank space at the end of paragraph **5.4.1**; otherwise the sentence should be crossed out and initialed by both parties. Note also that periodic adjustments in these items are expected (see ¶**5.3.4**).

12. **Insurance**—Paragraph 7.3 deals with general insurance coverages. Owners frequently require that their Engineers maintain professional liability insurance, and EJCDC considers it important that Engineers do so. Language addressing this matter is set forth below and should be added as additional subparagraphs to paragraph 7.3 (after the present paragraph 7.3 has been renumbered 7.3.1).

7.3.2. Also ENGINEER shall procure and maintain professional liability insurance for protection from claims arising out of performance of professional services caused by a negligent error, omission or act for which the insured is legally liable; such professional liability insurance will provide for coverage in such amounts, with such deductible provisions and for such period of time as set forth below, and certificates indicating that such insurance is in effect will be delivered to OWNER:

Amount ______________________________

Deductible ______________________________

Effective Through ______________________________

7.3.3. ENGINEER will also cause professional associates and consultants retained by ENGINEER for the Project to procure and maintain comparable professional liability insurance coverage.

13. **Paragraph 7.6—Arbitration.** This provides for compulsory and binding arbitration of all disputes between Owner and Engineer where the claimed amount at issue is not more than $200,000. It also restricts joinder in the arbitration proceedings of others who are not a party to the Agreement. The Engineers' Joint Contract Documents Committee believes that arbitration of such disputes under the Construction Industry Arbitration Rules of the American Arbitration Association is in the best interest of both Owner and Engineer, but also recognizes that others may differ, and the laws of all states are not similar. Accordingly, the form of Agreement has been prepared so that ¶**7.6** which deals with arbitration may be eliminated from the Agreement simply by tearing out the separate page (page 17) on which it is printed *and* completing the information at the bottom of the last page to indicate the page number as well as the total number of pages that make up the entire Agreement. The Table of Contents should also be modified.

14. **Paragraph 8.1** contains space for additional or special provisions required by Owner or Engineer. These may include

identification of the individuals in Engineer's office who will be in charge of the Project and identification of Engineer's professional associates and consultants who will be employed to assist in the performance of Basic Services.

It is the practice of some engineers to include in their agreements language limiting their liability for negligence to a fixed dollar amount or a certain percentage of total fee for services. Any such language should be specifically negotiated with Owner after consultation with insurance counsel as to recommended terms. Legal counsel should be consulted as to the enforceability of such language as well as its relationship with and bearing on any provisions requiring the maintenance of professional liability insurance. The language agreed to should be inserted as a subparagraph of ¶**8.1**.

15. **Paragraph 8.2** identifies two Exhibits: Exhibit A "Further Description of Basic Engineering Services and Related Matters" and Exhibit B "Duties, Responsibilities and Limitations of Authority of Resident Project Representative". The number of pages of each Exhibit should be inserted in the blank spaces. Other Exhibits should be listed in the space provided. If either Exhibit A or B is not to be attached, all reference to the document should be crossed out of the Agreement.

AGREEMENTS FOR LESS THAN CUSTOMARY FULL ENGINEERING SERVICES

There are an increasing number of engineering assignments in which Engineer is employed only to provide design services. This usually involves Study and Report, Preliminary Design and Final Design services. When such services are completed and the Project proceeds to construction, it is expected that Engineer who designed it will be employed to provide customary Bidding and Negotiating, Construction and Operational Phase services either under a separate agreement or as a supplement to the existing Agreement for design services only. The attached form of Agreement may be used as the basis for an agreement between Owner and Engineer for design services only and also as a basis for a supplement to or amendment of that agreement should Engineer be employed to provide the last three phases (construction related) of services.

A. **When design services only are to be provided** the entire standard form of Agreement should be reviewed and in particular the provisions of the following paragraphs should be modified as appropriate:

(1) If the Project Program or Report on which Engineer's design documentation will be based was not prepared by Engineer, that Program or Report should be identified. The extent to which Engineer may rely on it or is required to review and comment on it should be stated. Delete the entire ¶2.1, Study and Report Phase.

(2) It is customary for Engineer to be employed to provide services during the Bidding or Negotiating and Construction Phases in order to interpret and clarify ENGINEER's documentation, detect at an early date and correct imperfections in the documentation which must always be expected, issue addenda, review and approve shop drawings, determine the acceptability of substitute or or-equal items, prepare work directive changes and change orders and make recommendations for payment—all in order to give Owner some degree of assurance (but not a guarantee) that what Contractor produces is generally in accordance with the Contract Documents and assurance that the integrity of the design has been preserved. When Engineer is not permitted to provide such construction related services, Owner's interest may be prejudiced; and, in addition, Engineer may be denied the customary opportunity to discuss and analyze the documentation with Contractor (and, where appropriate, Owner) and avoid misunderstandings and difficulties that may develop during construction and afterwards. In such circumstances it is important for Owner and Engineer to recognize in writing the consequences of limiting Engineer to less than full services and the risks inherent in having a party other than the preparer interpret and clarify drawings and specifications, act on shop drawings and substitutions, issue addenda and change orders and provide other customary services. Suggested language for this purpose is set forth below and should be inserted at the end in the Basic Services. [¶¶**1.5**, **1.6** and **1.7** would obviously be eliminated.]

> "OWNER acknowledges that it is customary for the engineer who is responsible for the preparation and furnishing of Drawings and Specifications and other construction-related documents to be employed to provide professional services during the Bidding and Construction Phases of the Project (i) to interpret and clarify the documentation so furnished and to modify the same as circumstances revealed during bidding and construction may dictate, (ii) in connection with acceptance of substitute or or-equal items of materials and equipment proposed by bidders and Contractor(s), (iii) in connection with approval of shop drawings and sample submittals, and (iv) as a result of and in response to ENGINEER's detecting in advance of performance of affected work inconsistencies or irregularities in such documentation. OWNER agrees that if ENGINEER is not employed to provide such professional services during the Bidding (if the work is put out for bids) and the Construction Phases of the Project, ENGINEER will not be responsible for, and OWNER shall indemnify and hold ENGINEER (and ENGINEER's professional associates and consultants) harmless from, all claims, damages, losses and expenses including attorneys' fees arising out of, or resulting from, any interpretation, clarification, substitution acceptance, shop drawing or sample approval or modification of such documentation issued or carried out by OWNER or others. Nothing contained in this paragraph shall be construed to release ENGINEER (or ENGINEER's professional associates or consultants) from liability for failure to perform in accordance with professional standards any duty or responsibility which ENGINEER has undertaken or assumed under this Agreement."

The EJCDC cautions against providing less than whole Construction Phase services (such as shop drawing approval only) because of the risks and exposures involved when control of contract administration is out of the hands of the Engineer.

(3) Certain of the Additional Services should be modified or eliminated. See for instance ¶¶ 2.1.10, 2.1.12, 2.2.1, 2.2.4, 2.2.5, and 2.2.6.

(4) In Section 3, ¶ 3.5 would not be used and ¶ 3.9 would be modified to eliminate reference to inspections during construction. Paragraph 3.11 would also need to be changed to require Owner to advise Engineer how the construction coordination between the several prime Contractor(s) is to be addressed in the documentation prepared by Engineer. The last clause of ¶3.14 should be eliminated.

(5) Section 4 would require revision to eliminate reference to the Bidding and Negotiating, Construction and Post Construction Phases.

(6) Those portions of the Payment Section which pertain to services after completion of design would be eliminated.

(7) Section 6 dealing with the responsibility for control over the cost of the Project and the possible obligation to redesign should be rewritten after the issues involved have been discussed with and agreed to by Owner.

B. **When the Agreement for design services only is supplemented to include construction related services:** Since Engineer will have prepared and will be thoroughly familiar with the Drawings and Specifications and Contract Documents, it will be a relatively simple matter to prepare the supplement to the existing Agreement in order to add the three additional phases for construction related services. The language of ¶¶ 1.5, 1.6 and 1.7 of 1910-1 may be used almost verbatim in most cases and with only minor changes in others as noted below. There will, of course, be such additional changes as may be necessary to be responsive to the peculiarities of the individual assignment. It is likely that by the time the supplement is to be prepared, Engineer's design documentation will have been formally accepted by Owner and decisions will have been made on the matters of bidding vs. negotiating, whether or not there will be more than one prime Contractor for construction, whether or not there will be fast-tracking and if any of the construction contracts will contain a cost plus or incentive savings or similar provisions. While all of the provisions of ¶¶1.5, 1.6 and 1.7 should be reviewed carefully in each case, the following will probably require change:

(1) ¶1.5.1. should be changed to eliminate reference to Bidding or Negotiating—whichever approach will not be used.

(2) ¶1.6.1. may be changed to eliminate reference to the set of General Conditions to be included in the Contract Documents since this will already have been decided.

(3) Exhibit B "Duties, Responsibilities and Limitations of Authority of the Resident Project Representative" referred to in ¶1.6.2.2 will have to be prepared and attached as an Exhibit to the supplemental agreement.

(4) The ¶¶ in the Additional Services Section that were eliminated when the design services only agreement was prepared will probably be reinstated or the description of Basic Services during Phases 4, 5, and 6 expanded to include such Additional Services with an appropriate adjustment in Engineer's compensation for Basic Services.

(5) Since the number of prime Contractors will have been decided, ¶2.1.8 may be eliminated as well as the portions of ¶5.1.1 that deal with the same subject matter.

(6) In connection with Owner's obligations, the portions of ¶¶3.4 and 3.8 that were deleted in the design services only agreement should be included in the supplemental agreement. A decision on the matter of Owner's separate representative at the site will have been reached so that ¶3.10 can be changed, and, if such a person is to be employed, an appropriate exhibit prepared and attached to the supplemental agreement.

(7) ¶3.11 should be modified to reflect Owner's decisions on the matter of designating the party who will be responsible for coordination of the work of the various prime Contractors.

(8) As noted in paragraph 5 of the Guide Sheet, if a construction manager is to be employed by Owner extensive change in the language of ¶¶1.5, 1.6 and 1.7 will be necessary.

(9) The portions of Section 4 that pertain to the Bidding and Negotiating, Construction and Operational Phases, should be reinstated and included in the supplemental agreement, and consideration should be given to fixing an outside date for completion of construction.

(10) If the construction work is to be performed on a fast-track basis, the schedule called for in ¶4.12 will have to be developed and appropriate changes made in that paragraph.

(11) Changes in or supplements to the language of the Payment Section in the Agreement for design services only should be made. A method of payment different from that employed during the design of the project may be used in which case a great deal of the language contained in the applicable Method of Payment pages of 1910-1 should be included.

(12) A paragraph will have to be added to set forth the extent of Engineer's responsibilities for construction cost control during the last three phases of service.

C. **As indicated in the Guide Sheet**, the EJCDC does not recommend Engineer's undertaking less than full services during the Construction Phase, because the risks and legal exposure are quite serious when one is responsible for the administration of a construction contract while not having control over its administration. This amounts to divided responsibility that invariably entails serious legal exposure, and at times it is tantamount to responsibility without authority which can lead to worse consequences. Of particular concern are the relationships between the parties as to and responsibilities for quality control, review of applications for payment and coordination between interpretations, clarifications, work directive changes, change orders, shop drawing approvals and acceptance of substitutes. Also, the EJCDC suggests that Engineers carefully evaluate the legal and professional exposure involved before undertaking construction related services when the design documentation has been prepared by another party. Assignments of this nature have been successfully carried out, but the problems of reliance on the documentation of others with an undefined duty to detect errors and omissions can lead to misunderstandings and unacceptable legal exposure unless carefully discussed and resolved and contract language prepared after discussion with legal counsel. The language of the previously prepared Contract Documents will be determinative of Engineer's duties and responsibilities during the construction related phases, but it may be too late to change the wording and assigned responsibilities after the Contract Documents have been finalized by another design professional.

EXHIBIT A TO AGREEMENT BETWEEN OWNER AND ENGINEER FOR PROFESSIONAL SERVICES, dated ________ , 19__ (for use with No. 1910-1, 1984 Edition).

Further Description of Basic Engineering Services and Related Matters

1. This is an exhibit attached to, made a part of and incorporated by reference into the Agreement made on __ , 19____ between ____________________ (OWNER) and (ENGINEER) providing for professional engineering services. The Basic Services of Engineer as described in Section 1 of the Agreement are amended or supplemented as indicated below and the time periods for the performance of certain services as indicated in Section 4 of the Agreement are stipulated as indicated below.

2. During the Study and Report Phase ENGINEER shall:
(insert amendments or supplements to paragraph 1.2.)

The Study and Report Phase Services will be completed and the Report submitted within ________ calendar days following written authorization from OWNER to ENGINEER to proceed with that phase of services.

3. During the Preliminary Design Phase ENGINEER shall:
(insert amendments or supplements to paragraph 1.3.)

The Preliminary Design Phase Services will be completed and ENGINEER's documentation and opinion of costs submitted within ________ calendar days following written authorization from OWNER to ENGINEER to proceed with that phase of services.

4. During the Final Design Phase ENGINEER shall:
(insert amendments or supplements to paragraph 1.4.)

The Final Design Phase Services will be completed and Contract Documents and ENGINEER's opinion of costs submitted within ________ calendar days following written authorization from OWNER to ENGINEER to proceed with that phase of services.

[If the construction work is to be performed under more than one prime contract and the times of performance are to be staggered, the provisions of paragraph 4.12 will probably apply and it will be necessary to prepare a special schedule for the Final Design Phase Services.]

5. During the Bidding or Negotiating Phase ENGINEER shall:
(insert amendments or supplements to paragraph 1.5.)

[If the construction work is to be performed under more than one prime contract and the times of performance are to be staggered, the provisions of paragraph 4.12 will probably apply and it will be necessary to prepare a special schedule for the Bidding or Negotiating Phase Services.]

6. During the Construction Phase ENGINEER shall:
(insert amendments or supplements to paragraph 1.6.)

[If the construction work is to be performed under more than one prime contract and the times of performance are to be staggered, the provisions of paragraph 4.12 will probably apply and it will be necessary to prepare a special schedule for the Construction Phase Services.]

[If an outside date for completion of construction is to be added, it should be added here.]

Note: This change has to be coordinated with those in ¶¶4.1 and 4.7.

7. During the Operational Phase ENGINEER shall:
(insert amendments or supplements to paragraph 1.7)

[If an outside date for completion of all services is to be added, it should be added here.]
Note: This change is to be coordinated with those in ¶¶4.1 and 4.8.

vii

EXHIBIT B TO AGREEMENT BETWEEN OWNER AND ENGINEER FOR PROFESSIONAL SERVICES, dated ________ ,19__ (for use with No. 1910-1, 1984 Edition).

A LISTING OF THE DUTIES, RESPONSIBILITIES AND LIMITATIONS OF AUTHORITY OF THE RESIDENT PROJECT REPRESENTATIVE.

This is an Exhibit attached to, made a part of and incorporated by reference with the Agreement made on ____ , 19____ , between ____ (OWNER) and (ENGINEER) providing for professional engineering services.

ENGINEER shall furnish a Resident Project Representative (RPR), assistants and other field staff to assist ENGINEER in observing performance of the work of Contractor.

Through more extensive on-site observations of the work in progress and field checks of materials and equipment by the RPR and assistants, ENGINEER shall endeavor to provide further protection for OWNER against defects and deficiencies in the work of CONTRACTOR; but, the furnishing of such services will not make ENGINEER responsible for or give ENGINEER control over construction means, methods, techniques, sequences or procedures or for safety precautions or programs, or responsibility for CONTRACTOR's failure to perform the Work in accordance with the Contract Documents and in particular the specific limitations set forth in paragraph 1.6 of the Agreement are applicable.

The duties and responsibilities of the RPR are limited to those of ENGINEER in ENGINEER's agreement with the OWNER and in the construction Contract Documents, and are further limited and described as follows:

A. General

RPR is ENGINEER's agent at the site, will act as directed by and under the supervision of ENGINEER, and will confer with ENGINEER regarding RPR's actions. RPR's dealings in matters pertaining to the on-site work shall in general be with ENGINEER and CONTRACTOR keeping OWNER advised as necessary. RPR's dealings with subcontractors shall only be through or with the full knowledge and approval of CONTRACTOR. RPR shall generally communicate with OWNER with the knowledge of and under the direction of ENGINEER.

B. Duties and Responsibilities of RPR

1. *Schedules:* Review the progress schedule, schedule of Shop Drawing submittals and schedule of values prepared by CONTRACTOR and consult with ENGINEER concerning acceptability.
2. *Conferences and Meetings:* Attend meetings with CONTRACTOR, such as preconstruction conferences, progress meetings, job conferences and other project-related meetings, and prepare and circulate copies of minutes thereof.
3. *Liaison:*
   a. Serve as ENGINEER's liaison with CONTRACTOR, working principally through CONTRACTOR's superintendent and assist in understanding the intent of the Contract Documents; and assist ENGINEER in serving as OWNER's liaison with CONTRACTOR when CONTRACTOR's operations affect OWNER's on-site operations.
   b. Assist in obtaining from OWNER additional details or information, when required for proper execution of the Work.
4. *Shop Drawings and Samples:*
   a. Record date of receipt of Shop Drawings and samples.
   b. Receive samples which are furnished at the site by CONTRACTOR, and notify ENGINEER of availability of samples for examination.
   c. Advise ENGINEER and CONTRACTOR of the commencement of any Work requiring a Shop Drawing or sample if the submittal has not been approved by ENGINEER.
5. *Review of Work, Rejection of Defective Work, Inspections and Tests:*
   a. Conduct on-site observations of the Work in progress to assist ENGINEER in determining if the Work is in general proceeding in accordance with the Contract Documents.
   b. Report to ENGINEER whenever RPR believes that any Work is unsatisfactory, faulty or defective or does not conform to the Contract Documents, or has been damaged, or does not meet the requirements

of any inspection, test or approval required to be made; and advise ENGINEER of Work that RPR believes should be corrected or rejected or should be uncovered for observation, or requires special testing, inspection or approval.

c. Verify that tests, equipment and systems startups and operating and maintenance training are conducted in the presence of appropriate personnel, and that CONTRACTOR maintains adequate records thereof; and observe, record and report to ENGINEER appropriate details relative to the test procedures and startups.

d. Accompany visiting inspectors representing public or other agencies having jurisdiction over the Project, record the results of these inspections and report to ENGINEER.

6. *Interpretation of Contract Documents:* Report to ENGINEER when clarifications and interpretations of the Contract Documents are needed and transmit to CONTRACTOR clarifications and interpretations as issued by ENGINEER.

7. *Modifications:* Consider and evaluate CONTRACTOR's suggestions for modifications in Drawings or Specifications and report with RPR's recommendations to ENGINEER. Transmit to CONTRACTOR decisions as issued by ENGINEER.

8. *Records:*

   a. Maintain at the job site orderly files for correspondence, reports of job conferences, Shop Drawings and samples, reproductions of original Contract Documents including all Work Directive Changes, Addenda, Change Orders, Field Orders, additional Drawings issued subsequent to the execution of the Contract, ENGINEER's clarifications and interpretations of the Contract Documents, progress reports, and other Project related documents.

   b. Keep a diary or log book, recording CONTRACTOR hours on the job site, weather conditions, data relative to questions of Work Directive Changes, Change Orders or changed conditions, list of job site visitors, daily activities, decisions, observations in general, and specific observations in more detail as in the case of observing test procedures; and send copies to ENGINEER.

   c. Record names, addresses and telephone numbers of all CONTRACTORS, subcontractors and major suppliers of materials and equipment.

9. *Reports:*

   a. Furnish ENGINEER periodic reports as required of progress of the Work and of CONTRACTOR's compliance with the progress schedule and schedule of Shop Drawing and sample submittals.

   b. Consult with ENGINEER in advance of scheduled major tests, inspections or start of important phases of the Work.

   c. Draft proposed Change Orders and Work Directive Changes, obtaining backup material from CONTRACTOR and recommend to ENGINEER Change Orders, Work Directive Changes, and Field Orders.

   d. Report immediately to ENGINEER and OWNER upon the occurrence of any accident.

10. *Payment Requests:* Review applications for payment with CONTRACTOR for compliance with the established procedure for their submission and forward with recommendations to ENGINEER, noting particularly the relationship of the payment requested to the schedule of values, Work completed and materials and equipment delivered at the site but not incorporated in the Work.

11. *Certificates, Maintenance and Operation Manuals:* During the course of the Work, verify that certificates, maintenance and operation manuals and other data required to be assembled and furnished by CONTRACTOR are applicable to the items actually installed and in accordance with the Contract Documents, and have this material delivered to ENGINEER for review and forwarding to OWNER prior to final payment for the Work.

12. *Completion:*

   a. Before ENGINEER issues a Certificate of Substantial Completion, submit to CONTRACTOR a list of observed items requiring completion or correction.

   b. Conduct final inspection in the company of ENGINEER, OWNER, and CONTRACTOR and prepare a final list of items to be completed or corrected.

   c. Observe that all items on final list have been completed or corrected and make recommendations to ENGINEER concerning acceptance.

C. **Limitations of Authority**

Resident Project Representative:

1. Shall not authorize any deviation from the Contract Documents or substitution of materials or equipment, unless authorized by ENGINEER.
2. Shall not exceed limitations of ENGINEER's authority as set forth in the Agreement or the Contract Documents.
3. Shall not undertake any of the responsibilities of CONTRACTOR, subcontractors or CONTRACTOR's superintendent.
4. Shall not advise on, issue directions relative to or assume control over any aspect of the means, methods, techniques, sequences or procedures of construction unless such advice or directions are specifically required by the Contract Documents.
5. Shall not advise on, issue directions regarding or assume control over safety precautions and programs in connection with the Work.
6. Shall not accept Shop Drawing or sample submittals from anyone other than CONTRACTOR.
7. Shall not authorize OWNER to occupy the Project in whole or in part.
8. Shall not participate in specialized field or laboratory tests or inspections conducted by others except as specifically authorized by ENGINEER.

TABLE OF CONTENTS

This document has important legal consequences; consultation with an attorney is encouraged with respect to its completion or modification.

STANDARD FORM OF AGREEMENT BETWEEN OWNER AND ENGINEER FOR PROFESSIONAL SERVICES

THIS IS AN AGREEMENT made as of ______________________________, 19____

between ______________________________

______________________________ (OWNER) and

____

______________________________ (ENGINEER).

OWNER intends to ______________________________

____

____

____

____

______________________________ (hereinafter called the Project).

OWNER and ENGINEER in consideration of their mutual covenants herein agree in respect of the performance of professional engineering services by ENGINEER and the payment for those services by OWNER as set forth below.

SECTION 1—BASIC SERVICES OF ENGINEER

1.1. General.

1.1.1. ENGINEER shall provide for OWNER professional engineering services in all phases of the Project to which this Agreement applies as hereinafter provided. These services will include serving as OWNER's professional engineering representative for the Project, providing professional engineering consultation and advice and furnishing customary civil, structural, mechanical and electrical engineering services and customary architectural services incidental thereto.

1.2. Study and Report Phase.

After written authorization to proceed, ENGINEER shall:

1.2.1. Consult with OWNER to clarify and define OWNER's requirements for the Project and review available data.

1.2.2. Advise OWNER as to the necessity of OWNER's providing or obtaining from others data or services of the types described in paragraph 3.3, and assist OWNER in obtaining such data and services.

Page 1 of ____ pages

1.2.3. Identify and analyze requirements of governmental authorities having jurisdiction to approve the design of the Project and participate in consultations with such authorities.

1.2.4. Provide analyses of OWNER's needs, planning surveys, site evaluations and comparative studies of prospective sites and solutions.

1.2.5. Provide a general economic analysis of OWNER's requirements applicable to various alternatives.

1.2.6. Prepare a Report containing schematic layouts, sketches and conceptual design criteria with appropriate exhibits to indicate clearly the considerations involved (including applicable requirements of governmental authorities having jurisdiction as aforesaid) and the alternative solutions available to OWNER and setting forth ENGINEER's findings and recommendations. This Report will be accompanied by ENGINEER's opinion of probable costs for the Project, including the following which will be separately itemized: Construction Cost, allowance for engineering costs and contingencies, and (on the basis of information furnished by OWNER) allowances for such other items as charges of all other professionals and consultants, for the cost of land and rights-of-way, for compensation for or damages to properties, for interest and financing charges and for other services to be provided by others for Owner pursuant to paragraphs 3.7 through 3.11, inclusive. The total of all such costs, allowances, etc. are hereinafter called "Total Project Costs".

1.2.7. Furnish five copies of the Study and Report documents and review them in person with OWNER.

The duties and responsibilities of ENGINEER during the Study and Report Phase are amended and supplemented as indicated in paragraph 2 of Exhibit A "Further Descriptions of Basic Engineering Services and Related Matters".

1.3. Preliminary Design Phase.

After written authorization to proceed with the Preliminary Design Phase, ENGINEER shall:

1.3.1. In consultation with OWNER and on the basis of the accepted Study and Report documents, determine the general scope, extent and character of the Project.

1.3.2. Prepare Preliminary Design documents consisting of final design criteria, preliminary drawings, outline specifications and written descriptions of the Project.

1.3.3. Advise OWNER if additional data or services of the types described in paragraph 3.4 are necessary and assist OWNER in obtaining such data and services.

1.3.4. Based on the information contained in the preliminary design documents, submit a revised opinion of probable Total Project Costs.

1.3.5. Furnish five copies of the above Preliminary Design documents and present and review them in person with OWNER.

The duties and responsibilities of ENGINEER during the Preliminary Design Phase are amended and supplemented as indicated in paragraph 3 of Exhibit A "Further Description of Basic Engineering Services and Related Matters".

1.4. Final Design Phase.

After written authorization to proceed with the Final Design Phase, ENGINEER shall:

1.4.1. On the basis of the accepted Preliminary Design documents and the revised opinion of probable Total Project Costs prepare for incorporation in the Contract Documents final drawings to show the general scope, extent and character of the work to be furnished and performed by Contractor(s) (hereinafter called "Drawings") and Specifications (which will be prepared in conformance with the sixteen division format of the Construction Specifications Institute).

1.4.2. Provide technical criteria, written descriptions and design data for OWNER's use in filing applications for permits with or obtaining approvals of such governmental authorities as have jurisdiction to approve the design of the Project, and assist OWNER in consultations with appropriate authorities.

1.4.3. Advise OWNER of any adjustments to the latest opinion of probable Total Project Costs caused by changes in general scope, extent or character or design requirements of the Project or Construction Costs. Furnish to OWNER a revised opinion of probable Total Project Costs based on the Drawings and Specifications.

1.4.4. Prepare for review and approval by OWNER, its legal counsel and other advisors contract agreement forms, general conditions and supplementary conditions, and (where appropriate) bid forms, invitations to bid and instructions to bidders (all of which shall be consistent with the forms and pertinent guide sheets prepared by the Engineers Joint Contract Documents Committee), and assist in the preparation of other related documents.

1.4.5. Furnish five copies of the above documents and of the Drawings and Specifications and present and review them in person with OWNER.

The duties and responsibilities of ENGINEER during the Final Design Phase are amended and supplemented as indicated in paragraph 4 of Exhibit A "Further Description of Basic Engineering Services and Related Matters".

1.5. Bidding or Negotiating Phase.

After written authorization to proceed with the Bidding or Negotiating Phase, ENGINEER shall:

1.5.1. Assist OWNER in advertising for and obtaining bids or negotiating proposals for each separate prime contract for construction, materials, equipment and services; and, where applicable, maintain a record of prospective bidders to whom Bidding Documents have been issued, attend pre-bid conferences and receive and process deposits for Bidding Documents.

1.5.2. Issue addenda as appropriate to interpret, clarify or expand the Bidding Documents.

1.5.3. Consult with and advise OWNER as to the acceptability of subcontractors, suppliers and other persons and organizations proposed by the prime contractor(s) (herein called "Contractor(s)") for those portions of the work as to which such acceptability is required by the Bidding Documents.

1.5.4. Consult with OWNER concerning and determine the acceptability of substitute materials and equipment proposed by Contractor(s) when substitution prior to the award of contracts is allowed by the Bidding Documents.

1.5.5. Attend the bid opening, prepare bid tabulation sheets and assist OWNER in evaluating bids or proposals and in assembling and awarding contracts for construction, materials, equipment and services.

The duties and responsibilities of ENGINEER during the Bidding or Negotiating Phase are amended and supplemented as indicated in paragraph 5 of Exhibit A "Further Description of Basic Engineering Services and Related Matters".

1.6. Construction Phase.

During the Construction Phase:

1.6.1. *General Administration of Construction Contract.* ENGINEER shall consult with and advise OWNER and act as OWNER's representative as provided in Articles 1 through 17, inclusive, of the Standard General Conditions of the Construction Contract, No. 1910-8 (1983 edition) of the Engineers Joint Contract Documents Committee. The extent and limitations of the duties, responsibilities and authority of ENGINEER as assigned in said Standard General Conditions shall not be modified, except to the extent provided in paragraph 6 of Exhibit A "Further Description of Basic Engineering Services and Related Matters" and except as ENGINEER may otherwise agree in writing. All of OWNER's instructions to Contractor(s) will be issued through ENGINEER who will have authority to act on behalf of OWNER to the extent provided in said Standard General Conditions except as otherwise provided in writing.

1.6.2. *Visits to Site and Observation of Construction.* In connection with observations of the work of Contractor(s) while it is in progress:

1.6.2.1. ENGINEER shall make visits to the site at intervals appropriate to the various stages of construction as ENGINEER deems necessary in order to observe as an experienced and qualified design professional the progress and quality of the various aspects of Contractor(s)' work. In addition, ENGINEER shall provide the services of a Resident Project Representative (and assistants as agreed) at the site to assist ENGINEER and to provide more continuous observation of such work. Based on information obtained during such visits and on such observations, ENGINEER shall endeavor to determine in general if such work is proceeding in accordance with the Contract Documents and ENGINEER shall keep OWNER informed of the progress of the work.

1.6.2.2. The Resident Project Representative (and any assistants) will be ENGINEER's agent or employee and under ENGINEER's supervision. The duties and responsibilities of the Resident Project Representative (and assistants) are set forth in Exhibit B "Duties, Responsibilities and Limitation of Authority of Resident Project Representative".

1.6.2.3. The purpose of ENGINEER's visits to and representation by the Resident Project Representative (and assistants, if any) at the site will be to enable ENGINEER to better carry out the duties and responsibilities assigned to and undertaken by ENGINEER during the Construction Phase, and, in addition, by exercise of ENGINEER's efforts as an experienced and qualified design professional, to provide for OWNER a greater degree of confidence that the completed work of Contractor(s) will conform generally to the Contract Documents and that the integrity of the design concept as reflected in the Contract Documents has been implemented and preserved by Contractor(s). On the other hand, ENGINEER shall not, during such visits or as a result of such observations of Contractor(s)' work in progress, supervise, direct or have control over Contractor(s)' work nor shall ENGINEER have authority over or responsibility for the means, methods, techniques, sequences or procedures of construction selected by Contractor(s), for safety precautions and programs incident to the work of Contractor(s) or for any failure of Contractor(s) to comply with laws, rules, regulations, ordinances, codes or orders applicable to Contractor(s) furnishing and performing their work. Accordingly, ENGINEER can neither guarantee the performance of the construction contracts by Contractor(s) nor assume responsibility for Contractor(s)' failure to furnish and perform their work in accordance with the Contract Documents.

1.6.3. *Defective Work.* During such visits and on the basis of such observations, ENGINEER may disapprove of or reject Contractor(s)' work while it is in progress if ENGINEER believes that such work will not produce a completed Project that conforms generally to the Contract Documents or that it will prejudice the integrity of the design concept of the Project as reflected in the Contract Documents.

1.6.4. *Interpretations and Clarifications.* ENGINEER shall issue necessary interpretations and clarifications of the Contract Documents and in connection therewith prepare work directive changes and change orders as required.

1.6.5. *Shop Drawings.* ENGINEER shall review and approve (or take other appropriate action in respect of) Shop Drawings (as that term is defined in the aforesaid Standard General Conditions), samples and other data which Contractor(s) are required to submit, but only for conformance with the design concept of the Project and compliance with the information given in the Contract Documents. Such reviews and approvals or other action shall not extend to means, methods, techniques, sequences or procedures of construction or to safety precautions and programs incident thereto.

1.6.6. *Substitutes.* ENGINEER shall evaluate and determine the acceptability of substitute materials and equipment proposed by Contractor(s), but subject to the provision of paragraph 2.2.2.

1.6.7. *Inspections and Tests.* ENGINEER shall have authority, as OWNER's representative, to require special inspection or testing of the work, and shall receive and review all certificates of inspections, testings and approvals required by laws, rules, regulations, ordinances, codes, orders or the Contract Documents (but only to determine generally that their content complies with the requirements of, and the results certified indicate compliance with, the Contract Documents).

1.6.8. *Disputes between OWNER and Contractor.* ENGINEER shall act as initial interpreter of the requirements of the Contract Documents and judge of the acceptability of the work thereunder and make decisions on all claims of OWNER and Contractor(s) relating to the acceptability of the work or the interpretation of the requirements of the Contract Documents pertaining to the execution and progress of the work. ENGINEER shall not be liable for the results of any such interpretations or decisions rendered in good faith.

1.6.9. *Applications for Payment.* Based on ENGINEER's on-site observations as an experienced and qualified design professional, on information provided by the Resident Project Representative and on review of applications for payment and the accompanying data and schedules:

1.6.9.1. ENGINEER shall determine the amounts owing to Contractor(s) and recommend in writing payments to Contractor(s) in such amounts. Such recommendations of payment will constitute a representation to OWNER, based on such observations and review, that the work has progressed to the point indicated, and that, to the best of ENGINEER's knowledge, information and belief, the quality of such work is generally in accordance with the Contract Documents (subject to an evaluation of such work as a functioning whole prior to or upon Substantial Completion, to the results of any subsequent tests called for in the Contract Documents and to any other qualifications stated in the recommendation). In the case of unit price work, ENGINEER's recommendations of payment will include final determinations of quantities and classifications of such work (subject to any subsequent adjustments allowed by the Contract Documents).

1.6.9.2. By recommending any payment ENGINEER will not thereby be deemed to have represented that exhaustive, continuous or detailed reviews or examinations have been made by ENGINEER to check the quality or quantity of Contractor(s)' work

Page 4 of _____ pages

as it is furnished and performed beyond the responsibilities specifically assigned to ENGINEER in this Agreement and the Contract Documents. ENGINEER's review of Contractor(s)' work for the purposes of recommending payments will not impose on ENGINEER responsibility to supervise, direct or control such work or for the means, methods, techniques, sequences, or procedures of construction or safety precautions or programs incident thereto or Contractor(s) compliance with laws, rules, regulations, ordinances, codes or orders applicable to their furnishing and performing the work. It will also not impose responsibility on ENGINEER to make any examination to ascertain how or for what purposes any Contractor has used the moneys paid on account of the Contract Price, or to determine that title to any of the work, materials or equipment has passed to OWNER free and clear of any lien, claims, security interests or encumbrances, or that there may not be other matters at issue between OWNER and CONTRACTOR that might affect the amount that should be paid.

1.6.10. *Contractor(s)' Completion Documents*. ENGINEER shall receive and review maintenance and operating instructions, schedules, guarantees, bonds and certificates of inspection, tests and approvals which are to be assembled by Contractor(s) in accordance with the Contract Documents (but such review will only be to determine that their content complies with the requirements of, and in the case of certificates of inspection, tests and approvals the results certified indicate compliance with, the Contract Documents); and shall transmit them to OWNER with written comments.

1.6.11. *Inspections*. ENGINEER shall conduct an inspection to determine if the work is substantially complete and a final inspection to determine if the completed work is acceptable so that ENGINEER may recommend, in writing, final payment to Contractor(s) and may give written notice to OWNER and the Contractor(s) that the work is acceptable (subject to any conditions therein expressed), but any such recommendation and notice will be subject to the limitations expressed in paragraph 1.6.9.2.

1.6.12. *Limitation of Responsibilities*. ENGINEER shall not be responsible for the acts or omissions of any Contractor, or of any subcontractor or supplier, or any of the Contractor(s)' or subcontractor's or supplier's agents or employees or any other persons (except ENGINEER's own employees and agents) at the site or otherwise furnishing or performing any of the Contractor(s)' work; however, nothing contained in paragraphs 1.6.1 thru 1.6.11 inclusive, shall be construed to release ENGINEER from liability for failure to properly perform duties and responsibilities assumed by ENGINEER in the Contract Documents.

1.7. Operational Phase.

During the Operational Phase, ENGINEER shall, when requested by OWNER:

1.7.1. Provide assistance in the closing of any financial or related transaction for the Project.

1.7.2. Provide assistance in connection with the refining and adjusting of any equipment or system.

1.7.3. Assist OWNER in training OWNER's staff to operate and maintain the Project.

1.7.4. Assist OWNER in developing systems and procedures for control of the operation and maintenance of and record keeping for the Project.

1.7.5. Prepare a set of reproducible record prints of Drawings showing those changes made during the construction process, based on the marked-up prints, drawings and other data furnished by Contractor(s) to ENGINEER and which ENGINEER considers significant.

1.7.6. In company with OWNER, visit the Project to observe any apparent defects in the completed construction, assist OWNER in consultations and discussions with Contractor(s) concerning correction of such deficiencies, and make recommendations as to replacement or correction of defective work.

The duties and responsibilities of ENGINEER during the Operational Phase are amended and supplemented as indicated in paragraph 7 of Exhibit A "Further Description of Basic Engineering Services and Related Matters".

SECTION 2—ADDITIONAL SERVICES OF ENGINEER

2.1. Services Requiring Authorization in Advance.

If authorized in writing by OWNER, ENGINEER shall furnish or obtain from others Additional Services of the types listed in paragraphs 2.1.1 through 2.1.14, inclusive. These services are not included as part of Basic Services except to the extent provided otherwise in Exhibit A "Further Description of Basic Engineering Services and Related Matters"; these will be paid for by OWNER as indicated in Section 5.

2.1.1. Preparation of applications and supporting documents (in addition to those furnished under Basic Services) for private or governmental grants, loans or advances in connection with the Project; preparation or review of environmental assessments and impact statements; review and evaluation of the effect on the design requirements of the Project of any such statements and documents prepared by others; and assistance in obtaining approvals of authorities having jurisdiction over the anticipated environmental impact of the Project.

2.1.2. Services to make measured drawings of or to investigate existing conditions or facilities, or to verify the accuracy of drawings or other information furnished by OWNER.

2.1.3. Services resulting from significant changes in the general scope, extent or character of the Project or its design including, but not limited to, changes in size, complexity, OWNER's schedule, character of construction or method of financing; and revising previously accepted studies, reports, design documents or Contract Documents when such revisions are required by changes in laws, rules, regulations, ordinances, codes or orders enacted subsequent to the preparation of such studies, reports or documents, or are due to any other causes beyond ENGINEERS's control.

2.1.4. Providing renderings or models for OWNER's use.

2.1.5. Preparing documents for alternate bids requested by OWNER for Contractor(s)' work which is not executed or documents for out-of-sequence work.

2.1.6. Investigations and studies involving, but not limited to, detailed consideration of operations, maintenance and overhead expenses; providing value engineering during the course of design; the preparation of feasibility studies, cash flow and economic evaluations, rate schedules and appraisals; assistance in obtaining financing for the Project; evaluating processes available for licensing and assisting OWNER in obtaining process licensing; detailed quantity surveys of material, equipment and labor; and audits or inventories required in connection with construction performed by OWNER.

2.1.7. Furnishing services of independent professional associates and consultants for other than Basic Services (which include, but are not limited to, customary civil, structural, mechanical and electrical engineering and customary architectural design incidental thereto); and providing data or services of the types described in paragraph 3.4 when OWNER employs ENGINEER to provide such data or services in lieu of furnishing the same in accordance with paragraph 3.4.

2.1.8. If ENGINEER's compensation is on the basis of a lump sum or percentage of Construction Cost or cost-plus a fixed fee method of payment, services resulting from the award of more separate prime contracts for construction, materials or equipment for the Project than are contemplated by paragraph 5.1.1.2. If ENGINEER's compensation is on the basis of a percentage of Construction Cost and ENGINEER has been required to prepare Contract Documents on the assumption that more than one prime contract will be awarded for construction, materials and equipment, but only one prime contract is awarded for construction, materials and equipment for the Project, services attributable to the preparation of contract documentation that was rendered unuseable and any revisions or additions to contract documentation used that was necessitated by the award of only one prime contract.

2.1.9. Services during out-of-town travel required of ENGINEER other than visits to the site or OWNER's office as required by Section 1.

2.1.10. Assistance in connection with bid protests, rebidding or renegotiating contracts for construction, materials, equipment or services, except when such assistance is required to complete services called for in paragraph 6.2.2.5.

2.1.11. Providing any type of property surveys or related engineering services needed for the transfer of interests in real property and field surveys for design purposes and engineering surveys and staking to enable Contractor(s) to proceed with their work; and providing other special field surveys.

2.1.12. Preparation of operating, maintenance and staffing manuals to supplement Basic Services under paragraph 1.7.3.

2.1.13. Preparing to serve or serving as a consultant or witness for OWNER in any litigation, arbitration or other legal or administrative proceeding involving the Project (except for assistance in consultations which is included as part of Basic Services under paragraphs 1.2.3 and 1.4.2).

2.1.14. Additional services in connection with the Project, including services which are to be furnished by OWNER in accordance with Article 3, and services not otherwise provided for in this Agreement.

2.2. Required Additional Services.

When required by the Contract Documents in circumstances beyond ENGINEER's control, ENGINEER shall furnish or obtain from others, as circumstances require during construction and without waiting for specific authorization from OWNER, Additional Services of the types listed in paragraphs 2.2.1 through 2.2.6, inclusive (except to the extent otherwise provided in Exhibit A "Further Description of Basic Engineering Services and Related Matters"). These services are not included as part of Basic Services. ENGINEER shall advise OWNER promptly after starting any such Additional Services which will be paid for by OWNER as indicated in Section 5.

2.2.1. Services in connection with work directive changes and change orders to reflect changes requested by OWNER if the resulting change in compensation for Basic Services is not commensurate with the additional services rendered.

2.2.2. Services in making revisions to Drawings and Specifications occasioned by the acceptance of substitutions proposed by Contractor(s); and services after the award of each contract in evaluating and determining the acceptability of an unreasonable or excessive number of substitutions proposed by Contractor.

2.2.3. Services resulting from significant delays, changes or price increases occurring as a direct or indirect result of material, equipment or energy shortages.

2.2.4. Additional or extended services during construction made necessary by (1) work damaged by fire or other cause during construction, (2) a significant amount of defective or neglected work of any Contractor, (3) acceleration of the progress schedule involving services beyond normal working hours, and (4) default by any Contractor.

2.2.5. Services (other than Basic Services during the Operational Phase) in connection with any partial utilization of any part of the Project by OWNER prior to Substantial Completion.

2.2.6. Evaluating an unreasonable or extensive number of claims submitted by Contractor(s) or others in connection with the work.

SECTION 3—OWNER'S RESPONSIBILITIES

OWNER shall do the following in a timely manner so as not to delay the services of ENGINEER:

3.1. Designate in writing a person to act as OWNER's representative with respect to the services to be rendered under this Agreement. Such person shall have complete authority to transmit instructions, receive information, interpret and define OWNER's policies and decisions with respect to ENGINEER's services for the Project.

3.2. Provide all criteria and full information as to OWNER's requirements for the Project, including design objectives and constraints, space, capacity and performance requirements, flexibility and expandability, and any budgetary limitations; and furnish copies of all design and construction standards which OWNER will require to be included in the Drawings and Specifications.

3.3. Assist ENGINEER by placing at ENGINEER's disposal all available information pertinent to the Project including previous reports and any other data relative to design or construction of the Project.

3.4. Furnish to ENGINEER, as required for performance of ENGINEER's Basic Services (except to the extent provided otherwise in Exhibit A "Further Description of Basic Engineering Services and Related Matters"), the following:

3.4.1. data prepared by or services of others, including without limitation borings, probings and subsurface explorations, hydrographic surveys, laboratory tests and inspections of samples, materials and equipment;

3.4.2. appropriate professional interpretations of all of the foregoing;

3.4.3. environmental assessment and impact statements;

3.4.4. property, boundary, easement, right-of-way, topographic and utility surveys;

3.4.5. property descriptions;

3.4.6. zoning, deed and other land use restriction; and

3.4.7. other special data or consultations not covered in Section 2;

all of which ENGINEER may use and rely upon in performing services under this Agreement.

3.5. Provide engineering surveys to establish reference points for construction (except to the extent provided otherwise in Exhibit A "Further Description of Basic Engineering Services and Related Matters") to enable Contractor(s) to proceed with the layout of the work.

3.6. Arrange for access to and make all provisions for ENGINEER to enter upon public and private property as required for ENGINEER to perform services under this Agreement.

3.7. Examine all studies, reports, sketches, Drawings, Specifications, proposals and other documents presented by ENGINEER, obtain advice of an attorney, insurance counselor and other consultants as OWNER deems appropriate for such examination and render in writing decisions pertaining thereto within a reasonable time so as not to delay the services of ENGINEER.

3.8. Furnish approvals and permits from all governmental authorities having jurisdiction over the Project and such approvals and consents from others as may be necessary for completion of the Project.

3.9. Provide such accounting, independent cost estimating and insurance counseling services as may be required for the Project, such legal services as OWNER may require or ENGINEER may reasonably request with regard to legal issues pertaining to the Project including any that may be raised by Contractor(s), such auditing service as OWNER may require to ascertain how or for what purpose any Contractor has used the moneys paid under the construction contract, and such inspection services as OWNER may require to ascertain that Contractor(s) are complying with any law, rule, regulation, ordinance, code or order applicable to their furnishing and performing the work.

3.10. If OWNER designates a person to represent OWNER at the site who is not ENGINEER or ENGINEER's agent or employee, the duties, responsibilities and limitations of authority of such other person and the affect thereof on the duties and responsibilities of ENGINEER and the Resident Project Representative (and any assistants) will be set forth in an exhibit that is to be identified, attached to and made a part of this Agreement before such services begin.

3.11. If more than one prime contract is to be awarded for construction, materials, equipment and services for the entire Project, designate a person or organization to have authority and responsibility for coordinating the activities among the various prime contractors.

3.12. Furnish to ENGINEER data or estimated figures as to OWNER's anticipated costs for services to be provided by others for OWNER (such as services pursuant to paragraphs 3.7 through 3.11, inclusive and other costs of the types referred to in paragraph 1.2.6) so that ENGINEER may make the necessary findings to support opinions of probable Total Project Costs.

3.13. Attend the pre-bid conference, bid opening, preconstruction conferences, construction progress and other job related meetings and substantial completion inspections and final payment inspections.

3.14. Give prompt written notice to ENGINEER whenever OWNER observes or otherwise becomes aware of any development that affects the scope or timing of ENGINEER's services, or any defect or nonconformance in the work of any Contractor.

3.15. Furnish, or direct ENGINEER to provide, Additional Services as stipulated in paragraph 2.1 of this Agreement or other services as required.

3.16. Bear all costs incident to compliance with the requirements of this Section 3.

SECTION 4—PERIODS OF SERVICE

4.1. The provisions of this Section 4 and the various rates of compensation for ENGINEER's services provided for elsewhere in this Agreement have been agreed to in anticipation of the orderly and continuous progress of the Project through completion of the Construction Phase. ENGINEER's obligation to render services hereunder will extend for a period which may reasonably be required for the design, award of contracts, construction and initial operation of the Project including extra work and required extensions thereto. If in Exhibit A "Further Description of Basic Engineering Services and Related Matters" specific periods of time for rendering services are set forth or specific dates by which services are to be completed are provided and if such dates are exceeded through no fault of ENGINEER, all rates, measures and amounts of compensation provided herein shall be subject to equitable adjustment.

4.2. The services called for in the Study and Report Phase will be completed and the Report submitted within the stipulated period indicated in paragraph 2 of Exhibit A "Further Description of Basic Engineering Services and Related Matters" after written authorization to proceed with that phase of services which will be given by OWNER within thirty days after ENGINEER has signed this Agreement.

4.3. After acceptance by OWNER of the Study and Report Phase documents indicating any specific modifications or changes in the general scope, extent or character of the Project desired by OWNER, and upon written authorization from OWNER, ENGINEER shall proceed with the performance of the services called for in the Preliminary Design Phase, and shall submit preliminary design documents and a revised opinion of probable Total Project Costs within the stipulated period indicated in paragraph 3 of Exhibit A "Further Description of Basic Engineering Services and Related Matters".

4.4. After acceptance by OWNER of the Preliminary Design Phase documents and revised opinion of probable Total Project Costs, indicating any specific modifications or changes in the general scope, extent or character of the Project desired by OWNER, and upon written authorization from OWNER, ENGINEER shall proceed with the performance of the services called for in the Final Design Phase; and shall deliver Contract Documents and a revised opinion of probable Total Project Costs for all work of Contractor(s) on the Project within the stipulated period indicated in paragraph 4 of Exhibit A "Further Description of Basic Engineering Services and Related Matters".

4.5. ENGINEER's services under the Study and Report Phase, Preliminary Design Phase and Final Design Phase shall each be considered complete at the earlier of (1) the date when the submissions for that phase have been accepted by OWNER or (2) thirty days after the date when such submissions are delivered to OWNER for final acceptance, plus in each case such additional time as may be considered reasonable for obtaining approval of governmental authorities having jurisdiction to approve the design of the Project.

4.6. After acceptance by OWNER of the ENGINEER's Drawings, Specifications and other Final Design Phase documentation including the most recent opinion of probable Total Project Costs and upon written authorization to proceed, ENGINEER shall proceed with performance of the services called for in the Bidding or Negotiating Phase. This Phase shall terminate and the services to be rendered thereunder shall be considered complete upon commencement of the Construction Phase or upon cessation of negotiations with prospective Contractor(s) (except as may otherwise be required to complete the services called for in paragraph 6.2.2.5).

4.7. The Construction Phase will commence with the execution of the first prime contract to be executed for the work of the Project or any part thereof, and will terminate upon written recommendation by ENGINEER of final payment on the last prime contract to be completed. Construction Phase services may be rendered at different times in respect of separate prime contracts if the Project involves more than one prime contract.

4.8. The Operational Phase will commence during the Construction Phase and will terminate one year after the date of Substantial Completion of the last prime contract for construction, materials and equipment on which substantial completion is achieved.

4.9. If OWNER has requested significant modifications or changes in the general scope, extent or character of the Project, the time of performance of ENGINEER's services shall be adjusted equitably.

4.10. If OWNER fails to give prompt written authorization to proceed with any phase of services after completion of the immediately preceding phase, or if the Construction Phase has not commenced within _____ calendar days (plus such additional time as may be required to complete the services called for under paragraph 6.2.2.5) after completion of the Final Design Phase, ENGINEER may, after giving seven days' written notice to OWNER, suspend services under this Agreement.

4.11. If ENGINEER's services for design or during construction of the Project are delayed or suspended in whole or in part by OWNER for more than three months for reasons beyond ENGINEER's control, ENGINEER shall on written demand to OWNER (but without termination of this Agreement) be paid as provided in paragraph 5.3.2. If such delay or suspension extends for more than one year for reasons beyond ENGINEER's control, or if ENGINEER for any reason is required to render Construction Phase services in respect of any prime contract for construction, materials or equipment more than one year after Substantial Completion is achieved under that contract, the various rates of compensation provided for elsewhere in this Agreement shall be subject to equitable adjustment.

4.12. In the event that the work designed or specified by ENGINEER is to be furnished or performed under more than one prime contract, or if ENGINEER's services are to be separately sequenced with the work of one or more prime contractors (such as in the case of fast-tracking), OWNER and ENGINEER shall, prior to commencement of the Final Design Phase, develop a schedule for performance of ENGINEER's services during the Final Design, Bidding or Negotiating and Construction Phases in order to sequence and coordinate properly such services as are applicable to the work under such separate contracts. This schedule is to be prepared whether or not the work under such contracts is to proceed concurrently and is to be included in Exhibit A "Further Description of Basic Engineering Services and Related Matters", and the provisions of paragraphs 4.4 through 4.10 inclusive, will be modified accordingly.

LUMP SUM METHOD OF PAYMENT

SECTION 5—PAYMENTS TO ENGINEER

5.1. Methods of Payment for Services and Expenses of ENGINEER.

5.1.1. *For Basic Services.* OWNER shall pay ENGINEER for Basic Services rendered under Section 1 (as amended and supplemented by Exhibit A "Further Description of Basic Engineering Services and Related Matters") as follows:

5.1.1.1. *One Prime Contract.* If only one prime contract is awarded for construction, materials and equipment for the Project, a lump sum fee of $ ______________ for all Basic Services (except services of ENGINEER's Resident Project Representative (and assistants) furnished under paragraph 1.6.2.1 and Operational Phase services furnished under paragraph 1.7); but, if the prime contract contains cost-plus or incentive savings provisions for Contractor's basic compensation, a lump sum fee of $ ______________ for such services.

[Delete inapplicable language and initial]

5.1.1.2. *Several Prime Contracts.* If more than one but less than ________ separate prime contracts are awarded for construction, materials and equipment for the Project, a lump sum fee of $ ____________ for all Basic Services (except services of ENGINEER's Resident Project Representative (and assistants) furnished under paragraph 1.6.2.1 and Operational Phase services furnished under paragraph 1.7): but, if any prime contract contains cost-plus or incentive savings provisions for Contractor's basic compensation, a lump sum fee of $ ____________ for such services.

5.1.1.3. *Resident Project Services.* For services of ENGINEER's Resident Project Representative (and assistants) furnished under paragraph 1.6.2.1, on the basis of Salary Costs times a factor of ________ for services rendered by principals and employees assigned to resident Project representation.

5.1.1.4. *Operational Phase Services.* For Operational Phase services furnished under paragraph 1.7, an amount equal to ENGINEER's Salary Costs times a factor of ________ for services rendered by principals and employees engaged directly on the Project.

5.1.2. *For Additional Services.* OWNER shall pay ENGINEER for Additional Services rendered under Section 2 as follows:

5.1.2.1. *General.* For Additional Services of ENGINEER's principals and employees engaged directly on the Project and rendered pursuant to paragraph 2.1 or 2.2 (except services as a consultant or witness under paragraph 2.1.13), on the basis of ENGINEER's Salary Costs times a factor of ________.

5.1.2.2. *Professional Associates and Consultants.* For services and Reimbursable Expenses of independent professional associates and consultants employed by ENGINEER to render Additional Services pursuant to paragraph 2.1 or 2.2, the amount billed to ENGINEER therefor times a factor of ________.

5.1.2.3. *Serving as a Witness.* For services rendered by ENGINEER's principals and employees as consultants or witnesses in any litigation, arbitration or other legal or administrative proceeding in accordance with paragraph 2.1.13, at the rate of $ ________ per day or any portion thereof (but compensation for time spent in preparing to appear in any such litigation, arbitration or proceeding will be on the basis provided in paragraph 5.1.2.1). Compensation for ENGINEER's independent professional associates and consultants will be on the basis provided in paragraph 5.1.2.2.

5.1.3. *For Reimbursable Expenses.* In addition to payments provided for in paragraphs 5.1.1 and 5.1.2, OWNER shall pay ENGINEER the actual costs of all Reimbursable Expenses incurred in connection with all Basic and Additional Services.

5.1.4. The terms "Salary Costs" and "Reimbursable Expenses" have the meanings assigned to them in paragraph 5.4.

5.2. Times of Payments.

5.2.1. ENGINEER shall submit monthly statements for Basic and Additional Services rendered and for Reimbursable Expenses incurred. The statements will be based upon ENGINEER's estimate of the proportion of the total services actually completed at the time of billing. OWNER shall make prompt monthly payments in response to ENGINEER's monthly statements.

Page 11 of _____ pages

5.3. Other Provisions Concerning Payments.

5.3.1. If OWNER fails to make any payment due ENGINEER for services and expenses within thirty days after receipt of ENGINEER's statement therefor, the amounts due ENGINEER will be increased at the rate of 1% per month from said thirtieth day, and in addition, ENGINEER may, after giving seven days' written notice to OWNER, suspend services under this Agreement until ENGINEER has been paid in full all amounts due for services, expenses and charges.

5.3.2. In the event of termination by OWNER under paragraph 7.1 upon the completion of any phase of the Basic Services, progress payments due ENGINEER for services rendered through such phase shall constitute total payment for such services. In the event of such termination by OWNER during any phase of the Basic Services, ENGINEER will be paid for services rendered during that phase on the basis of ENGINEER's Salary Costs times a factor of ________ for services rendered during that phase to date of termination by ENGINEER's principals and employees engaged directly on the Project. In the event of any such termination, ENGINEER also will be reimbursed for the charges of independent professional associates and consultants employed by ENGINEER to render Basic Services, and paid for all unpaid Additional Services and unpaid Reimbursable Expenses, plus all termination expenses. Termination expenses mean Reimbursable Expenses directly attributable to termination, which, if termination is at OWNER's convenience, shall include an amount computed as a percentage of total compensation for Basic Services earned by ENGINEER to the date of termination, as follows:

> 20% if termination occurs after commencement of the Preliminary Design Phase but prior to commencement of the Final Design Phase; or
>
> 10% if termination occurs after commencement of the Final Design Phase.

5.3.3. Records of ENGINEER's Salary Costs pertinent to ENGINEER's compensation under this Agreement will be kept in accordance with generally accepted accounting practices. Copies will be made available to OWNER at cost on request prior to final payment for ENGINEER's services.

5.3.4. Whenever a factor is applied to Salary Costs in determining compensation payable to ENGINEER that factor will be adjusted periodically and equitably to reflect changes in the various elements that comprise such factor. All such adjustments will be in accordance with generally accepted accounting practices as applied on a consistent basis by ENGINEER and consistent with ENGINEER's overall compensation practices and procedures.

5.4. Definitions.

5.4.1. The Salary Costs used as a basis for payment mean salaries and wages (basic and incentive) paid to all ENGINEER's personnel engaged directly on the Project, including, but not limited to, engineers, architects, surveyors, designers, draftsmen, specification writers, estimators, other technical and business personnel; plus the cost of customary and statutory benefits including, but not limited to, social security contributions, unemployment, excise and payroll taxes, workers' compensation, health and retirement benefits, sick leave, vacation and holiday pay and other group benefits. For the purposes of this Agreement, the principals of ENGINEER and their current hourly Salary Costs are:

____

____

The hourly Salary Costs of principals of ENGINEER will be adjusted equitably to reflect changes in personnel and in ENGINEER's overall compensation procedures and practices.

The amount of customary and statutory benefits of all other personnel of ENGINEER will be considered equal to ________ % of salaries and wages, subject to equitable adjustment to reflect changes in ENGINEER's overall compensation procedures and practices.

5.4.2. Reimbursable Expenses mean the actual expenses incurred by ENGINEER or ENGINEER's independent professional associates or consultants, directly or indirectly in connection with the Project, such as expenses for: transportation and subsistence incidental thereto; obtaining bids or proposals from Contractor(s); providing and maintaining field office facilities including furnishings and utilities; subsistence and transportation of Resident Project Representatives and their assistants; toll telephone calls and telegrams; reproduction of reports, Drawings, Specifications, Bidding Documents, and similar Project-related items in addition to those required under Section 1; and, if authorized in advance by OWNER, overtime work requiring higher than regular rates.

Page 12 of _____ pages

PERCENTAGE METHOD OF PAYMENT

SECTION 5—PAYMENTS TO ENGINEER

5.1. Methods of Payment for Services and Expenses of ENGINEER.

5.1.1. *For Basic Services.* OWNER shall pay ENGINEER for Basic Services rendered under Section 1 (as amended and supplemented by Exhibit A "Further Description of Basic Engineering Services and Related Matters") as follows:

5.1.1.1. *One Prime Contract.* If only one prime contract is awarded for construction, materials and equipment for the Project, an amount equal to ________ % of the Construction Cost for all Basic Services (except services of ENGINEER's Resident Project Representative (and assistants) furnished under paragraph 1.6.2.1 and Operational Phase services furnished under paragraph 1.7); but, if the prime contract contains cost-plus or incentive savings provisions for the Contractor's basic compensation, an amount equal to ________ % of the Construction Cost for such services.

[Delete inapplicable language and initial]

5.1.1.2. *Several Prime Contracts.* If more than one but less than ________ separate prime contracts are awarded for construction, materials and equipment for the Project, an amount equal to ________ % of the Construction Cost for all Basic Services (except services of ENGINEER's Resident Project Representative (and assistants) furnished under paragraph 1.6.2.1 and Operational Phase services furnished under paragraph 1.7); but, if any prime contract contains cost-plus or incentive savings provisions for Contractor's basic compensation, ________ % of the Construction Cost for such services.

5.1.1.3. *Resident Project Services.* For services of ENGINEER's Resident Project Representative (and assistants) furnished under paragraph 1.6.2.1, on the basis of Salary Costs times a factor of ________ for services rendered by principals and employees assigned to resident Project representation.

5.1.1.4. *Operational Phase Services.* For Operational Phase services furnished under paragraph 1.7, an amount equal to ENGINEER's Salary Costs times a factor of ________ for services rendered by principals and employees engaged directly on the Project.

5.1.2 *For Additional Services.* OWNER shall pay ENGINEER for Additional Services rendered under Section 2 as follows:

5.1.2.1. *General.* For Additional Services of ENGINEER's principals and employees engaged directly on the Project and rendered pursuant to paragraph 2.1 or 2.2 (except services as a consultant or witness under paragraph 2.1.13), on the basis of ENGINEER's Salary Costs times a factor of ________ .

5.1.2.2. *Professional Associates and Consultants.* For services and Reimbursable Expenses of independent professional associates and consultants employed by ENGINEER to render Additional Services pursuant to paragraph 2.1 or 2.2, the amount billed to ENGINEER therefor times a factor of ________ .

5.1.2.3. *Serving as a Witness.* For services rendered by ENGINEER's principals and employees as consultants or witnesses in any litigation, arbitration or other legal or administrative proceeding in accordance with paragraph 2.1.13, at the rate of $ ______ per day or any portion thereof (but compensation for time spent in preparing to appear in any such litigation, arbitration or proceeding will be on the basis provided in paragraph 5.1.2.1). Compensation for ENGINEER's independent professional associates and consultants will be on the basis provided in paragraph 5.1.2.2.

5.1.3. *For Reimbursable Expenses.* In addition to payments provided for in paragraphs 5.1.1 and 5.1.2, OWNER shall pay ENGINEER the actual costs of all Reimbursable Expenses incurred in connection with all Basic and Additional Services.

5.1.4. As used in this paragraph 5.1, the terms "Salary Costs" and "Reimbursable Expenses" have the meanings assigned to them in paragraph 5.4, and the term "Construction Cost" has the meaning assigned to it in paragraph 6.1. When Construction Cost is used as a basis for payment it will be based on one of the following sources with precedence in the order listed for work designed or specified by ENGINEER:

5.1.4.1. For completed construction work, the total cost of all work performed as designed or specified by ENGINEER.

5.1.4.2. For work designed or specified but not constructed, the lowest bona fide bid received from a qualified bidder for such work; or, if the work is not bid, the lowest bona fide negotiated proposal for such work.

Page 11 of ______ pages

5.1.4.3. For work designed or specified but not constructed upon which no such bid or proposal is received, the most recent estimate of Construction Cost, or, if none is available, ENGINEER's most recent opinion of probable Construction Cost.

Labor furnished by OWNER for the Project will be included in the Construction Cost at current market rates including a reasonable allowance for overhead and profit. Materials and equipment furnished by OWNER will be included at current market prices. No deduction is to be made from Construction Costs on account of any penalty, liquidated damages, or other amounts withheld from payments to Contractor(s).

5.2. Times of Payments.

5.2.1. ENGINEER shall submit monthly statements for Basic and Additional Services rendered and for Reimbursable Expenses incurred. The statements will be based upon ENGINEER's estimate of the proportion of the total services actually completed at the time of billing. OWNER shall make prompt monthly payments in response to ENGINEER's monthly statements.

5.2.2. Upon conclusion of each phase of Basic Services, OWNER shall pay such additional amount, if any, as may be necessary to bring total compensation paid on account of such phase to the following percentages of total compensation payable for all phases of Basic Services:

Phase	*Insert Actual Percentage*
Study and Report	________ %
Preliminary Design	________ %
Final Design	________ %
Bidding or Negotiating	________ %
Construction	________ %
Operational	________ %
	100 %

5.3. Other Provisions Concerning Payments.

5.3.1. If OWNER fails to make any payment due ENGINEER for services and expenses within thirty days after receipt of ENGINEER's statement therefor, the amounts due ENGINEER will be increased at the rate of 1% per month from said thirtieth day, and in addition, ENGINEER may, after giving seven days' written notice to OWNER, suspend services under this Agreement until ENGINEER has been paid in full all amounts due for services, expenses and charges.

5.3.2. In the event of termination by OWNER under paragraph 7.1 upon the completion of any phase of the Basic Services, progress payments due ENGINEER for services rendered through such phase shall constitute total payment for such services. In the event of such termination by OWNER during any phase of the Basic Services, ENGINEER will be paid for services rendered during that phase on the basis of ENGINEER's Salary Costs times a factor of ________ for services rendered during that phase to date of termination by ENGINEER's principals and employees engaged directly on the Project. In the event of any such termination, ENGINEER also will be reimbursed for the charges of independent professional associates and consultants employed by ENGINEER to render Basic Services, and paid for all unpaid Additional Services and unpaid Reimbursable Expenses, plus all termination expenses. Termination expenses mean additional Reimbursable Expenses directly attributable to termination, which, if termination is at Owner's convenience, shall include an amount computed as a percentage of total compensation for Basic Services earned by ENGINEER to the date of termination, as follows:

20% if termination occurs after commencement of the Preliminary Design Phase but prior to commencement of the Final Design Phase; or

10% if termination occurs after commencement of the Final Design Phase.

5.3.3. Records of ENGINEER's Salary Costs pertinent to ENGINEER's compensation under this Agreement will be kept in accordance with generally accepted accounting principles. Copies will be made available to OWNER at cost on request prior to final payment for ENGINEER's services.

5.3.4. Whenever a factor is applied to Salary Costs in determining compensation payable to ENGINEER that factor will be adjusted periodically and equitably to reflect changes in the various elements that comprise such factor. All such adjustments will be in accordance with generally accepted accounting practices as applied on a consistent basis by ENGINEER and consistent with ENGINEER's overall compensation practices and procedures.

5.4. Definitions.

5.4.1. The Salary Costs used as a basis for payment mean salaries and wages (basic and incentive) paid to all ENGINEER's personnel engaged directly on the Project, including, but not limited to, engineers, architects, surveyors, designers, draftsmen, specification writers, estimators, other technical and business per-

Page 12 of ____ pages

sonnel; plus the cost of customary and statutory benefits including, but not limited to, social security contributions, unemployment, excise and payroll taxes, workers' compensation, health and retirement benefits, sick leave, vacation and holiday and other group benefits. For the purposes of this Agreement, the principals of ENGINEER and their current hourly Salary Costs are:

____

____

The hourly Salary Costs of principals of ENGINEER will be adjusted equitably to reflect changes in personnel and in ENGINEER's overall compensation procedures and practices.

The amount of customary and statutory benefits of all other personnel of ENGINEER will be considered equal to ______ % of salaries and wages, subject to equitable adjustment to reflect changes in ENGINEER's overall compensation procedures and practices.

5.4.2. Reimbursable Expenses mean the actual expenses incurred by ENGINEER or ENGINEER's independent professional associates or consultants, directly or indirectly in connection with the Project, such as expenses for: transportation and subsistence incidental thereto; obtaining bids or proposals from Contractor(s); providing and maintaining field office facilities including furnishings and utilities; subsistence and transportation of Resident Project Representatives and their assistants; toll telephone calls and telegrams; reproduction of reports, Drawings, Specifications, Bidding Documents, and similar Project-related items in addition to those required under Section 1; and, if authorized in advance by OWNER, overtime work requiring higher than regular rates.

Page 13 of _____ pages

Page 14 of ______ pages

COST PLUS A FIXED FEE METHOD OF PAYMENT

SECTION 5—PAYMENTS TO ENGINEER

5.1. Methods of Payment for Services and Expenses of ENGINEER.

5.1.1. *For Basic Services.* OWNER shall pay ENGINEER for Basic Services rendered under Section 1 (as amended and supplemented by Exhibit A "Further Description of Basic Engineering Services and Related Matters") as follows:

5.1.1.1. *One Prime Contract.* If only one prime contract is awarded for construction, materials and equipment for the Project, a fixed fee of $ ____________ plus an amount equal to ENGINEER's Direct Labor Cost times a factor of ______ for all Basic Services rendered by principals and employees engaged directly on the Project (except for services of ENGINEER's Resident Project Representative (and assistants) furnished under paragraph 1.6.2.1 and for Operational Phase services furnished under paragraph 1.7); but, if the prime contract contains cost-plus or incentive savings provisions for Contractor's basic compensation, the fixed fee will be $ ____________ plus the aforesaid amount based on ENGINEER's Direct Labor Costs for such services.

[Delete inapplicable language and initial]

5.1.1.2. *Several Prime Contracts.* If more than one, but less than ______ separate prime contracts are awarded for construction, materials and equipment for the Project, a fixed fee of $ ____________ plus an amount equal to ENGINEER's Direct Labor Costs times a factor of ______ for all Basic Services rendered by principals and employees engaged directly on the Project (except for services of ENGINEER's Resident Project Representative (and assistants) furnished under paragraph 1.6.2.1 and Operational Phase services furnished under paragraph 1.7); but, if any prime contract contains cost-plus or incentive savings provisions for Contractor's basic compensation, the fixed fee will be $ ____________ plus the aforesaid amount based on ENGINEER's Direct Labor Costs for such services.

5.1.1.3. *Resident Project Services.* For services of ENGINEER's Resident Project Representative (and assistants) furnished under paragraph 1.6.2.1, on the basis of Direct Labor Costs times a factor of ______ for services rendered by principals and employees assigned to resident Project representation.

5.1.1.4. *Professional Associates and Consultants.* For services and Reimbursable Expenses of independent professional associates and consultants employed by ENGINEER to render Basic Services, the amount billed to ENGINEER therefor times a factor of ______ .

5.1.1.5. *Operational Phase Services.* For Operational Phase services furnished under paragraph 1.7, an amount equal to ENGINEER's Direct Labor Costs times a factor of ______ for services rendered by principals and employees engaged directly on the Project.

5.1.1.6. *Notice of Relationship of Payments to Estimated Total Compensation.* If it becomes apparent to ENGINEER at any time before Basic Services to be rendered under this Agreement have been about eighty percent completed that the total amount of compensation to be paid to ENGINEER on account of Basic Services pursuant to paragraphs 5.1.1.1, 5.1.1.3, 5.1.1.4 and 5.1.1.5 will exceed $ ____________ or pursuant to paragraphs 5.1.1.2 through 5.1.1.5 inclusive will exceed $ __________ (which figures are ENGINEER's estimates of the amounts that will become payable for such services), ENGINEER shall give OWNER written notice thereof. Promptly thereafter OWNER and ENGINEER shall review the matter of compensation for such Basic Services, and either OWNER shall accede to such compensation exceeding said estimated amounts or OWNER and ENGINEER shall agree to a reduction in the remaining services to be rendered by ENGINEER under this Agreement so that total compensation for such Basic Services will not exceed such estimated amounts when such services are completed.

5.1.2. *Additional Services.* OWNER shall pay ENGINEER for Additional Services rendered under Section 2 as follows:

5.1.2.1. *General.* For Additional Services of ENGINEER's principals and employees engaged directly on the Project and rendered pursuant to paragraph 2.1 or 2.2 (except services as a consultant or witness under paragraph 2.1.13), on the basis of ENGINEER's Direct Labor Costs times a factor of ______ .

5.1.2.2. *Professional Associates and Consultants.* For services and Reimbursable Expenses of independent professional associates and consultants employed by ENGINEER to render Additional Services pursuant to paragraph 2.1 or 2.2, the amount billed to ENGINEER therefor times a factor of ______ .

5.1.2.3. *Serving as a Witness.* For services rendered by ENGINEER's principals and employees as consultants or witnesses in any litigation, arbitration or other legal or administrative proceeding in accordance with paragraph 2.1.13, at the rate of $______ per day or any portion thereof (but compensation for time spent in preparing to appear in any such litigation, arbitration or proceeding will be on the basis provided in paragraph 5.1.2.1). Compensation for ENGINEER's independent professional associates and consultants will be on the basis provided in paragraph 5.1.2.2.

5.1.3. *Reimbursable Expenses.* In addition to payments provided for in paragraphs 5.1.1 and 5.1.2, OWNER shall pay ENGINEER the actual costs (except where specifically provided otherwise) of all Reimbursable Expenses incurred in connection with all Basic and Additional Services.

5.1.4. The "Direct Labor Costs" and "Reimbursable Expenses" have the meanings assigned to them in paragraph 5.4.

5.2. Times of Payments.

5.2.1. ENGINEER shall submit monthly statements for Basic and Additional Services rendered and for Reimbursable Expenses incurred. The charge on account of the fixed fee will be based upon ENGINEER's estimate of the proportion of the total services actually completed at the time of billing. OWNER shall make prompt monthly payments in response to ENGINEER's monthly statements.

5.3. Other Provisions Concerning Payments.

5.3.1. If OWNER fails to make any payment due ENGINEER for services and expenses within thirty days after receipt of ENGINEER's statement therefor, the amounts due ENGINEER will be increased at the rate of 1% per month from said thirtieth day, and in addition, ENGINEER may, after giving seven days' written notice to OWNER, suspend services under this Agreement until ENGINEER has been paid in full all amounts due for services, expenses and charges.

5.3.2. In the event of termination by OWNER under paragraph 7.1 upon the completion of any phase of the Basic Services, progress payments due ENGINEER for services rendered through such phase shall constitute total payment for such services. In the event of such termination by OWNER during any phase of the Basic Services, ENGINEER also will be reimbursed for the charges of independent professional associates and consultants, employed by ENGINEER to render Basic Services, and paid for services rendered during that phase on the basis of ENGINEER's Direct Labor Costs times a factor of ______ for services rendered during that phase to date of termination by ENGINEER's principals and employees engaged directly on the Project. In the event of any such termination, ENGINEER will be paid for all unpaid Additional Services and unpaid Reimbursable Expenses, plus all termination expenses. Termination expenses mean additional Reimbursable Expenses directly attributable to termination, which, if termination is at Owner's convenience, shall include an amount computed as a percentage of total compensation for Basic Services earned by ENGINEER to the date of termination, as follows:

20% if termination occurs after commencement of the Preliminary Design Phase but prior to commencement of the Final Design Phase; or

10% if termination occurs after commencement of the Final Design Phase.

5.3.3. Records of ENGINEER's Direct Labor Costs pertinent to ENGINEER's compensation under this Agreement will be kept in accordance with generally accepted accounting practices. Copies will be made available to OWNER at cost on request prior to final payment for ENGINEER's services.

5.3.4. Whenever a factor is applied to Direct Labor Costs in determining compensation payable to ENGINEER, that factor will be adjusted periodically and equitably to reflect changes in the various elements that comprise such factor. All such adjustments will be in accordance with generally accepted accounting practices as applied on a consistent basis by ENGINEER and consistent with ENGINEER's overall compensation practices and procedures.

5.4. Definitions.

5.4.1. Direct Labor Costs used as a basis for payment mean salaries and wages (basic and incentive) paid to all ENGINEER's personnel engaged directly on the Project, including, but not limited to, engineers, architects, surveyors, designers, draftsmen, specification writers, estimators, other technical and business personnel; but does not include indirect payroll related costs or fringe benefits. For the purposes of this Agreement the principals of ENGINEER and their current hourly Direct Labor Costs are:

____

____

The hourly Direct Labor Costs of principals of ENGINEER will be adjusted equitably to reflect changes in personnel and in ENGINEER's overall compensation procedures and practices.

5.4.2. Reimbursable Expenses mean the actual expenses incurred by ENGINEER or ENGINEER's independent professional associates or consultants, directly or indirectly in connection with the Project, such as expenses for: transportation and subsistence incidental thereto; obtaining bids or proposals from Contractor(s); providing and maintaining field office facilities including furnishings and utilities; subsistence and transportation of Resident Project Representatives and their assistants; toll telephone calls and telegrams; reproduction of reports, Drawings, Specifications, Bidding Documents, and similar Project-related items in addition to those required under Section 1; and, if authorized in advance by OWNER, overtime work requiring higher than regular rates. In addition, if authorized in advance by OWNER, Reimbursable Expenses will also include expenses incurred for computer time and other highly specialized equipment, including an appropriate charge for previously established programs and expenses of photographic production techniques times a factor of _______ .

Page 13 of _____ pages

Page 14 of _____ pages

DIRECT LABOR COSTS METHOD OF PAYMENT

SECTION 5—PAYMENTS TO ENGINEER

5.1. Methods of Payments for Services and Expenses of ENGINEER.

5.1.1. *For Basic Services.* OWNER shall pay ENGINEER for Basic Services rendered under Section 1 (as amended and supplemented by Exhibit A "Further Description of Basic Engineering Services and Related Matters") as follows:

5.1.1.1. An amount equal to ENGINEER's Direct Labor Costs times a factor of ______ for all Basic Services rendered by principals and employees engaged directly on the Project (except for services of ENGINEER's Resident Project Representative (and assistants) furnished under paragraph 1.6.2.1 and for Operational Phase services furnished under paragraph 1.7).

5.1.1.2. *Resident Project Services.* For services of ENGINEER's Resident Project Representative (and assistants) furnished under paragraph 1.6.2.1, on the basis of Direct Labor Costs times a factor of ______ for services rendered by principals and employees assigned to resident Project representation.

5.1.1.3. *Professional Associates and Consultants.* For services and Reimbursable Expenses of independent professional associates and consultants employed by ENGINEER to render Basic Services, the amount billed to ENGINEER therefor times a factor of ______ .

5.1.1.4. *Operational Phase Services.* For Operational Phase services furnished under paragraph 1.7, an amount equal to ENGINEER's Direct Labor Costs times a factor of ______ for services rendered by principals and employees engaged directly on the Project.

5.1.1.5. *Notice of Relationship of Payments to Estimated Total Compensation.* If it becomes apparent to ENGINEER at any time before Basic Services to be rendered under this Agreement are about eighty percent completed that the total amount of compensation to be paid to ENGINEER on account of Basic Services pursuant to paragraph 5.1.1 will exceed ENGINEER's estimate of the amount set forth below which is applicable to the services rendered, ENGINEER shall give OWNER written notice thereof. Promptly thereafter OWNER and ENGINEER shall review the matter of compensation for such Basic Services, and either OWNER shall accede to such compensation exceeding said estimated amounts or OWNER and ENGINEER shall agree to a reduction in the remaining services to be rendered by ENGINEER under this Agreement so that total compensation for such Basic Services will not exceed said estimated amount when such services are completed.

5.1.1.5.1. If only one prime contract is awarded for construction, materials, and equipment for the Project, ENGINEER's estimate of the amount that will become payable for Basic Services pursuant to paragraphs 5.1.1.1 through 5.1.1.4 inclusive is $ ______________ ; but, if the prime contract contains cost-plus or incentive savings provisions for Contractor's basic compensation, the estimated amount will be $ ______________ .

5.1.1.5.2. If more than one but less than ______ separate prime contracts are awarded for construction, materials and equipment for the Project, ENGINEER's estimate of the amount that will become payable for Basic Services pursuant to paragraphs 5.1.1.1 through 5.1.1.4 inclusive is $ ______________ ; but, if any prime contract contains cost-plus or incentive savings provisions for Contractor's basic compensation, the estimated amount will be $ ______________ .

5.1.2. *Additional Services.* OWNER shall pay ENGINEER for Additional Services rendered under Section 2 as follows:

5.1.2.1. *General.* For Additional Services of ENGINEER's principals and employees engaged directly on the Project and rendered pursuant to paragraph 2.1 or 2.2 (except services as a consultant or witness under paragraph 2.1.13), on the basis of ENGINEER's Direct Labor Costs times a factor of ______ .

5.1.2.2. *Professional Associates and Consultants.* For services and Reimbursable Expenses of independent professional associates and consultants employed by ENGINEER to render Additional Services pursuant to paragraph 2.1 or 2.2, the amount billed to ENGINEER therefor times a factor of ______ .

5.1.2.3. *Serving as a Witness.* For services rendered by ENGINEER's principals and employees as consultants or witnesses in any litigation, arbitration or other legal or administrative proceeding in accordance with paragraph 2.1.13, at the rate of $ _____ per day or any portion thereof (but compensation for time spent in preparing to appear in any such litigation, arbitration or proceeding will be on the basis provided in paragraph 5.1.2.1). Compensation for ENGINEER's independent professional associates and consultants will be on the basis provided in paragraph 5.1.2.2.

5.1.3. *For Reimbursable Expenses*. In addition to payments provided for in paragraphs 5.1.1 and 5.1.2, OWNER shall pay ENGINEER the actual costs (except where specifically provided otherwise) of all Reimbursable Expenses incurred in connection with all Basic and Additional Services.

5.1.4. The terms "Direct Labor Costs" and "Reimbursable Expenses" have the meanings assigned to them in paragraph 5.4.

5.2. Times of Payments.

5.2.1. ENGINEER shall submit monthly statements for Basic and Additional Services rendered and for Reimbursable Expenses incurred. OWNER shall make prompt monthly payments in response to ENGINEER's monthly statements.

5.3. Other Provisions Concerning Payments.

5.3.1. If OWNER fails to make any payment due ENGINEER for services and expenses within thirty days after receipt of ENGINEER's statement therefor, the amounts due ENGINEER will be increased at the rate of 1% per month from said thirtieth day, and in addition, ENGINEER may, after giving seven days' written notice to OWNER, suspend services under this Agreement until ENGINEER has been paid in full all amounts due for services, expenses and charges.

5.3.2. In the event of termination by OWNER under paragraph 7.1 upon the completion of any phase of the Basic Services, progress payments due ENGINEER for services rendered through such phase shall constitute total payment for such services. In the event of such termination by OWNER during any phase of the Basic Services, ENGINEER also will be reimbursed for the charges of independent professional associates and consultants employed by ENGINEER to render Basic Services, and paid for services rendered during that phase on the basis of ENGINEER's Direct Labor Costs times a factor of _____ for services rendered during that phase to date of termination by ENGINEER's principals and employees engaged directly on the Project. In the event of any such termination, ENGINEER will be paid for all unpaid Additional Services and unpaid Reimbursable Expenses, plus all termination expenses. Termination expenses mean additional Reimbursable Expenses directly attributable to termination, which, if termination is at OWNER's convenience, shall include an amount computed as a percentage of total compensation for Basic Services earned by ENGINEER to the date of termination, as follows:

20% if termination occurs after commencement of the Preliminary Design Phase but prior to commencement of the Final Design Phase; or

10% if termination occurs after commencement of the Final Design Phase.

5.3.3. Records of ENGINEER's Direct Labor Costs pertinent to ENGINEER's compensation under this Agreement will be kept in accordance with generally accepted accounting practices. Copies will be made available to OWNER at cost on request prior to final payment for ENGINEER's services.

5.3.4. Whenever a factor is applied to Direct Labor Costs in determining compensation payable to ENGINEER, that factor will be adjusted periodically and equitably to reflect changes in various elements that comprise such factor. All such adjustments will be in accordance with generally accepted accounting practices as applied on a consistent basis by ENGINEER and consistent with ENGINEER's overall compensation practices and procedures.

5.4. Definitions.

5.4.1. Direct Labor Costs used as a basis for payment mean salaries and wages (basic and incentive) paid to all ENGINEER's personnel engaged directly on the Project, including, but not limited to, engineers, architects, surveyors, designers, draftsmen, specification writers, estimators, other technical and business personnel; but does not include indirect payroll related costs or fringe benefits. For the purposes of this Agreement the principals of ENGINEER and their current hourly Direct Labor Costs are:

____

____

The hourly Direct Labor Costs of principals of ENGINEER will be adjusted equitably to reflect changes in

Page 12 of _____ pages

personnel and in ENGINEER's overall compensation procedures and practices.

5.4.2. Reimbursable Expenses mean the actual expenses incurred by ENGINEER or ENGINEER's independent professional associates or consultants, directly or indirectly in connection with the Project, such as expenses for: transportation and subsistence incidental thereto; obtaining bids or proposals from Contractor(s); providing and maintaining field office facilities including furnishings and utilities; subsistence and transportation of Resident Project Representatives and their assistants; toll telephone calls and telegrams; reproduction of reports, Drawings, Specifications, Bidding Documents, and similar Project-related items in addition to those required under Section 1; and, if authorized in advance by OWNER, overtime work requiring higher than regular rates. In addition, if authorized in advance by OWNER, Reimbursable Expenses will also include expenses incurred for computer time and other highly specialized equipment, including an appropriate charge for previously established programs and expenses of photographic production techniques times a factor of ____________ .

Page 14 of _____ pages

SALARY COSTS METHOD OF PAYMENT

SECTION 5—PAYMENTS TO ENGINEER

5.1. Methods of Payment for Services and Expenses of ENGINEER.

5.1.1. *For Basic Services.* OWNER shall pay ENGINEER for Basic Services rendered under Section 1 (as amended and supplemented by Exhibit A "Further Description of Basic Engineering Services and Related Matters") an amount as follows:

5.1.1.1. An amount equal to ENGINEER's Salary Cost times a factor of ______ for all Basic Services rendered by principals and employees engaged directly on the Project (except for services of ENGINEER's Resident Project Representative (and assistants) furnished under paragraph 1.6.2.1. and for Operational Phase Services furnished under paragraph 1.7).

5.1.1.2. *Resident Project Services.* For services of ENGINEER's Resident Project Representative (and assistants) furnished under pragraph 1.6.2.1, on the basis of Salary Costs times a factor of ______ for services rendered by principals and employees assigned to resident Project representation.

5.1.1.3. *Professional Associates and Consultants.* For services and Reimbursable Expenses of independent professional associates and consultants employed by ENGINEER to render Basic Services, the amount billed to ENGINEER therefor times a factor of ______ .

5.1.1.4. *Operational Phase Services.* For Operational Phase services furnished under paragraph 1.7, an amount equal to ENGINEER's Salary Costs times a factor of ______ for services rendered by principals and employees engaged directly on the Project.

5.1.1.5. *Notice of Relationship of Payments to Estimated Total Compensation.* If it becomes apparent to ENGINEER at any time before Basic Services to be rendered under this Agreement are about eighty percent completed that the total amount of compensation to be paid to ENGINEER on account of Basic Services pursuant to paragraph 5.1.1 will exceed ENGINEER's estimate of the amount set forth below which is applicable to the services rendered, ENGINEER shall give OWNER written notice thereof. Promptly thereafter OWNER and ENGINEER shall review the matter of compensation for such Basic Services, and either OWNER shall accede to such compensation exceeding said estimated amounts or OWNER and ENGINEER shall agree to a reduction in the remaining services to be rendered by ENGINEER under this Agreement so that total compensation for such Basic Services will not exceed said estimated amount when such services are completed.

5.1.1.5.1. If only one prime contract is awarded for construction, materials and equipment for the Project, ENGINEER's estimate of the amount that will become payable for Basic Services pursuant to paragraphs 5.1.1.1 through 5.1.1.4 inclusive is $ ______ ; but, if the prime contract contains cost-plus or incentive savings provisions for Contractor's basic compensation, the estimated amount will be $ ______ .

5.1.1.5.2. If more than one but less than ______ separate prime contracts are awarded for construction, materials and equipment for the Project, ENGINEER's estimate of the amount that will become payable for Basic Services pursuant to paragraphs 5.1.1.1 through 5.1.1.4 inclusive is $ ______ ; but, if any prime contract contains cost-plus or incentive savings provisions for Contractor's basic compensation, the estimated amount will be $ ______ .

5.1.2. *For Additional Services.* OWNER shall pay ENGINEER for Additional Services rendered under Section 2 as follows:

5.1.2.1. *General.* For Additional Services of ENGINEER's principals and employees engaged directly on the Project and rendered pursuant to paragraph 2.1 or 2.2 (except services as a consultant or witness under paragraph 2.1.13), on the basis of ENGINEER's Salary Costs times a factor of ______ .

5.1.2.2. *Professional Associates and Consultants.* For services and Reimbursable Expenses of independent professional associates and consultants employed by ENGINEER to render Additional Services pursuant to paragraph 2.1 or 2.2, the amount billed to ENGINEER therefor times a factor of ______ .

5.1.2.3. *Serving as a Witness.* For services rendered by ENGINEER's principals and employees as consultants or witnesses in any litigation, arbitration or other legal or administrative proceeding in accordance with paragraph 2.1.13, at the rate of $ ______

Page 11 of ______ pages

per day or any portion thereof (but compensation for time spent in preparing to appear in any such litigation, arbitration or proceeding will be on the basis provided in paragraph 5.1.2.1). Compensation for ENGINEER's independent professional associates and consultants will be on the basis provided in paragraph 5.1.2.2.

5.1.3. *For Reimbursable Expenses.* In addition to payments provided for in paragraphs 5.1.1 and 5.1.2, OWNER shall pay ENGINEER the actual costs (except where specifically provided otherwise) of all Reimbursable Expenses incurred in connection with all Basic and Additional Services.

5.1.4. The terms "Salary Costs" and "Reimbursable Expenses" have the meanings assigned to them in paragraph 5.4.

5.2. Times of Payments.

5.2.1. ENGINEER shall submit monthly statements for Basic and Additional Services rendered and for Reimbursable Expenses incurred. OWNER shall make prompt monthly payments in response to ENGINEER's monthly statements.

5.3. Other Provisions Concerning Payments.

5.3.1. If OWNER fails to make any payment due ENGINEER for services and expenses within thirty days after receipt of ENGINEER's statement therefor, the amounts due ENGINEER will be increased at the rate of 1% per month from said thirtieth day, and in addition, ENGINEER may, after giving seven days' written notice to OWNER, suspend services under this Agreement until ENGINEER has been paid in full all amounts due for services, expenses and charges.

5.3.2. In the event of termination by OWNER under paragraph 7.1 upon the completion of any phase of the Basic Services, progress payments due ENGINEER for services rendered through such phase shall constitute total payment for such services. In the event of such termination by OWNER during any phase of the Basic Services, ENGINEER also will be reimbursed for the charges of independent professional associates and consultants employed by ENGINEER to render Basic Services, and paid for services rendered during that phase on the basis of ENGINEER's Salary Costs times a factor of ______ for services rendered during that phase to date of termination by ENGINEER's principals and employees engaged directly on the Project. In the event of any such termination, ENGINEER will be paid for all unpaid Additional Services and unpaid Reimbursable Expenses, plus all termination expenses. Termination expenses mean additional Reimbursable Expenses directly attributable to termination, which, if termination is at Owner's convenience, shall include an amount computed as a percentage of total compensation for Basic Services earned by ENGINEER to the date of termination, as follows:

20% if termination occurs after commencement of the Preliminary Design Phase but prior to commencement of the Final Design Phase; or

10% if termination occurs after commencement of the Final Design Phase.

5.3.3. Records of ENGINEER's Salary Costs pertinent to ENGINEER's compensation under this Agreement will be kept in accordance with generally accepted accounting practices. Copies will be made available to OWNER at cost on request prior to final payment for ENGINEER's services.

5.3.4. Whenever a factor is applied to Salary Costs in determining compensation payable to ENGINEER, that factor will be adjusted periodically and equitably to reflect changes in the various elements that comprise such factor. All such adjustments will be in accordance with generally accepted accounting practices as applied on a consistent basis by ENGINEER and consistent with ENGINEER's overall compensation practices and procedures.

5.4. Definitions.

5.4.1. The Salary Costs used as a basis for payment mean salaries and wages (basic and incentive) paid to all ENGINEER's personnel engaged directly on the Project, including, but not limited to, engineers, architects, surveyors, designers, draftsmen, specification writers, estimators, other technical and business personnel; plus the cost of customary and statutory benefits including, but not limited to, social security contributions, unemployment, excise and payroll taxes, workers' compensation, health and retirement benefits, sick leave, vacation and holiday pay and other group benefits. For the purposes of this Agreement, the principals of ENGINEER and their current hourly Salary Costs are:

____

____

The hourly Salary Costs of principals of ENGINEER will be adjusted equitably to reflect changes in personnel and in ENGINEER's overall compensation procedures and practices.

The amount of customary and statutory benefits of all other personnel of ENGINEER will be considered equal to ______ % of salaries and wages, subject to equitable adjustment to reflect changes in ENGINEER's overall compensation procedures and practices.

5.4.2. Reimbursable Expenses mean the actual expenses incurred by ENGINEER or ENGINEER's independent professional associates or consultants directly or indirectly in connection with the Project, such as expenses for: transportation and subsistence incidental thereto; obtaining bids or proposals from Contractor(s); providing and maintaining field office facilities including furnishings and utilities; subsistence and transportation of Resident Project Representatives and their assistants; toll telephone calls and telegrams; reproduction of reports, Drawings, Specifications, Bidding Documents and similar Project-related items in addition to those required under Section 1; and if authorized in advance by OWNER, overtime work requiring higher than regular rates. In addition, if authorized in advance by OWNER, Reimbursable Expenses will also include expenses incurred for computer time and other highly specialized equipment, including an appropriate charge for previously established programs and expenses of photographic production techniques times a factor of ______ .

Page 14 of _____ pages

SECTION 6—CONSTRUCTION COST AND OPINIONS OF COST

6.1. Construction Cost.

The construction cost of the entire Project (herein referred to as "Construction Cost") means the total cost to OWNER of those portions of the entire Project designed and specified by ENGINEER, but it will not include ENGINEER's compensation and expenses, the cost of land, rights-of-way, or compensation for or damages to, properties unless this Agreement so specifies, nor will it include OWNER's legal, accounting, insurance counseling or auditing services, or interest and financing charges incurred in connection with the Project or the cost of other services to be provided by others to OWNER pursuant to Section 3, as applicable. [Construction Cost is one of the items comprising Total Project Costs which is defined in paragraph 1.2.6.]

6.2. Opinions of Cost.

6.2.1. Since ENGINEER has no control over the cost of labor, materials, equipment or services furnished by others, or over the Contractor(s)' methods of determining prices, or over competitive bidding or market conditions, ENGINEER's opinions of probable Total Project Costs and Construction Cost provided for herein are to be made on the basis of ENGINEER's experience and qualifications and represent ENGINEER's best judgment as an experienced and qualified professional engineer, familiar with the construction industry; but ENGINEER cannot and does not guarantee that proposals, bids or actual Total Project or Construction Costs will not vary from opinions of probable cost prepared by ENGINEER. If prior to the Bidding or Negotiating Phase OWNER wishes greater assurance as to Total Project or Construction Costs, OWNER shall employ an independent cost estimator as provided in paragraph 3.9.

6.2.2. If a Construction Cost limit is established by written agreement between OWNER and ENGINEER and specifically set forth in this Agreement as a condition thereto, the following will apply:

6.2.2.1. The acceptance by OWNER at any time during the Basic Services of a revised opinion of probable Total Project or Construction Costs in excess of the then established cost limit will constitute a corresponding revision in the Construction Cost limit to the extent indicated in such revised opinion.

6.2.2.2. Any Construction Cost limit so established will include a contingency of ten percent unless another amount is agreed upon in writing.

6.2.2.3. ENGINEER will be permitted to determine what types of materials, equipment and component systems are to be included in the Drawings and Specifications and to make reasonable adjustments in the general scope, extent and character of the Project to bring it within the cost limit.

6.2.2.4. If the Bidding or Negotiating Phase has not commenced within six months after completion of the Final Design Phase, the established Construction Cost limit will not be binding on ENGINEER, and OWNER shall consent to an adjustment in such cost limit commensurate with any applicable change in the general level of prices in the construction industry between the date of completion of the Final Design Phase and the date on which proposals or bids are sought.

6.2.2.5. If the lowest bona fide proposal or bid exceeds the established Construction Cost limit, OWNER shall (1) give written approval to increase such cost limit, (2) authorize negotiating or rebidding the Project within a reasonable time, or (3) cooperate in revising the Project's general scope, extent or character to the extent consistent with the Project's requirements and with sound engineering practices. In the case of (3), ENGINEER shall modify the Contract Documents as necessary to bring the Construction Cost within the cost limit. In lieu of other compensation for services in making such modifications, OWNER shall pay ENGINEER, ENGINEER's cost of such services, all overhead expenses reasonably related thereto and Reimbursable Expenses, but without profit to ENGINEER on account of such services. The providing of such service will be the limit of ENGINEER's responsibility in this regard and, having done so, ENGINEER shall be entitled to payment for services in accordance with this Agreement and will not otherwise be liable for damages attributable to the lowest bona fide proposal or bid exceeding the established Construction Cost.

SECTION 7—GENERAL CONSIDERATION

7.1. Termination.

The obligation to provide further services under this Agreement may be terminated by either party upon thirty days' written notice in the event of substantial failure by the other party to perform in accordance with the terms hereof through no fault of the terminating party.

7.2. Reuse of Documents.

All documents including Drawings and Specifications prepared or furnished by ENGINEER (and ENGINEER's independent professional associates and consultants) pursuant to this Agreement are instruments of service in respect of the Project and ENGINEER shall retain an ownership and property interest therein whether or not the Project is completed. OWNER may make and retain copies for information and reference in connection with the use and occupancy of the Project by OWNER and others; however, such documents are not intended or represented to be suitable for reuse by OWNER or others on extensions of the Project or on any other project. Any reuse without written verification or adaptation by ENGINEER for the specific purpose intended will be at OWNER's sole risk and without liability or legal exposure to ENGINEER, or to ENGINEER's independent professional associates or consultants, and OWNER shall indemnify and hold harmless ENGINEER and ENGINEER's independent professional associates and consultants from all claims, damages, losses and expenses including attorneys' fees arising out of or resulting therefrom. Any such verification or adaptation will entitle ENGINEER to further compensation at rates to be agreed upon by OWNER and ENGINEER.

7.3. Insurance.

7.3.1. ENGINEER shall procure and maintain insurance for protection from claims under workers' compensation acts, claims for damages because of bodily injury including personal injury, sickness or disease or death of any and all employees or of any person other than such employees, and from claims or damages because of injury to or destruction of property including loss of use resulting therefrom.

7.4. Controlling Law.

This Agreement is to be governed by the law of the principal place of business of ENGINEER.

7.5. Successors and Assigns.

7.5.1. OWNER and ENGINEER each is hereby bound and the partners, successors, executors, administrators and legal representatives of OWNER and ENGINEER (and to the extent permitted by paragraph 7.5.2 the assigns of OWNER and ENGINEER) are hereby bound to the other party to this Agreement and to the partners, successors, executors, administrators and legal representatives (and said assigns) of such other party, in respect of all covenants, agreements and obligations of this Agreement.

7.5.2. Neither OWNER nor ENGINEER shall assign, sublet or transfer any rights under or interest in (including, but without limitation, moneys that may become due or moneys that are due) this Agreement without the written consent of the other, except to the extent that any assignment, subletting or transfer is mandated by law or the effect of this limitation may be restricted by law. Unless specifically stated to the contrary in any written consent to an assignment, no assignment will release or discharge the assignor from any duty or responsibility under this Agreement. Nothing contained in this paragraph shall prevent ENGINEER from employing such independent professional associates and consultants as ENGINEER may deem appropriate to assist in the performance of services hereunder.

7.5.3. Nothing under this Agreement shall be construed to give any rights or benefits in this Agreement to anyone other than OWNER and ENGINEER, and all duties and responsibilities undertaken pursuant to this Agreement will be for the sole and exclusive benefit of OWNER and ENGINEER and not for the benefit of any other party.

7.6. Arbitration.

7.6.1. All claims, counterclaims, disputes and other matters in question between the parties hereto arising out of or relating to this Agreement or the breach thereof will be decided by arbitration in accordance with the Construction Industry Arbitration Rules of the American Arbitration Association then obtaining, subject to the limitations and restrictions stated in paragraphs 7.6.3 and 7.6.4 below. This Agreement so to arbitrate and any other agreement or consent to arbitrate entered into in accordance herewith as provided in this paragraph 7.6 will be specifically enforceable under the prevailing law of any court having jurisdiction.

7.6.2. Notice of demand for arbitration must be filed in writing with the other parties to this Agreement and with the American Arbitration Association. The demand must be made within a reasonable time after the claim, dispute or other matter in question has arisen. In no event may the demand for arbitration be made after the date when institution of legal or equitable proceedings based on such claim, dispute or other matter in question would be barred by the applicable statute of limitations.

7.6.3. All demands for arbitration and all answering statements thereto which include any monetary claim must contain a statement that the total sum or value in controversy as alleged by the party making such demand or answering statement is not more than $200,000 (exclusive of interest and costs) and the arbitrators will not have jurisdiction, power or authority to render a monetary award in response thereto against any party which totals more than $200,000 (exclusive of interest and costs). The arbitrators will not have jurisdiction, power or authority to consider, or make findings (except in denial of their own jurisdiction) concerning any claim, counterclaim, dispute or other matter in question where the amount in controversy of any such claim, counterclaim, dispute or matter is more than $200,000 (exclusive of interest and costs).

7.6.4. No arbitration arising out of, or relating to, this Agreement may include, by consolidation, joinder or in any other manner, any person or entity who is not a party to this Agreement.

7.6.5. By written consent signed by all the parties to this Agreement and containing a specific reference hereto, the limitations and restrictions contained in paragraphs 7.6.3 and 7.6.4 may be waived in whole or in part as to any claim, counterclaim, dispute or other matter specifically described in such consent. No consent to arbitration in respect of a specifically described claim, counterclaim, dispute or other matter in question will constitute consent to arbitrate any other claim, counterclaim, dispute or other matter in question which is not specifically described in such consent or in which the sum or value in controversy exceeds $200,000 (exclusive of interest and costs) or which is with any party not specifically described therein.

7.6.6. The award rendered by the arbitrators will be final, judgment may be entered upon it in any court having jurisdiction thereof, and will not be subject to modification or appeal except to the extent permitted by Sections 10 and 11 of the Federal Arbitration Act (9 U.S.C. §§10, 11).

Page 17 of ____ pages

Page 18 of _____ pages

SECTION 8—SPECIAL PROVISIONS, EXHIBITS and SCHEDULES.

8.1. This Agreement is subject to the following special provisions.

8.1.1.

8.2. The following Exhibits are attached to and made a part of this Agreement:

8.2.1. Exhibit A "Further Description of Basic Engineering Services and Related Matters" consisting of ______ pages.

8.2.2. Exhibit B "Duties, Responsibilities and Limitations of Authority of Resident Project Representative" consisting of ________ pages.

8.2.3.

8.3. This Agreement (consisting of pages 1 to ________ , inclusive) together with the Exhibits and schedules identified above constitute the entire agreement between OWNER and ENGINEER and supersede all prior written or oral understandings. This Agreement and said Exhibits and schedules may only be amended, supplemented, modified or cancelled by a duly executed written instrument.

IN WITNESS WHEREOF, the parties hereto have made and executed this Agreement as of the day and year first above written.

OWNER:	ENGINEER:
______________________________	______________________________
______________________________	______________________________
______________________________	______________________________
Address for giving notices:	Address for giving notices:
______________________________	______________________________
______________________________	______________________________

Page ___ of _____ pages

This document has important legal consequences: consultation with an attorney is encouraged with respect to its completion or modification.

STANDARD GENERAL CONDITIONS OF THE CONSTRUCTION CONTRACT

Prepared by

Engineers Joint Contract Documents Committee

and

Issued and Published Jointly By

PROFESSIONAL ENGINEERS IN PRIVATE PRACTICE
A practice division of the
NATIONAL SOCIETY OF PROFESSIONAL ENGINEERS

AMERICAN CONSULTING ENGINEERS COUNCIL

AMERICAN SOCIETY OF CIVIL ENGINEERS

CONSTRUCTION SPECIFICATIONS INSTITUTE

This document has been approved and endorsed by

The Associated General Contractors of America

These General Conditions have been prepared for use with the Owner-Contractor Agreements (No. 1910-8-A-1 or 1910-8-A-2, 1983 editions). Their provisions are interrelated and a change in one may necessitate a change in the others. Comments concerning their usage are contained in the Commentary on Agreements for Engineering Services and Contract Documents, No. 1910-9, 1981 edition. For guidance in the preparation of Supplementary Conditions, see Guide to the Preparation of Supplementary Conditions (No. 1910-17, 1983 edition). When bidding is involved, the Standard Form of Instructions to Bidders (No. 1910-12, 1983 edition) may be used.

EJCDC No. 1910-8 (1983 Edition)
Reprinted 11/88

American Consulting Engineers Council
1015 15th Street, N.W., Washington, D.C. 20005

American Society of Civil Engineers
345 East 47th Street, New York, NY 10017

Construction Specifications Institute
601 Madison St., Alexandria, VA 22314

TABLE OF CONTENTS OF GENERAL CONDITIONS

INDEX TO GENERAL CONDITIONS

Article or Paragraph Number

GENERAL CONDITIONS

ARTICLE 1—DEFINITIONS

Wherever used in these General Conditions or in the other Contract Documents the following terms have the meanings indicated which are applicable to both the singular and plural thereof:

Addenda—Written or graphic instruments issued prior to the opening of Bids which clarify, correct or change the bidding documents or the Contract Documents.

Agreement—The written agreement between OWNER and CONTRACTOR covering the Work to be performed; other Contract Documents are attached to the Agreement and made a part thereof as provided therein.

Application for Payment—The form accepted by ENGINEER which is to be used by CONTRACTOR in requesting progress or final payments and which is to include such supporting documentation as is required by the Contract Documents.

Bid—The offer or proposal of the bidder submitted on the prescribed form setting forth the prices for the Work to be performed.

Bonds—Bid, performance and payment bonds and other instruments of security.

Change Order—A document recommended by ENGINEER, which is signed by CONTRACTOR and OWNER and authorizes an addition, deletion or revision in the Work, or an adjustment in the Contract Price or the Contract Time, issued on or after the Effective Date of the Agreement.

Contract Documents—The Agreement, Addenda (which pertain to the Contract Documents), CONTRACTOR's Bid (including documentation accompanying the Bid and any post-Bid documentation submitted prior to the Notice of Award) when attached as an exhibit to the Agreement, the Bonds, these General Conditions, the Supplementary Conditions, the Specifications and the Drawings as the same are more specifically identified in the Agreement, together with all amendments, modifications and supplements issued pursuant to paragraphs 3.4 and 3.5 on or after the Effective Date of the Agreement.

Contract Price—The moneys payable by OWNER to CONTRACTOR under the Contract Documents as stated in the Agreement (subject to the provisions of paragraph 11.9.1 in the case of Unit Price Work).

Contract Time—The number of days (computed as provided in paragraph 17.2) or the date stated in the Agreement for the completion of the Work.

CONTRACTOR—The person, firm or corporation with whom OWNER has entered into the Agreement.

defective—An adjective which when modifying the word Work refers to Work that is unsatisfactory, faulty or deficient, or does not conform to the Contract Documents, or does not meet the requirements of any inspection, reference standard, test or approval referred to in the Contract Documents, or has been damaged prior to ENGINEER's recommendation of final payment (unless responsibility for the protection thereof has been assumed by OWNER at Substantial Completion in accordance with paragraph 14.8 or 14.10).

Drawings—The drawings which show the character and scope of the Work to be performed and which have been prepared or approved by ENGINEER and are referred to in the Contract Documents.

Effective Date of the Agreement—The date indicated in the Agreement on which it becomes effective, but if no such date is indicated it means the date on which the Agreement is signed and delivered by the last of the two parties to sign and deliver.

ENGINEER—The person, firm or corporation named as such in the Agreement.

Field Order—A written order issued by ENGINEER which orders minor changes in the Work in accordance with paragraph 9.5 but which does not involve a change in the Contract Price or the Contract Time.

General Requirements—Sections of Division 1 of the Specifications.

Laws and Regulations; Laws or Regulations—Laws, rules, regulations, ordinances, codes and/or orders.

Notice of Award—The written notice by OWNER to the apparent successful bidder stating that upon compliance by the apparent successful bidder with the conditions precedent enumerated therein, within the time specified, OWNER will sign and deliver the Agreement.

Notice to Proceed—A written notice given by OWNER to CONTRACTOR (with a copy to ENGINEER) fixing the date on which the Contract Time will commence to run and on which CONTRACTOR shall start to perform CONTRACTOR'S obligations under the Contract Documents.

OWNER—The public body or authority, corporation, association, firm or person with whom CONTRACTOR has entered into the Agreement and for whom the Work is to be provided.

Partial Utilization—Placing a portion of the Work in service for the purpose for which it is intended (or a related purpose) before reaching Substantial Completion for all the Work.

Project—The total construction of which the Work to be provided under the Contract Documents may be the whole, or a part as indicated elsewhere in the Contract Documents.

Resident Project Representative—The authorized representative of ENGINEER who is assigned to the site or any part thereof.

Shop Drawings—All drawings, diagrams, illustrations, schedules and other data which are specifically prepared by or for CONTRACTOR to illustrate some portion of the Work and all illustrations, brochures, standard schedules, performance charts, instructions, diagrams and other information prepared by a Supplier and submitted by CONTRACTOR to illustrate material or equipment for some portion of the Work.

Specifications—Those portions of the Contract Documents consisting of written technical descriptions of materials, equipment, construction systems, standards and workmanship as applied to the Work and certain administrative details applicable thereto.

Subcontractor—An individual, firm or corporation having a direct contract with CONTRACTOR or with any other Subcontractor for the performance of a part of the Work at the site.

Substantial Completion—The Work (or a specified part thereof) has progressed to the point where, in the opinion of ENGINEER as evidenced by ENGINEER's definitive certificate of Substantial Completion, it is sufficiently complete, in accordance with the Contract Documents, so that the Work (or specified part) can be utilized for the purposes for which it is intended; or if there be no such certificate issued, when final payment is due in accordance with paragraph 14.13. The terms "substantially complete" and "substantially completed" as applied to any Work refer to Substantial Completion thereof.

Supplementary Conditions—The part of the Contract Documents which amends or supplements these General Conditions.

Supplier—A manufacturer, fabricator, supplier, distributor, materialman or vendor.

Underground Facilities—All pipelines, conduits, ducts, cables, wires, manholes, vaults, tanks, tunnels or other such facilities or attachments, and any encasements containing such facilities which have been installed underground to furnish any of the following services or materials: electricity, gases, steam, liquid petroleum products, telephone or other communications, cable television, sewage and drainage removal, traffic or other control systems or water.

Unit Price Work—Work to be paid for on the basis of unit prices.

Work—The entire completed construction or the various separately identifiable parts thereof required to be furnished under the Contract Documents. Work is the result of performing services, furnishing labor and furnishing and incorporating materials and equipment into the construction, all as required by the Contract Documents.

Work Directive Change—A written directive to CONTRACTOR, issued on or after the Effective Date of the Agreement and signed by OWNER and recommended by ENGINEER, ordering an addition, deletion or revision in the Work, or responding to differing or unforeseen physical conditions under which the Work is to be performed as provided in paragraph 4.2 or 4.3 or to emergencies under paragraph 6.22. A Work Directive Change may not change the Contract Price or the Contract Time, but is evidence that the parties expect that the change directed or documented by a Work Directive Change will be incorporated in a subsequently issued Change Order following negotiations by the parties as to its effect, if any, on the Contract Price or Contract Time as provided in paragraph 10.2.

Written Amendment—A written amendment of the Contract Documents, signed by OWNER and CONTRACTOR on or after the Effective Date of the Agreement and normally dealing with the nonengineering or nontechnical rather than strictly Work-related aspects of the Contract Documents.

ARTICLE 2—PRELIMINARY MATTERS

Delivery of Bonds:

2.1. When CONTRACTOR delivers the executed Agreements to OWNER, CONTRACTOR shall also deliver to OWNER such Bonds as CONTRACTOR may be required to furnish in accordance with paragraph 5.1.

Copies of Documents:

2.2. OWNER shall furnish to CONTRACTOR up to ten copies (unless otherwise specified in the Supplementary Conditions) of the Contract Documents as are reasonably necessary for the execution of the Work. Additional copies will be furnished, upon request, at the cost of reproduction.

Commencement of Contract Time; Notice to Proceed:

2.3. The Contract Time will commence to run on the thirtieth day after the Effective Date of the Agreement, or, if a Notice to Proceed is given, on the day indicated in the Notice to Proceed. A Notice to Proceed may be given at any time within thirty days after the Effective Date of the Agreement. In no event will the Contract Time commence to run later than the seventy-fifth day after the day of Bid opening or the thirtieth day after the Effective Date of the Agreement, whichever date is earlier.

Starting the Project:

2.4. CONTRACTOR shall start to perform the Work on the date when the Contract Time commences to run, but no Work shall be done at the site prior to the date on which the Contract Time commences to run.

Before Starting Construction:

2.5. Before undertaking each part of the Work, CONTRACTOR shall carefully study and compare the Contract Documents and check and verify pertinent figures shown thereon and all applicable field measurements. CONTRACTOR shall promptly report in writing to ENGINEER any conflict, error or discrepancy which CONTRACTOR may discover and shall obtain a written interpretation or clarification from ENGINEER before proceeding with any Work affected thereby; however, CONTRACTOR shall not be liable to OWNER or ENGINEER for failure to report any conflict, error or discrepancy in the Contract Documents, unless CONTRACTOR had actual knowledge thereof or should reasonably have known thereof.

2.6. Within ten days after the Effective Date of the Agreement (unless otherwise specified in the General Requirements), CONTRACTOR shall submit to ENGINEER for review:

2.6.1. an estimated progress schedule indicating the starting and completion dates of the various stages of the Work;

2.6.2. a preliminary schedule of Shop Drawing submissions; and

2.6.3. a preliminary schedule of values for all of the Work which will include quantities and prices of items aggregating the Contract Price and will subdivide the Work into component parts in sufficient detail to serve as the basis for progress payments during construction. Such prices will include an appropriate amount of overhead and profit applicable to each item of Work which will be confirmed in writing by CONTRACTOR at the time of submission.

2.7. Before any Work at the site is started, CONTRACTOR shall deliver to OWNER, with a copy to ENGINEER, certificates (and other evidence of insurance requested by OWNER) which CONTRACTOR is required to purchase and maintain in accordance with paragraphs 5.3 and 5.4, and OWNER shall deliver to CONTRACTOR certificates (and other evidence of insurance requested by CONTRACTOR) which OWNER is required to purchase and maintain in accordance with paragraphs 5.6 and 5.7.

Preconstruction Conference:

2.8. Within twenty days after the Effective Date of the Agreement, but before CONTRACTOR starts the Work at the site, a conference attended by CONTRACTOR, ENGINEER and others as appropriate will be held to discuss the schedules referred to in paragraph 2.6, to discuss procedures for handling Shop Drawings and other submittals and for processing Applications for Payment, and to establish a working understanding among the parties as to the Work.

Finalizing Schedules:

2.9. At least ten days before submission of the first Application for Payment a conference attended by CONTRACTOR, ENGINEER and others as appropriate will be held to finalize the schedules submitted in accordance with paragraph 2.6. The finalized progress schedule will be acceptable to ENGINEER as providing an orderly progression of the Work to completion within the Contract Time, but such acceptance will neither impose on ENGINEER responsibility for the progress or scheduling of the Work nor relieve CONTRACTOR from full responsibility therefor. The finalized schedule of Shop Drawing submissions will be acceptable to ENGINEER as providing a workable arrangement for processing the submissions. The finalized schedule of values will be acceptable to ENGINEER as to form and substance.

ARTICLE 3—CONTRACT DOCUMENTS: INTENT, AMENDING, REUSE

Intent:

3.1. The Contract Documents comprise the entire agreement between OWNER and CONTRACTOR concerning the Work. The Contract Documents are complementary; what is called for by one is as binding as if called for by all. The Contract Documents will be construed in accordance with the law of the place of the Project.

3.2. It is the intent of the Contract Documents to describe a functionally complete Project (or part thereof) to be constructed in accordance with the Contract Documents. Any Work, materials or equipment that may reasonably be inferred from the Contract Documents as being required to produce the intended result will be supplied whether or not specifically called for. When words which have a well-known technical or trade meaning are used to describe Work, materials or equipment such words shall be interpreted in accordance with that meaning. Reference to standard specifications, manuals or codes of any technical society, organization or association, or to the Laws or Regulations of any governmental authority, whether such reference be specific or by implication, shall mean the latest standard specification, manual, code or Laws or Regulations in effect at the time of opening of Bids (or, on the Effective Date of the Agreement if there were no Bids), except as may be otherwise specifically stated. However, no provision of any referenced standard specification, manual or code (whether or not specifically incorporated by reference in the Contract Documents) shall be effective to change the duties and responsibilities of OWNER, CONTRACTOR or ENGINEER, or any of their consultants, agents or employees from those set forth in the Contract Documents, nor shall it be effective to assign to ENGINEER, or any of ENGINEER's consultants, agents or employees, any duty or authority to supervise or direct the furnishing or performance of the Work or any duty or authority to undertake responsibility contrary to the provisions of paragraph 9.15 or 9.16. Clarifications and interpretations of the Contract Documents shall be issued by ENGINEER as provided in paragraph 9.4.

3.3. If, during the performance of the Work, CONTRACTOR finds a conflict, error or discrepancy in the Contract Documents, CONTRACTOR shall so report to ENGINEER in writing at once and before proceeding with the Work affected thereby shall obtain a written interpretation or clarification from ENGINEER; however, CONTRACTOR shall not be liable to OWNER or ENGINEER for failure to report any conflict, error or discrepancy in the Contract Documents unless CONTRACTOR had actual knowledge thereof or should reasonably have known thereof.

Amending and Supplementing Contract Documents:

3.4. The Contract Documents may be amended to provide for additions, deletions and revisions in the Work or to modify the terms and conditions thereof in one or more of the following ways:

3.4.1. a formal Written Amendment,

3.4.2. a Change Order (pursuant to paragraph 10.4), or

3.4.3. a Work Directive Change (pursuant to paragraph 10.1).

As indicated in paragraphs 11.2 and 12.1, Contract Price and Contract Time may only be changed by a Change Order or a Written Amendment.

3.5. In addition, the requirements of the Contract Documents may be supplemented, and minor variations and deviations in the Work may be authorized, in one or more of the following ways:

3.5.1. a Field Order (pursuant to paragraph 9.5),

3.5.2. ENGINEER's approval of a Shop Drawing or sample (pursuant to paragraphs 6.26 and 6.27), or

3.5.3. ENGINEER's written interpretation or clarification (pursuant to paragraph 9.4).

Reuse of Documents:

3.6. Neither CONTRACTOR nor any Subcontractor or Supplier or other person or organization performing or furnishing any of the Work under a direct or indirect contract with OWNER shall have or acquire any title to or ownership rights in any of the Drawings, Specifications or other documents (or copies of any thereof) prepared by or bearing the seal of ENGINEER; and they shall not reuse any of them on extensions of the Project or any other project without written consent of OWNER and ENGINEER and specific written verification or adaptation by ENGINEER.

ARTICLE 4—AVAILABILITY OF LANDS; PHYSICAL CONDITIONS; REFERENCE POINTS

Availability of Lands:

4.1. OWNER shall furnish, as indicated in the Contract Documents, the lands upon which the Work is to be performed, rights-of-way and easements for access thereto, and such other lands which are designated for the use of CONTRACTOR. Easements for permanent structures or permanent changes in existing facilities will be obtained and paid for by OWNER, unless otherwise provided in the Contract Documents. If CONTRACTOR believes that any delay in OWNER's furnishing these lands, rights-of-way or easements entitles CONTRACTOR to an extension of the Contract Time, CONTRACTOR may make a claim therefor as provided in Article 12. CONTRACTOR shall provide for all additional lands and access thereto that may be required for temporary construction facilities or storage of materials and equipment.

Physical Conditions:

4.2.1. *Explorations and Reports:* Reference is made to the Supplementary Conditions for identification of those reports of explorations and tests of subsurface conditions at the site that have been utilized by ENGINEER in preparation of the Contract Documents. CONTRACTOR may rely upon the accuracy of the technical data contained in such reports, but not upon nontechnical data, interpretations or opinions contained therein or for the completeness thereof for CONTRACTOR's purposes. Except as indicated in the immediately preceding sentence and in paragraph 4.2.6, CONTRACTOR shall have full responsibility with respect to subsurface conditions at the site.

4.2.2. *Existing Structures:* Reference is made to the Supplementary Conditions for identification of those drawings of physical conditions in or relating to existing surface and subsurface structures (except Underground Facilities referred to in paragraph 4.3) which are at or contiguous to the site that have been utilized by ENGINEER in preparation of the Contract Documents. CONTRACTOR may rely upon the accuracy of the technical data contained in such drawings, but not for the completeness thereof for CONTRACTOR's purposes. Except as indicated in the immediately preceding sentence and in paragraph 4.2.6, CONTRACTOR shall have full responsibility with respect to physical conditions in or relating to such structures.

4.2.3. *Report of Differing Conditions:* If CONTRACTOR believes that:

4.2.3.1. any technical data on which CONTRACTOR is entitled to rely as provided in paragraphs 4.2.1 and 4.2.2 is inaccurate, or

4.2.3.2. any physical condition uncovered or revealed at the site differs materially from that indicated, reflected or referred to in the Contract Documents,

CONTRACTOR shall, promptly after becoming aware thereof and before performing any Work in connection therewith (except in an emergency as permitted by paragraph 6.22), notify OWNER and ENGINEER in writing about the inaccuracy or difference.

4.2.4. *ENGINEER's Review:* ENGINEER will promptly review the pertinent conditions, determine the necessity of obtaining additional explorations or tests with respect thereto and advise OWNER in writing (with a copy to CONTRACTOR) of ENGINEER's findings and conclusions.

4.2.5. *Possible Document Change:* If ENGINEER concludes that there is a material error in the Contract Documents or that because of newly discovered conditions a change in the Contract Documents is required, a Work Directive Change or a Change Order will be issued as provided in Article 10 to reflect and document the consequences of the inaccuracy or difference.

4.2.6. *Possible Price and Time Adjustments:* In each such case, an increase or decrease in the Contract Price or an extension or shortening of the Contract Time, or any combination thereof, will be allowable to the extent that they are attributable to any such inaccuracy or difference. If OWNER and CONTRACTOR are unable to agree as to the amount or length thereof, a claim may be made therefor as provided in Articles 11 and 12.

Physical Conditions—Underground Facilities:

4.3.1. *Shown or Indicated:* The information and data shown or indicated in the Contract Documents with respect to existing Underground Facilities at or contiguous to the site is based on information and data furnished to OWNER or ENGINEER by the owners of such Underground Facilities or by others. Unless it is otherwise expressly provided in the Supplementary Conditions:

4.3.1.1. OWNER and ENGINEER shall not be responsible for the accuracy or completeness of any such information or data; and,

4.3.1.2. CONTRACTOR shall have full responsibility for reviewing and checking all such information and data, for locating all Underground Facilities shown or indicated in the Contract Documents, for coordination of the Work with the owners of such Underground Facilities during construction, for the safety and protection thereof as provided in paragraph 6.20 and repairing any damage thereto resulting from the Work, the cost of all of which will be considered as having been included in the Contract Price.

4.3.2. *Not Shown or Indicated.* If an Underground Facility is uncovered or revealed at or contiguous to the site which was not shown or indicated in the Contract Documents and which CONTRACTOR could not reasonably have been expected to be aware of, CONTRACTOR shall, promptly after becoming aware thereof and before performing any Work affected thereby (except in an emergency as permitted by paragraph 6.22), identify the owner of such Underground Facility and give written notice thereof to that owner and to OWNER and ENGINEER. ENGINEER will promptly review the Underground Facility to determine the extent to which the Contract Documents should be modified to reflect and document the consequences of the existence of the Underground Facility, and the Contract Documents will be amended or supplemented to the extent necessary. During such time, CONTRACTOR shall be responsible for the safety and protection of such Underground Facility as provided in paragraph 6.20. CONTRACTOR shall be allowed an increase in the Contract Price or an extension of the Contract Time, or both, to the extent that they are attributable to the existence of any Underground Facility that was not shown or indicated in the Contract Documents and which CONTRACTOR could not reasonably have been expected to be aware of. If the parties are unable to agree as to the amount or length thereof, CONTRACTOR may make a claim therefor as provided in Articles 11 and 12.

Reference Points:

4.4. OWNER shall provide engineering surveys to establish reference points for construction which in ENGINEER's judgment are necessary to enable CONTRACTOR to proceed with the Work. CONTRACTOR shall be responsible for laying out the Work (unless otherwise specified in the General Requirements), shall protect and preserve the established reference points and shall make no changes or relocations without the prior written approval of OWNER. CONTRACTOR shall report to ENGINEER whenever any reference point is lost or destroyed or requires relocation because of necessary changes in grades or locations, and shall be responsible for the accurate replacement or relocation of such reference points by professionally qualified personnel.

ARTICLE 5—BONDS AND INSURANCE

Performance and Other Bonds:

5.1. CONTRACTOR shall furnish performance and payment Bonds, each in an amount at least equal to the Contract Price as security for the faithful performance and payment of all CONTRACTOR's obligations under the Contract Documents. These Bonds shall remain in effect at least until one year after the date when final payment becomes due, except as otherwise provided by Law or Regulation or by the Contract Documents. CONTRACTOR shall also furnish such other Bonds as are required by the Supplementary Conditions. All Bonds shall be in the forms prescribed by Law or Regulation or by the Contract Documents and be executed by such sureties as are named in the current list of "Companies Holding Certificates of Authority as Acceptable Sureties on Federal Bonds and as Acceptable Reinsuring Companies" as published in Circular 570 (amended) by the Audit Staff Bureau of Accounts, U.S. Treasury Department. All Bonds signed by an agent must be accompanied by a certified copy of the authority to act.

5.2. If the surety on any Bond furnished by CONTRACTOR is declared a bankrupt or becomes insolvent or its right to do business is terminated in any state where any part of the Project is located or it ceases to meet the requirements of paragraph 5.1, CONTRACTOR shall within five days thereafter substitute another Bond and Surety, both of which must be acceptable to OWNER.

Contractor's Liability Insurance:

5.3. CONTRACTOR shall purchase and maintain such comprehensive general liability and other insurance as is appropriate for the Work being performed and furnished and as will provide protection from claims set forth below which may arise out of or result from CONTRACTOR's performance and furnishing of the Work and CONTRACTOR's other obligations under the Contract Documents, whether it is to be performed or furnished by CONTRACTOR, by any Subcontractor, by anyone directly or indirectly employed by any of them to perform or furnish any of the Work, or by anyone for whose acts any of them may be liable:

5.3.1. Claims under workers' or workmen's compensation, disability benefits and other similar employee benefit acts;

5.3.2. Claims for damages because of bodily injury, occupational sickness or disease, or death of CONTRACTOR's employees;

5.3.3. Claims for damages because of bodily injury, sickness or disease, or death of any person other than CONTRACTOR's employees;

5.3.4. Claims for damages insured by personal injury liability coverage which are sustained (a) by any person as a result of an offense directly or indirectly related to the employment of such person by CONTRACTOR, or (b) by any other person for any other reason;

5.3.5. Claims for damages, other than to the Work itself, because of injury to or destruction of tangible property wherever located, including loss of use resulting therefrom;

5.3.6. Claims arising out of operation of Laws or Regulations for damages because of bodily injury or death of any person or for damage to property; and

5.3.7. Claims for damages because of bodily injury or death of any person or property damage arising out of the ownership, maintenance or use of any motor vehicle.

The insurance required by this paragraph 5.3 shall include the specific coverages and be written for not less than the limits of liability and coverages provided in the Supplementary Conditions, or required by law, whichever is greater. The comprehensive general liability insurance shall include completed operations insurance. All of the policies of insurance so required to be purchased and maintained (or the certificates or other evidence thereof) shall contain a provision or endorsement that the coverage afforded will not be cancelled, materially changed or renewal refused until at least thirty days' prior written notice has been given to OWNER and ENGINEER by certified mail. All such insurance shall remain in effect until final payment and at all times thereafter when CONTRACTOR may be correcting, removing or replacing *defective* Work in accordance with paragraph 13.12. In addition, CONTRACTOR shall maintain such completed operations insurance for at least two years after final payment and furnish OWNER with evidence of continuation of such insurance at final payment and one year thereafter.

Contractual Liability Insurance:

5.4. The comprehensive general liability insurance required by paragraph 5.3 will include contractual liability insurance applicable to CONTRACTOR's obligations under paragraphs 6.30 and 6.31.

Owner's Liability Insurance:

5.5. OWNER shall be responsible for purchasing and maintaining OWNER's own liability insurance and, at OWNER's option, may purchase and maintain such insurance as will protect OWNER against claims which may arise from operations under the Contract Documents.

Property Insurance:

5.6. Unless otherwise provided in the Supplementary Conditions, OWNER shall purchase and maintain property insurance upon the Work at the site to the full insurable value thereof (subject to such deductible amounts as may be provided in the Supplementary Conditions or required by Laws and Regulations). This insurance shall include the interests of OWNER, CONTRACTOR, Subcontractors, ENGINEER and ENGINEER's consultants in the Work, all of whom shall be listed as insureds or additional insured parties, shall insure against the perils of fire and extended coverage and shall include "all risk" insurance for physical loss and damage including theft, vandalism and malicious mischief, collapse and water damage. and such other perils as may be provided in the Supplementary Conditions, and shall include damages, losses and expenses arising out of or resulting from any insured loss or incurred in the repair or replacement of any insured property (including but not limited to fees and charges of engineers, architects, attorneys and other professionals). If not covered under the "all risk" insurance or otherwise provided in the Supplementary Conditions, CONTRACTOR shall purchase and maintain similar property insurance on portions of the Work stored on and off the site or in transit when such portions of the Work are to be included in an Application for Payment.

5.7. OWNER shall purchase and maintain such boiler and machinery insurance or additional property insurance as may be required by the Supplementary Conditions or Laws and Regulations which will include the interests of OWNER, CONTRACTOR, Subcontractors, ENGINEER AND ENGINEER's consultants in the Work, all of whom shall be listed as insured or additional insured parties.

5.8. All the policies of insurance (or the certificates or other evidence thereof) required to be purchased and maintained by OWNER in accordance with paragraphs 5.6 and 5.7 will contain a provision or endorsement that the coverage afforded will not be cancelled or materially changed or renewal refused until at least thirty days' prior written notice has been given to CONTRACTOR by certified mail and will contain waiver provisions in accordance with paragraph 5.11.2.

5.9. OWNER shall not be responsible for purchasing and maintaining any property insurance to protect the interests of CONTRACTOR, Subcontractors or others in the Work to the extent of any deductible amounts that are provided in the Supplementary Conditions. The risk of loss within the deductible amount, will be borne by CONTRACTOR, Subcontractor or others suffering any such loss and if any of them wishes property insurance coverage within the limits of such amounts, each may purchase and maintain it at the purchaser's own expense.

5.10. If CONTRACTOR requests in writing that other special insurance be included in the property insurance policy, OWNER shall, if possible, include such insurance, and the cost thereof will be charged to CONTRACTOR by appropriate Change Order or Written Amendment. Prior to commencement of the Work at the site, OWNER shall in writing advise CONTRACTOR whether or not such other insurance has been procured by OWNER.

Waiver of Rights:

5.11.1. OWNER and CONTRACTOR waive all rights against each other for all losses and damages caused by any of the perils covered by the policies of insurance provided in response to paragraphs 5.6 and 5.7 and any other property insurance applicable to the Work, and also waive all such rights against the Subcontractors, ENGINEER, ENGINEER's consultants and all other parties named as insureds in such policies for losses and damages so caused. As required by paragraph 6.11, each subcontract between CONTRACTOR and a Subcontractor will contain similar waiver provisions by the Subcontractor in favor of OWNER, CONTRACTOR, ENGINEER, ENGINEER's consultants and all other parties named as insureds. None of the above waivers shall extend to the rights that any of the insured parties may have to the proceeds of insurance held by OWNER as trustee or otherwise payable under any policy so issued.

5.11.2. OWNER and CONTRACTOR intend that any policies provided in response to paragraphs 5.6 and 5.7 shall protect all of the parties insured and provide primary coverage for all losses and damages caused by the perils covered thereby. Accordingly, all such policies shall contain provisions to the effect that in the event of payment of any loss or damage the insurer will have no rights of recovery against any of the parties named as insureds or additional insureds, and if the insurers require separate waiver forms to be signed by ENGINEER or ENGINEER's consultant OWNER will obtain the same, and if such waiver forms are required of any Subcontractor, CONTRACTOR will obtain the same.

Receipt and Application of Proceeds:

5.12. Any insured loss under the policies of insurance required by paragraphs 5.6 and 5.7 will be adjusted with OWNER and made payable to OWNER as trustee for the insureds, as their interests may appear, subject to the requirements of any applicable mortgage clause and of paragraph 5.13. OWNER shall deposit in a separate account any money so received, and shall distribute it in accordance with such agreement as the parties in interest may reach. If no other special agreement is reached the damaged Work shall be repaired or replaced, the moneys so received applied on account thereof and the Work and the cost thereof covered by an appropriate Change Order or Written Amendment.

5.13. OWNER as trustee shall have power to adjust and settle any loss with the insurers unless one of the parties in interest shall object in writing within fifteen days after the occurrence of loss to OWNER's exercise of this power. If such objection be made, OWNER as trustee shall make settlement with the insurers in accordance with such agreement as the parties in interest may reach. If required in writing by any party in interest, OWNER as trustee shall, upon the occurrence of an insured loss, give bond for the proper performance of such duties.

Acceptance of Insurance:

5.14. If OWNER has any objection to the coverage afforded by or other provisions of the insurance required to be purchased and maintained by CONTRACTOR in accordance with paragraphs 5.3 and 5.4 on the basis of its not complying with the Contract Documents, OWNER shall notify CONTRACTOR in writing thereof within ten days of the date of delivery of such certificates to OWNER in accordance with paragraph 2.7. If CONTRACTOR has any objection to the coverage afforded by or other provisions of the policies of insurance required to be purchased and maintained by OWNER in accordance with paragraphs 5.6 and 5.7 on the basis of their not complying with the Contract Documents, CONTRACTOR shall notify OWNER in writing thereof within ten days of the date of delivery of such certificates to CONTRACTOR in accordance with paragraph 2.7. OWNER and CONTRACTOR shall each provide to the other such additional information in respect of insurance provided by each as the other may reasonably request. Failure by OWNER or CONTRACTOR to give any such notice of objection within the time provided shall constitute acceptance of such insurance purchased by the other as complying with the Contract Documents.

Partial Utilization—Property Insurance:

5.15. If OWNER finds it necessary to occupy or use a portion or portions of the Work prior to Substantial Completion of all the Work, such use or occupancy may be accomplished in accordance with paragraph 14.10; provided that no such use or occupancy shall commence before the insurers providing the property insurance have acknowledged notice thereof and in writing effected the changes in coverage necessitated thereby. The insurers providing the property insurance shall consent by endorsement on the policy or policies, but the property insurance shall not be cancelled or lapse on account of any such partial use or occupancy.

ARTICLE 6—CONTRACTOR'S RESPONSIBILITIES

Supervision and Superintendence:

6.1. CONTRACTOR shall supervise and direct the Work competently and efficiently, devoting such attention thereto and applying such skills and expertise as may be necessary to perform the Work in accordance with the Contract Documents. CONTRACTOR shall be solely responsible for the means, methods, techniques, sequences and procedures of construction, but CONTRACTOR shall not be responsible for the negligence of others in the design or selection of a specific means, method, technique, sequence or procedure of construction which is indicated in and required by the Contract Documents. CONTRACTOR shall be responsible to see that the finished Work complies accurately with the Contract Documents.

6.2. CONTRACTOR shall keep on the Work at all times during its progress a competent resident superintendent, who shall not be replaced without written notice to OWNER and ENGINEER except under extraordinary circumstances. The superintendent will be CONTRACTOR's representative at the site and shall have authority to act on behalf of CONTRACTOR. All communications given to the superintendent shall be as binding as if given to CONTRACTOR.

Labor, Materials and Equipment:

6.3. CONTRACTOR shall provide competent, suitably qualified personnel to survey and lay out the Work and perform construction as required by the Contract Documents. CONTRACTOR shall at all times maintain good discipline and order at the site. Except in connection with the safety or protection of persons or the Work or property at the site or adjacent thereto, and except as otherwise indicated in the Contract Documents, all Work at the site shall be performed during regular working hours, and CONTRACTOR will not permit overtime work or the performance of Work on Saturday, Sunday or any legal holiday without OWNER's written consent given after prior written notice to ENGINEER.

6.4. Unless otherwise specified in the General Requirements, CONTRACTOR shall furnish and assume full responsibility for all materials, equipment, labor, transportation, construction equipment and machinery, tools, appliances, fuel, power, light, heat, telephone, water, sanitary facilities, temporary facilities and all other facilities and incidentals necessary for the furnishing, performance, testing, start-up and completion of the Work.

6.5. All materials and equipment shall be of good quality and new, except as otherwise provided in the Contract Documents. If required by ENGINEER, CONTRACTOR shall furnish satisfactory evidence (including reports of required tests) as to the kind and quality of materials and equipment. All materials and equipment shall be applied, installed, connected, erected, used, cleaned and conditioned in accordance with the instructions of the applicable Supplier except as otherwise provided in the Contract Documents; but no provision of any such instructions will be effective to assign to ENGINEER, or any of ENGINEER's consultants, agents or employees, any duty or authority to supervise or direct the furnishing or performance of the Work or any duty or authority to undertake responsibility contrary to the provisions of paragraph 9.15 or 9.16.

Adjusting Progress Schedule:

6.6. CONTRACTOR shall submit to ENGINEER for acceptance (to the extent indicated in paragraph 2.9) adjustments in the progress schedule to reflect the impact thereon of new developments; these will conform generally to the progress schedule then in effect and additionally will comply with any provisions of the General Requirements applicable thereto.

Substitutes or "Or-Equal" Items:

6.7.1. Whenever materials or equipment are specified or described in the Contract Documents by using the name of a proprietary item or the name of a particular Supplier the naming of the item is intended to establish the type, function and quality required. Unless the name is followed by words indicating that no substitution is permitted, materials or equipment of other Suppliers may be accepted by ENGINEER if sufficient information is submitted by CONTRACTOR to allow ENGINEER to determine that the material or equipment proposed is equivalent or equal to that named. The procedure for review by ENGINEER will include the following as supplemented in the General Requirements. Requests for review of substitute items of material and equipment will not be accepted by ENGINEER from anyone other than CONTRACTOR. If CONTRACTOR wishes to furnish or use a substitute item of material or equipment, CONTRACTOR shall make written application to ENGINEER for acceptance thereof, certifying that the proposed substitute will perform adequately the functions and achieve the results called for by the general design, be similar and of equal substance to that specified and be suited to the same use as that specified. The application will state that the evaluation and acceptance of the proposed substitute will not prejudice CONTRACTOR's achievement of Substantial Completion on time, whether or not acceptance of the substitute for use in the Work will require a change in any of the Contract Documents (or in the provisions of any other direct contract with OWNER for work on the Project) to adapt the design to the proposed substitute and whether or not incorporation or use of the substitute in connection with the Work is subject to payment of any license fee or royalty. All variations of the proposed substitute from that specified will be identified in the application and available maintenance, repair and replacement service will be indicated. The application will also contain an itemized estimate of all costs that will result directly or indirectly from acceptance of such substitute, including costs of redesign and claims of other contractors affected by the resulting change, all of which shall be considered by ENGINEER in evaluating the proposed substitute. ENGINEER may require CONTRACTOR to furnish at CONTRACTOR's expense additional data about the proposed substitute.

6.7.2. If a specific means, method, technique, sequence or procedure of construction is indicated in or required by the Contract Documents, CONTRACTOR may furnish or utilize a substitute means, method, sequence, technique or procedure of construction acceptable to ENGINEER, if CONTRACTOR submits sufficient information to allow ENGINEER to determine that the substitute proposed is equivalent to that indicated or required by the Contract Documents. The procedure for review by ENGINEER will be similar to that provided in paragraph 6.7.1 as applied by ENGINEER and as may be supplemented in the General Requirements.

6.7.3. ENGINEER will be allowed a reasonable time within which to evaluate each proposed substitute. ENGINEER will be the sole judge of acceptability, and no substitute will be ordered, installed or utilized without ENGINEER's prior written acceptance which will be evidenced by either a Change Order or an approved Shop Drawing. OWNER may require CONTRACTOR to furnish at CONTRACTOR's expense a special performance guarantee or other surety with respect to any substitute. ENGINEER will record time required by ENGINEER and ENGINEER's consultants in evaluating substitutions proposed by CONTRACTOR and in making changes in the Contract Documents occasioned thereby. Whether or not ENGINEER accepts a proposed substitute, CONTRACTOR shall reimburse OWNER for the charges of ENGINEER and ENGINEER's consultants for evaluating each proposed substitute.

Concerning Subcontractors, Suppliers and Others:

6.8.1. CONTRACTOR shall not employ any Subcontractor, Supplier or other person or organization (including those acceptable to OWNER and ENGINEER as indicated in paragraph 6.8.2), whether initially or as a substitute, against whom OWNER or ENGINEER may have reasonable objection. CONTRACTOR shall not be required to employ any Subcontractor, Supplier or other person or organization to furnish or perform any of the Work against whom CONTRACTOR has reasonable objection.

6.8.2. If the Supplementary Conditions require the identity of certain Subcontractors, Suppliers or other persons or organizations (including those who are to furnish the principal items of materials and equipment) to be submitted to OWNER in advance of the specified date prior to the Effective Date of the Agreement for acceptance by OWNER and ENGINEER and if CONTRACTOR has submitted a list thereof in accordance with the Supplementary Conditions, OWNER's or ENGINEER's acceptance (either in writing or by failing to make written objection thereto by the date indicated for acceptance or objection in the bidding documents or the Contract Documents) of any such Subcontractor, Supplier or other person or organization so identified may be revoked on the basis of reasonable objection after due investigation, in which case CONTRACTOR shall submit an acceptable substitute, the Contract Price will be increased by the difference in the cost occasioned by such substitution and an appropriate Change Order will be issued or Written Amendment signed. No acceptance by OWNER or ENGINEER of any such Subcontractor, Supplier or other person or organization shall constitute a waiver of any right of OWNER or ENGINEER to reject *defective* Work.

6.9. CONTRACTOR shall be fully responsible to OWNER and ENGINEER for all acts and omissions of the Subcontractors, Suppliers and other persons and organizations performing or furnishing any of the Work under a direct or indirect contract with CONTRACTOR just as CONTRACTOR is responsible for CONTRACTOR's own acts and omissions. Nothing in the Contract Documents shall create any contractual relationship between OWNER or ENGINEER and any such Subcontractor, Supplier or other person or organization, nor shall it create any obligation on the part of OWNER or ENGINEER to pay or to see to the payment of any moneys due any such Subcontractor, Supplier or other person or organization except as may otherwise be required by Laws and Regulations.

6.10. The divisions and sections of the Specifications and the identifications of any Drawings shall not control CONTRACTOR in dividing the Work among Subcontractors or Suppliers or delineating the Work to be performed by any specific trade.

6.11. All Work performed for CONTRACTOR by a Subcontractor will be pursuant to an appropriate agreement between CONTRACTOR and the Subcontractor which specifically binds the Subcontractor to the applicable terms and conditions of the Contract Documents for the benefit of OWNER and ENGINEER and contains waiver provisions as required by paragraph 5.11. CONTRACTOR shall pay each Subcontractor a just share of any insurance moneys received by CONTRACTOR on account of losses under policies issued pursuant to paragraphs 5.6 and 5.7.

Patent Fees and Royalties:

6.12. CONTRACTOR shall pay all license fees and royalties and assume all costs incident to the use in the performance of the Work or the incorporation in the Work of any invention, design, process, product or device which is the subject of patent rights or copyrights held by others. If a particular invention, design, process, product or device is specified in the Contract Documents for use in the performance of the Work and if to the actual knowledge of OWNER or ENGINEER its use is subject to patent rights or copyrights calling for the payment of any license fee or royalty to others, the existence of such rights shall be disclosed by OWNER in the Contract Documents. CONTRACTOR shall indemnify and hold harmless OWNER and ENGINEER and anyone directly or indirectly employed by either of them from and against all claims, damages, losses and expenses (including attorneys' fees and court and arbitration costs) arising out of any infringement of patent rights or copyrights incident to the use in the performance of the Work or resulting from the incorporation in the Work of any invention, design, process, product or device not specified in the Contract Documents, and shall defend all such claims in connection with any alleged infringement of such rights.

Permits:

6.13. Unless otherwise provided in the Supplementary Conditions, CONTRACTOR shall obtain and pay for all construction permits and licenses. OWNER shall assist CONTRACTOR, when necessary, in obtaining such permits and licenses. CONTRACTOR shall pay all governmental charges and inspection fees necessary for the prosecution of the Work, which are applicable at the time of opening of Bids, or if there are no Bids on the Effective Date of the Agreement. CONTRACTOR shall pay all charges of utility owners for connections to the Work, and OWNER shall pay all charges of such utility owners for capital costs related thereto such as plant investment fees.

Laws and Regulations:

6.14.1. CONTRACTOR shall give all notices and comply with all Laws and Regulations applicable to furnishing and performance of the Work. Except where otherwise expressly required by applicable Laws and Regulations, neither OWNER nor ENGINEER shall be responsible for monitoring CONTRACTOR's compliance with any Laws or Regulations.

6.14.2. If CONTRACTOR observes that the Specifications or Drawings are at variance with any Laws or Regulations, CONTRACTOR shall give ENGINEER prompt written notice thereof, and any necessary changes will be authorized by one of the methods indicated in paragraph 3.4. If CONTRACTOR performs any Work knowing or having reason to know that it is contrary to such Laws or Regulations, and without such notice to ENGINEER, CONTRACTOR shall bear all costs arising therefrom; however, it shall not be CONTRACTOR's primary responsibility to make certain that the Specifications and Drawings are in accordance with such Laws and Regulations.

Taxes:

6.15. CONTRACTOR shall pay all sales, consumer, use and other similar taxes required to be paid by CONTRACTOR in accordance with the Laws and Regulations of the place of the Project which are applicable during the performance of the Work.

Use of Premises:

6.16. CONTRACTOR shall confine construction equipment, the storage of materials and equipment and the operations of workers to the Project site and land and areas identified in and permitted by the Contract Documents and other land and areas permitted by Laws and Regulations, rights-of-way, permits and easements, and shall not unreasonably encumber the premises with construction equipment or other materials or equipment. CONTRACTOR shall assume full responsibility for any damage to any such land or area, or to the owner or occupant thereof or of any land or areas contiguous thereto, resulting from the performance of the Work. Should any claim be made against OWNER or ENGINEER by any such owner or occupant because of the performance of the Work, CONTRACTOR shall promptly attempt to settle with such other party by agreement or otherwise resolve the claim by arbitration or at law. CONTRACTOR shall, to the fullest extent permitted by Laws and Regulations, indemnify and hold OWNER and ENGINEER harmless from and against all claims, damages, losses and expenses (including, but not limited to, fees of engineers, architects, attorneys and other professionals and court and arbitration costs) arising directly, indirectly or consequentially out of any action, legal or equitable, brought by any such other party against OWNER or ENGINEER to the extent based on a claim arising out of CONTRACTOR's performance of the Work.

6.17. During the progress of the Work, CONTRACTOR shall keep the premises free from accumulations of waste materials, rubbish and other debris resulting from the Work. At the completion of the Work CONTRACTOR shall remove all waste materials, rubbish and debris from and about the premises as well as all tools, appliances, construction equipment and machinery, and surplus materials, and shall leave the site clean and ready for occupancy by OWNER. CONTRACTOR shall restore to original condition all property not designated for alteration by the Contract Documents.

6.18. CONTRACTOR shall not load nor permit any part of any structure to be loaded in any manner that will endanger the structure, nor shall CONTRACTOR subject any part of the Work or adjacent property to stresses or pressures that will endanger it.

Record Documents:

6.19. CONTRACTOR shall maintain in a safe place at the site one record copy of all Drawings, Specifications, Addenda, Written Amendments, Change Orders, Work Directive Changes, Field Orders and written interpretations and clarifications (issued pursuant to paragraph 9.4) in good order and annotated to show all changes made during construction. These record documents together with all approved samples and a counterpart of all approved Shop Drawings will be available to ENGINEER for reference. Upon completion of the Work, these record documents, samples and Shop Drawings will be delivered to ENGINEER for OWNER.

Safety and Protection:

6.20. CONTRACTOR shall be responsible for initiating, maintaining and supervising all safety precautions and programs in connection with the Work. CONTRACTOR shall take all necessary precautions for the safety of, and shall provide the necessary protection to prevent damage, injury or loss to:

6.20.1. all employees on the Work and other persons and organizations who may be affected thereby;

6.20.2. all the Work and materials and equipment to be incorporated therein, whether in storage on or off the site; and

6.20.3. other property at the site or adjacent thereto, including trees, shrubs, lawns, walks, pavements, roadways, structures, utilities and Underground Facilities not designated for removal, relocation or replacement in the course of construction.

CONTRACTOR shall comply with all applicable Laws and Regulations of any public body having jurisdiction for the safety of persons or property or to protect them from damage, injury or loss; and shall erect and maintain all necessary safeguards for such safety and protection. CONTRACTOR shall notify owners of adjacent property and of Underground Facilities and utility owners when prosecution of the Work may affect them, and shall cooperate with them in the protection, removal, relocation and replacement of their property. All damage, injury or loss to any property referred to in paragraph 6.20.2 or 6.20.3 caused, directly or indirectly, in whole or in part, by CONTRACTOR, any Subcontractor, Supplier or any other person or organization directly or indirectly employed by any of them to perform or furnish any of the Work or anyone for whose acts any of them may be liable, shall be remedied by CONTRACTOR (except damage or loss attributable to the fault of Drawings or Specifications or to the acts or omissions of OWNER or ENGINEER or anyone employed by either of them or anyone for whose acts either of them may be liable, and not attributable, directly or indirectly, in whole or in part, to the fault or negligence of CONTRACTOR). CONTRACTOR's duties and responsibilities for the safety and protection of the Work shall continue until such time as all the Work is completed and ENGINEER has issued a notice to OWNER and CONTRACTOR in accordance with paragraph 14.13 that the Work is acceptable (except as otherwise expressly provided in connection with Substantial Completion).

6.21. CONTRACTOR shall designate a responsible representative at the site whose duty shall be the prevention of accidents. This person shall be CONTRACTOR's superintendent unless otherwise designated in writing by CONTRACTOR to OWNER.

Emergencies:

6.22. In emergencies affecting the safety or protection of persons or the Work or property at the site or adjacent thereto, CONTRACTOR, without special instruction or authorization from ENGINEER or OWNER, is obligated to act to prevent threatened damage, injury or loss. CONTRACTOR shall give ENGINEER prompt written notice if CONTRACTOR believes that any significant changes in the Work or variations from the Contract Documents have been caused thereby. If ENGINEER determines that a change in the Contract Documents is required because of the action taken in response to an emergency, a Work Directive Change or Change Order will be issued to document the consequences of the changes or variations.

Shop Drawings and Samples:

6.23. After checking and verifying all field measurements and after complying with applicable procedures specified in the General Requirements, CONTRACTOR shall submit to ENGINEER for review and approval in accordance with the accepted schedule of Shop Drawing submissions (see paragraph 2.9), or for other appropriate action if so indicated in the Supplementary Conditions, five copies (unless otherwise specified in the General Requirements) of all Shop Drawings, which will bear a stamp or specific written indication that CONTRACTOR has satisfied CONTRACTOR's responsibilities under the Contract Documents with respect to the review of the submission. All submissions will be identified as ENGINEER may require. The data shown on the Shop Drawings will be complete with respect to quantities, dimensions, specified performance and design criteria, materials and similar data to enable ENGINEER to review the information as required.

6.24. CONTRACTOR shall also submit to ENGINEER for review and approval with such promptness as to cause no delay in Work, all samples required by the Contract Documents. All samples will have been checked by and accompanied by a specific written indication that CONTRACTOR has satisfied CONTRACTOR's responsibilities under the Contract Documents with respect to the review of the submission and will be identified clearly as to material, Supplier, pertinent data such as catalog numbers and the use for which intended.

6.25.1. Before submission of each Shop Drawing or sample CONTRACTOR shall have determined and verified all quantities, dimensions, specified performance criteria, installation requirements, materials, catalog numbers and similar data with respect thereto and reviewed or coordinated each Shop Drawing or sample with other Shop Drawings and samples and with the requirements of the Work and the Contract Documents.

6.25.2. At the time of each submission, CONTRACTOR shall give ENGINEER specific written notice of each variation that the Shop Drawings or samples may have from the requirements of the Contract Documents, and, in addition, shall cause a specific notation to be made on each Shop Drawing submitted to ENGINEER for review and approval of each such variation.

6.26. ENGINEER will review and approve with reasonable promptness Shop Drawings and samples, but ENGINEER's review and approval will be only for conformance with the design concept of the Project and for compliance with the information given in the Contract Documents and shall not extend to means, methods, techniques, sequences or procedures of construction (except where a specific means, method, technique, sequence or procedure of construction is indicated in or required by the Contract Documents) or to safety precautions or programs incident thereto. The review and approval of a separate item as such will not indicate approval of the assembly in which the item functions. CONTRACTOR shall make corrections required by ENGINEER, and shall return the required number of corrected copies of Shop Drawings and submit as required new samples for review and approval. CONTRACTOR shall direct specific attention in writing to revisions other than the corrections called for by ENGINEER on previous submittals.

6.27. ENGINEER's review and approval of Shop Drawings or samples shall not relieve CONTRACTOR from responsibility for any variation from the requirements of the Contract Documents unless CONTRACTOR has in writing called ENGINEER's attention to each such variation at the time of submission as required by paragraph 6.25.2 and ENGINEER has given written approval of each such variation by a specific written notation thereof incorporated in or accompanying the Shop Drawing or sample approval; nor will any approval by ENGINEER relieve CONTRACTOR from responsibility for errors or omissions in the Shop Drawings or from responsibility for having complied with the provisions of paragraph 6.25.1.

6.28. Where a Shop Drawing or sample is required by the Specifications, any related Work performed prior to ENGINEER's review and approval of the pertinent submission will be the sole expense and responsibility of CONTRACTOR.

Continuing the Work:

6.29. CONTRACTOR shall carry on the Work and adhere to the progress schedule during all disputes or disagreements with OWNER. No Work shall be delayed or postponed pending resolution of any disputes or disagreements, except as permitted by paragraph 15.5 or as CONTRACTOR and OWNER may otherwise agree in writing.

Indemnification:

6.30. To the fullest extent permitted by Laws and Regulations CONTRACTOR shall indemnify and hold harmless OWNER and ENGINEER and their consultants, agents and employees from and against all claims, damages, losses and expenses, direct, indirect or consequential (including but not limited to fees and charges of engineers, architects, attorneys and other professionals and court and arbitration costs) arising out of or resulting from the performance of the Work, provided that any such claim, damage, loss or expense (a) is attributable to bodily injury, sickness, disease or death, or to injury to or destruction of tangible property (other than the Work itself) including the loss of use resulting therefrom and (b) is caused in whole or in part by any negligent act or omission of CONTRACTOR, any Subcontractor, any person or organization directly or indirectly employed by any of them to perform or furnish any of the Work or anyone for whose acts any of them may be liable, regardless of whether or not it is caused in part by a party indemnified hereunder or arises by or is imposed by Law and Regulations regardless of the negligence of any such party.

6.31. In any and all claims against OWNER or ENGINEER or any of their consultants, agents or employees by any employee of CONTRACTOR, any Subcontractor, any person or organization directly or indirectly employed by any of them to perform or furnish any of the Work or anyone for whose acts any of them may be liable, the indemnification obligation under paragraph 6.30 shall not be limited in any way by any limitation on the amount or type of damages, compensation or benefits payable by or for CONTRACTOR or any such Subcontractor or other person or organization under workers' or workmen's compensation acts, disability benefit acts or other employee benefit acts.

6.32. The obligations of CONTRACTOR under paragraph 6.30 shall not extend to the liability of ENGINEER, ENGINEER's consultants, agents or employees arising out of the preparation or approval of maps, drawings, opinions, reports, surveys, Change Orders, designs or specifications.

ARTICLE 7—OTHER WORK

Related Work at Site:

7.1. OWNER may perform other work related to the Project at the site by OWNER's own forces, have other work performed by utility owners or let other direct contracts therefor which shall contain General Conditions similar to these. If the fact that such other work is to be performed was not noted in the Contract Documents, written notice thereof will be given to CONTRACTOR prior to starting any such other work; and, if CONTRACTOR believes that such performance will involve additional expense to CONTRACTOR or requires additional time and the parties are unable to agree as to the extent thereof, CONTRACTOR may make a claim therefor as provided in Articles 11 and 12.

7.2. CONTRACTOR shall afford each utility owner and other contractor who is a party to such a direct contract (or OWNER, if OWNER is performing the additional work with OWNER's employees) proper and safe access to the site and a reasonable opportunity for the introduction and storage of materials and equipment and the execution of such work, and shall properly connect and coordinate the Work with theirs. CONTRACTOR shall do all cutting, fitting and patching of the Work that may be required to make its several parts come together properly and integrate with such other work. CON-

TRACTOR shall not endanger any work of others by cutting, excavating or otherwise altering their work and will only cut or alter their work with the written consent of ENGINEER and the others whose work will be affected. The duties and responsibilities of CONTRACTOR under this paragraph are for the benefit of such utility owners and other contractors to the extent that there are comparable provisions for the benefit of CONTRACTOR in said direct contracts between OWNER and such utility owners and other contractors.

7.3. If any part of CONTRACTOR's Work depends for proper execution or results upon the work of any such other contractor or utility owner (or OWNER), CONTRACTOR shall inspect and promptly report to ENGINEER in writing any delays, defects or deficiencies in such work that render it unavailable or unsuitable for such proper execution and results. CONTRACTOR's failure so to report will constitute an acceptance of the other work as fit and proper for integration with CONTRACTOR's Work except for latent or non-apparent defects and deficiencies in the other work.

Coordination:

7.4. If OWNER contracts with others for the performance of other work on the Project at the site, the person or organization who will have authority and responsibility for coordination of the activities among the various prime contractors will be identified in the Supplementary Conditions, and the specific matters to be covered by such authority and responsibility will be itemized, and the extent of such authority and responsibilities will be provided, in the Supplementary Conditions. Unless otherwise provided in the Supplementary Conditions, neither OWNER nor ENGINEER shall have any authority or responsibility in respect of such coordination.

ARTICLE 8—OWNER'S RESPONSIBILITIES

8.1. OWNER shall issue all communications to CONTRACTOR through ENGINEER.

8.2. In case of termination of the employment of ENGINEER, OWNER shall appoint an engineer against whom CONTRACTOR makes no reasonable objection, whose status under the Contract Documents shall be that of the former ENGINEER. Any dispute in connection with such appointment shall be subject to arbitration.

8.3. OWNER shall furnish the data required of OWNER under the Contract Documents promptly and shall make payments to CONTRACTOR promptly after they are due as provided in paragraphs 14.4 and 14.13.

8.4. OWNER's duties in respect of providing lands and easements and providing engineering surveys to establish reference points are set forth in paragraphs 4.1 and 4.4. Paragraph 4.2 refers to OWNER's identifying and making available to CONTRACTOR copies of reports of explorations and tests of subsurface conditions at the site and in existing structures which have been utilized by ENGINEER in preparing the Drawings and Specifications.

8.5. OWNER's responsibilities in respect of purchasing and maintaining liability and property insurance are set forth in paragraphs 5.5 through 5.8.

8.6. OWNER is obligated to execute Change Orders as indicated in paragraph 10.4.

8.7. OWNER's responsibility in respect of certain inspections, tests and approvals is set forth in paragraph 13.4.

8.8. In connection with OWNER's right to stop Work or suspend Work, see paragraphs 13.10 and 15.1. Paragraph 15.2 deals with OWNER's right to terminate services of CONTRACTOR under certain circumstances.

ARTICLE 9—ENGINEER'S STATUS DURING CONSTRUCTION

Owner's Representative:

9.1. ENGINEER will be OWNER's representative during the construction period. The duties and responsibilities and the limitations of authority of ENGINEER as OWNER's representative during construction are set forth in the Contract Documents and shall not be extended without written consent of OWNER and ENGINEER.

Visits to Site:

9.2. ENGINEER will make visits to the site at intervals appropriate to the various stages of construction to observe the progress and quality of the executed Work and to determine, in general, if the Work is proceeding in accordance with the Contract Documents. ENGINEER will not be required to make exhaustive or continuous on-site inspections to check the quality or quantity of the Work. ENGINEER's efforts will be directed toward providing for OWNER a greater degree of confidence that the completed Work will conform to the Contract Documents. On the basis of such visits and on-site observations as an experienced and qualified design professional, ENGINEER will keep OWNER informed of the progress of the Work and will endeavor to guard OWNER against defects and deficiencies in the Work.

Project Representation:

9.3. If OWNER and ENGINEER agree, ENGINEER will furnish a Resident Project Representative to assist ENGINEER in observing the performance of the Work. The duties, responsibilities and limitations of authority of any such Resident Project Representative and assistants will be as provided in the Supplementary Conditions. If OWNER designates another agent to represent OWNER at the site who is not ENGINEER's agent or employee, the duties, responsibilities and limitations of authority of such other person will be as provided in the Supplementary Conditions.

Clarifications and Interpretations:

9.4. ENGINEER will issue with reasonable promptness such written clarifications or interpretations of the requirements of the Contract Documents (in the form of Drawings or otherwise) as ENGINEER may determine necessary, which shall be consistent with or reasonably inferable from the overall intent of the Contract Documents. If CONTRACTOR believes that a written clarification or interpretation justifies an increase in the Contract Price or an extension of the Contract Time and the parties are unable to agree to the amount or extent thereof, CONTRACTOR may make a claim therefor as provided in Article 11 or Article 12.

Authorized Variations in Work:

9.5. ENGINEER may authorize minor variations in the Work from the requirements of the Contract Documents which do not involve an adjustment in the Contract Price or the Contract Time and are consistent with the overall intent of the Contract Documents. These may be accomplished by a Field Order and will be binding on OWNER, and also on CONTRACTOR who shall perform the Work involved promptly. If CONTRACTOR believes that a Field Order justifies an increase in the Contract Price or an extension of the Contract Time and the parties are unable to agree as to the amount or extent thereof, CONTRACTOR may make a claim therefor as provided in Article 11 or 12.

Rejecting Defective Work:

9.6. ENGINEER will have authority to disapprove or reject Work which ENGINEER believes to be *defective,* and will also have authority to require special inspection or testing of the Work as provided in paragraph 13.9, whether or not the Work is fabricated, installed or completed.

Shop Drawings, Change Orders and Payments:

9.7 In connection with ENGINEER's responsibility for Shop Drawings and samples, see paragraphs 6.23 through 6.28 inclusive.

9.8. In connection with ENGINEER's responsibilities as to Change Orders, see Articles 10, 11 and 12.

9.9. In connection with ENGINEER's responsibilities in respect of Applications for Payment, etc., see Article 14.

Determinations for Unit Prices:

9.10. ENGINEER will determine the actual quantities and classifications of Unit Price Work performed by CONTRACTOR. ENGINEER will review with CONTRACTOR ENGINEER's preliminary determinations on such matters before rendering a written decision thereon (by recommendation of an Application for Payment or otherwise). ENGINEER's written decisions thereon will be final and binding upon OWNER and CONTRACTOR, unless, within ten days after the date of any such decision, either OWNER or CONTRACTOR delivers to the other party to the Agreement and to ENGINEER written notice of intention to appeal from such a decision.

Decisions on Disputes:

9.11. ENGINEER will be the initial interpreter of the requirements of the Contract Documents and judge of the acceptability of the Work thereunder. Claims, disputes and other matters relating to the acceptability of the Work or the interpretation of the requirements of the Contract Documents pertaining to the performance and furnishing of the Work and claims under Articles 11 and 12 in respect of changes in the Contract Price or Contract Time will be referred initially to ENGINEER in writing with a request for a formal decision in accordance with this paragraph, which ENGINEER will render in writing within a reasonable time. Written notice of each such claim, dispute and other matter will be delivered by the claimant to ENGINEER and the other party to the Agreement promptly (but in no event later than thirty days) after the occurrence of the event giving rise thereto, and written supporting data will be submitted to ENGINEER and the other party within sixty days after such occurrence unless ENGINEER allows an additional period of time to ascertain more accurate data in support of the claim.

9.12. When functioning as interpreter and judge under paragraphs 9.10 and 9.11, ENGINEER will not show partiality to OWNER or CONTRACTOR and will not be liable in connection with any interpretation or decision rendered in good faith in such capacity. The rendering of a decision by ENGINEER pursuant to paragraphs 9.10 and 9.11 with respect to any such claim, dispute or other matter (except any which have been waived by the making or acceptance of final payment as provided in paragraph 14.16) will be a condition precedent to any exercise by OWNER or CONTRACTOR of such rights or remedies as either may otherwise have under the Contract Documents or by Laws or Regulations in respect of any such claim, dispute or other matter.

Limitations on ENGINEER's Responsibilities:

9.13. Neither ENGINEER's authority to act under this Article 9 or elsewhere in the Contract Documents nor any decision made by ENGINEER in good faith either to exercise or not exercise such authority shall give rise to any duty or responsibility of ENGINEER to CONTRACTOR, any Subcontractor, any Supplier, or any other person or organization performing any of the Work, or to any surety for any of them.

9.14. Whenever in the Contract Documents the terms "as ordered", "as directed", "as required", "as allowed", "as approved" or terms of like effect or import are used, or the adjectives "reasonable", "suitable", "acceptable", "proper" or "satisfactory" or adjectives of like effect or import are used to describe a requirement, direction, review or judgment of ENGINEER as to the Work, it is intended that such requirement, direction, review or judgment will be solely to evaluate the Work for compliance with the Contract Documents (unless there is a specific statement indicating otherwise). The use of any such term or adjective shall not be effective to assign to ENGINEER any duty or authority to supervise or direct the furnishing or performance of the Work or any duty or authority to undertake responsibility contrary to the provisions of paragraph 9.15 or 9.16.

9.15. ENGINEER will not be responsible for CONTRACTOR's means, methods, techniques, sequences or procedures of construction, or the safety precautions and programs incident thereto, and ENGINEER will not be responsible for CONTRACTOR's failure to perform or furnish the Work in accordance with the Contract Documents.

9.16. ENGINEER will not be responsible for the acts or omissions of CONTRACTOR or of any Subcontractor, any Supplier, or of any other person or organization performing or furnishing any of the Work.

ARTICLE 10—CHANGES IN THE WORK

10.1. Without invalidating the Agreement and without notice to any surety, OWNER may, at any time or from time to time, order additions, deletions or revisions in the Work; these will be authorized by a Written Amendment, a Change Order, or a Work Directive Change. Upon receipt of any such document, CONTRACTOR shall promptly proceed with the Work involved which will be performed under the applicable conditions of the Contract Documents (except as otherwise specifically provided).

10.2. If OWNER and CONTRACTOR are unable to agree as to the extent, if any, of an increase or decrease in the Contract Price or an extension or shortening of the Contract Time that should be allowed as a result of a Work Directive Change, a claim may be made therefor as provided in Article 11 or Article 12.

10.3. CONTRACTOR shall not be entitled to an increase in the Contract Price or an extension of the Contract Time with respect to any Work performed that is not required by the Contract Documents as amended, modified and supplemented as provided in paragraphs 3.4 and 3.5, except in the case of an emergency as provided in paragraph 6.22 and except in the case of uncovering Work as provided in paragraph 13.9.

10.4. OWNER and CONTRACTOR shall execute appropriate Change Orders (or Written Amendments) covering:

10.4.1. changes in the Work which are ordered by OWNER pursuant to paragraph 10.1, are required because of acceptance of *defective* Work under paragraph 13.13 or correcting *defective* Work under paragraph 13.14, or are agreed to by the parties;

10.4.2. changes in the Contract Price or Contract Time which are agreed to by the parties; and

10.4.3. changes in the Contract Price or Contract Time which embody the substance of any written decision rendered by ENGINEER pursuant to paragraph 9.11;

provided that, in lieu of executing any such Change Order, an appeal may be taken from any such decision in accordance with the provisions of the Contract Documents and applicable Laws and Regulations, but during any such appeal, CONTRACTOR shall carry on the Work and adhere to the progress schedule as provided in paragraph 6.29.

10.5. If notice of any change affecting the general scope of the Work or the provisions of the Contract Documents (including, but not limited to, Contract Price or Contract Time) is required by the provisions of any Bond to be given to a surety, the giving of any such notice will be CONTRACTOR's responsibility, and the amount of each applicable Bond will be adjusted accordingly.

ARTICLE 11—CHANGE OF CONTRACT PRICE

11.1. The Contract Price constitutes the total compensation (subject to authorized adjustments) payable to CONTRACTOR for performing the Work. All duties, responsibilities and obligations assigned to or undertaken by CONTRACTOR shall be at his expense without change in the Contract Price.

11.2. The Contract Price may only be changed by a Change Order or by a Written Amendment. Any claim for an increase or decrease in the Contract Price shall be based on written notice delivered by the party making the claim to the other party and to ENGINEER promptly (but in no event later than thirty days) after the occurrence of the event giving rise to the claim and stating the general nature of the claim. Notice of the amount of the claim with supporting data shall be delivered within sixty days after such occurrence (unless ENGINEER allows an additional period of time to ascertain more accurate data in support of the claim) and shall be accompanied by claimant's written statement that the amount claimed covers all known amounts (direct, indirect and consequential) to which the claimant is entitled as a result of the occurrence of said event. All claims for adjustment in the Contract Price shall be determined by ENGINEER in accordance with paragraph 9.11 if OWNER and CONTRACTOR cannot otherwise agree on the amount involved. No claim for an adjustment in the Contract Price will be valid if not submitted in accordance with this paragraph 11.2.

11.3. The value of any Work covered by a Change Order or of any claim for an increase or decrease in the Contract Price shall be determined in one of the following ways:

11.3.1. Where the Work involved is covered by unit prices contained in the Contract Documents, by application of unit prices to the quantities of the items involved (subject to the provisions of paragraphs 11.9.1. through 11.9.3, inclusive).

11.3.2. By mutual acceptance of a lump sum (which may include an allowance for overhead and profit not necessarily in accordance with paragraph 11.6.2.1).

11.3.3. On the basis of the Cost of the Work (determined as provided in paragraphs 11.4 and 11.5) plus a CONTRACTOR's Fee for overhead and profit (determined as provided in paragraphs 11.6 and 11.7).

Cost of the Work:

11.4. The term Cost of the Work means the sum of all costs necessarily incurred and paid by CONTRACTOR in the proper performance of the Work. Except as otherwise may be agreed to in writing by OWNER, such costs shall be in amounts no higher than those prevailing in the locality of the Project, shall include only the following items and shall not include any of the costs itemized in paragraph 11.5:

11.4.1. Payroll costs for employees in the direct employ of CONTRACTOR in the performance of the Work under schedules of job classifications agreed upon by OWNER and CONTRACTOR. Payroll costs for employees not employed full time on the Work shall be apportioned on the basis of their time spent on the Work. Payroll costs shall include, but not be limited to, salaries and wages plus the cost of fringe benefits which shall include social security contributions, unemployment, excise and payroll taxes, workers' or workmen's compensation, health and retirement benefits, bonuses, sick leave, vacation and holiday pay applicable thereto. Such employees shall include superintendents and foremen at the site. The expenses of performing Work after regular working hours, on Saturday, Sunday or legal holidays, shall be included in the above to the extent authorized by OWNER.

11.4.2. Cost of all materials and equipment furnished and incorporated in the Work, including costs of transportation and storage thereof, and Suppliers' field services required in connection therewith. All cash discounts shall accrue to CONTRACTOR unless OWNER deposits funds with CONTRACTOR with which to make payments, in which case the cash discounts shall accrue to OWNER. All trade discounts, rebates and refunds and all returns from sale of surplus materials and equipment shall accrue to OWNER, and CONTRACTOR shall make provisions so that they may be obtained.

11.4.3. Payments made by CONTRACTOR to the Subcontractors for Work performed by Subcontractors. If required by OWNER, CONTRACTOR shall obtain competitive bids from Subcontractors acceptable to CONTRACTOR and shall deliver such bids to OWNER who will then determine, with the advice of ENGINEER, which bids will be accepted. If a subcontract provides that the Subcontractor is to be paid on the basis of Cost of the Work Plus a Fee, the Subcontractor's Cost of the Work shall be determined in the same manner as CONTRACTOR's Cost of the Work. All subcontracts shall be subject to the other provisions of the Contract Documents insofar as applicable.

11.4.4. Costs of special consultants (including but not limited to engineers, architects, testing laboratories, surveyors, attorneys and accountants) employed for services specifically related to the Work.

11.4.5. Supplemental costs including the following:

11.4.5.1. The proportion of necessary transportation, travel and subsistence expenses of CONTRACTOR's employees incurred in discharge of duties connected with the Work.

11.4.5.2. Cost, including transportation and maintenance, of all materials, supplies, equipment, machinery, appliances, office and temporary facilities at the site and hand tools not owned by the workers, which are consumed in the performance of the Work, and cost less market value of such items used but not consumed which remain the property of CONTRACTOR.

11.4.5.3. Rentals of all construction equipment and machinery and the parts thereof whether rented from CONTRACTOR or others in accordance with rental agreements approved by OWNER with the advice of ENGINEER, and the costs of transportation, loading, unloading, installation, dismantling and removal thereof—all in accordance with terms of said rental agreements. The rental of any such equipment, machinery or parts shall cease when the use thereof is no longer necessary for the Work.

11.4.5.4. Sales, consumer, use or similar taxes related to the Work, and for which CONTRACTOR is liable, imposed by Laws and Regulations.

11.4.5.5. Deposits lost for causes other than negligence of CONTRACTOR, any Subcontractor or anyone directly or indirectly employed by any of them or for whose acts any of them may be liable, and royalty payments and fees for permits and licenses.

11.4.5.6. Losses and damages (and related expenses), not compensated by insurance or otherwise, to the Work or otherwise sustained by CONTRACTOR in connection with the performance and furnishing of the Work (except losses and damages within the deductible amounts of property insurance established by OWNER in accordance with paragraph 5.9), provided they have resulted from causes other than the negligence of CONTRACTOR, any Subcontractor, or anyone directly or indirectly employed by any of them or for whose acts any of them may be liable. Such losses shall include settlements made with the written consent and approval of OWNER. No such losses, damages and expenses shall be included in the Cost of the Work for the purpose of determining CONTRACTOR's Fee. If, however, any such loss or damage requires reconstruction and CONTRACTOR is placed in charge thereof, CONTRACTOR shall be paid for services a fee proportionate to that stated in paragraph 11.6.2.

11.4.5.7. The cost of utilities, fuel and sanitary facilities at the site.

11.4.5.8. Minor expenses such as telegrams, long distance telephone calls, telephone service at the site, expressage and similar petty cash items in connection with the Work.

11.4.5.9. Cost of premiums for additional Bonds and insurance required because of changes in the Work and premiums for property insurance coverage within the limits of the deductible amounts established by OWNER in accordance with paragraph 5.9.

11.5. The term Cost of the Work shall not include any of the following:

11.5.1. Payroll costs and other compensation of CONTRACTOR's officers, executives, principals (of partnership and sole proprietorships), general managers, engineers, architects, estimators, attorneys, auditors, accountants, purchasing and contracting agents, expeditors, timekeepers, clerks and other personnel employed by CONTRACTOR whether at the site or in CONTRACTOR's principal or a branch office for general administration of the Work and not specifically included in the agreed upon schedule of job classifications referred to in paragraph 11.4.1 or specifically covered by paragraph 11.4.4—all of which are to be considered administrative costs covered by the CONTRACTOR's Fee.

11.5.2. Expenses of CONTRACTOR's principal and branch offices other than CONTRACTOR's office at the site.

11.5.3. Any part of CONTRACTOR's capital expenses, including interest on CONTRACTOR'S capital employed for the Work and charges against CONTRACTOR for delinquent payments.

11.5.4. Cost of premiums for all Bonds and for all insurance whether or not CONTRACTOR is required by the Contract Documents to purchase and maintain the same (except for the cost of premiums covered by subparagraph 11.4.5.9 above).

11.5.5. Costs due to the negligence of CONTRACTOR, any Subcontractor, or anyone directly or indirectly employed by any of them or for whose acts any of them may be liable, including but not limited to, the correction of *defective* Work, disposal of materials or equipment wrongly supplied and making good any damage to property.

11.5.6. Other overhead or general expense costs of any kind and the costs of any item not specifically and expressly included in paragraph 11.4.

CONTRACTOR's Fee:

11.6. The CONTRACTOR's Fee allowed to CONTRACTOR for overhead and profit shall be determined as follows:

11.6.1. a mutually acceptable fixed fee; or if none can be agreed upon,

11.6.2. a fee based on the following percentages of the various portions of the Cost of the Work:

11.6.2.1. for costs incurred under paragraphs 11.4.1 and 11.4.2, the CONTRACTOR's Fee shall be fifteen percent;

11.6.2.2. for costs incurred under paragraph 11.4.3, the CONTRACTOR's Fee shall be five percent; and if a subcontract is on the basis of Cost of the Work Plus a Fee, the maximum allowable to CONTRACTOR on account of overhead and profit of all Subcontractors shall be fifteen percent;

11.6.2.3. no fee shall be payable on the basis of costs itemized under paragraphs 11.4.4, 11.4.5 and 11.5;

11.6.2.4. the amount of credit to be allowed by CONTRACTOR to OWNER for any such change which results in a net decrease in cost will be the amount of the actual net decrease plus a deduction in CONTRACTOR's Fee by an amount equal to ten percent of the net decrease; and

11.6.2.5. when both additions and credits are involved in any one change, the adjustment in CONTRACTOR's Fee shall be computed on the basis of the net change in accordance with paragraphs 11.6.2.1 through 11.6.2.4, inclusive.

11.7. Whenever the cost of any Work is to be determined pursuant to paragraph 11.4 or 11.5, CONTRACTOR will submit in form acceptable to ENGINEER an itemized cost breakdown together with supporting data.

Cash Allowances:

11.8. It is understood that CONTRACTOR has included in the Contract Price all allowances so named in the Contract Documents and shall cause the Work so covered to be done by such Subcontractors or Suppliers and for such sums within the limit of the allowances as may be acceptable to ENGINEER. CONTRACTOR agrees that:

11.8.1. The allowances include the cost to CONTRACTOR (less any applicable trade discounts) of materials and equipment required by the allowances to be delivered at the site, and all applicable taxes; and

11.8.2. CONTRACTOR's costs for unloading and handling on the site, labor, installation costs, overhead, profit and other expenses contemplated for the allowances have been included in the Contract Price and not in the allowances. No demand for additional payment on account of any thereof will be valid.

Prior to final payment, an appropriate Change Order will be issued as recommended by ENGINEER to reflect actual amounts due CONTRACTOR on account of Work covered by allowances, and the Contract Price shall be correspondingly adjusted.

Unit Price Work:

11.9.1. Where the Contract Documents provide that all or part of the Work is to be Unit Price Work, initially the Contract Price will be deemed to include for all Unit Price Work an amount equal to the sum of the established unit prices for each separately identified item of Unit Price Work times the estimated quantity of each item as indicated in the Agreement. The estimated quantities of items of Unit Price Work are not guaranteed and are solely for the purpose of comparison of Bids and determining an initial Contract Price. Determinations of the actual quantities and classifications of Unit Price Work performed by CONTRACTOR will be made by ENGINEER in accordance with Paragraph 9.10.

11.9.2. Each unit price will be deemed to include an amount considered by CONTRACTOR to be adequate to cover CONTRACTOR's overhead and profit for each separately identified item.

11.9.3. Where the quantity of any item of Unit Price Work performed by CONTRACTOR differs materially and significantly from the estimated quantity of such item indicated in the Agreement and there is no corresponding adjustment with respect to any other item of Work and if CONTRACTOR believes that CONTRACTOR has incurred additional expense as a result thereof, CONTRACTOR may make a claim for an increase in the Contract Price in accordance with Article 11 if the parties are unable to agree as to the amount of any such increase.

ARTICLE 12—CHANGE OF CONTRACT TIME

12.1. The Contract Time may only be changed by a Change Order or a Written Amendment. Any claim for an extension or shortening of the Contract Time shall be based on written notice delivered by the party making the claim to the other party and to ENGINEER promptly (but in no event later than thirty days) after the occurrence of the event giving rise to the claim and stating the general nature of the claim. Notice of the extent of the claim with supporting data shall be delivered within sixty days after such occurrence (unless ENGINEER allows an additional period of time to ascertain more accurate data in support of the claim) and shall be accompanied by the claimant's written statement that the adjustment claimed is the entire adjustment to which the claimant has reason to believe it is entitled as a result of the occurrence of said event. All claims for adjustment in the Contract Time shall be determined by ENGINEER in accordance with paragraph 9.11 if OWNER and CONTRACTOR cannot otherwise agree. No claim for an adjustment in the Contract Time will be valid if not submitted in accordance with the requirements of this paragraph 12.1.

12.2. The Contract Time will be extended in an amount equal to time lost due to delays beyond the control of CONTRACTOR if a claim is made therefor as provided in paragraph 12.1. Such delays shall include, but not be limited to, acts or neglect by OWNER or others performing additional work as contemplated by Article 7, or to fires, floods, labor disputes, epidemics, abnormal weather conditions or acts of God.

12.3. All time limits stated in the Contract Documents are of the essence of the Agreement. The provisions of this Article 12 shall not exclude recovery for damages (including but not limited to fees and charges of engineers, architects, attorneys and other professionals and court and arbitration costs) for delay by either party.

ARTICLE 13—WARRANTY AND GUARANTEE; TESTS AND INSPECTIONS; CORRECTION, REMOVAL OR ACCEPTANCE OF DEFECTIVE WORK

Warranty and Guarantee:

13.1. CONTRACTOR warrants and guarantees to OWNER and ENGINEER that all Work will be in accordance with the Contract Documents and will not be *defective*. Prompt notice of all defects shall be given to CONTRACTOR. All *defective* Work, whether or not in place, may be rejected, corrected or accepted as provided in this Article 13.

Access to Work:

13.2. ENGINEER and ENGINEER's representatives, other representatives of OWNER, testing agencies and governmental agencies with jurisdictional interests will have access to the Work at reasonable times for their observation, inspecting and testing. CONTRACTOR shall provide proper and safe conditions for such access.

Tests and Inspections:

13.3. CONTRACTOR shall give ENGINEER timely notice of readiness of the Work for all required inspections, tests or approvals.

13.4. If Laws or Regulations of any public body having jurisdiction require any Work (or part thereof) to specifically be inspected, tested or approved, CONTRACTOR shall assume full responsibility therefor, pay all costs in connection therewith and furnish ENGINEER the required certificates of inspection, testing or approval. CONTRACTOR shall also be responsible for and shall pay all costs in connection with any inspection or testing required in connection with OWNER's or ENGINEER's acceptance of a Supplier of materials or equipment proposed to be incorporated in the Work, or of materials or equipment submitted for approval prior to CONTRACTOR's purchase thereof for incorporation in the Work. The cost of all inspections, tests and approvals in addition to the above which are required by the Contract Documents shall be paid by OWNER (unless otherwise specified).

13.5. All inspections, tests or approvals other than those required by Laws or Regulations of any public body having jurisdiction shall be performed by organizations acceptable to OWNER and CONTRACTOR (or by ENGINEER if so specified).

13.6. If any Work (including the work of others) that is to be inspected, tested or approved is covered without written concurrence of ENGINEER, it must, if requested by ENGINEER, be uncovered for observation. Such uncovering shall be at CONTRACTOR's expense unless CONTRACTOR has given ENGINEER timely notice of CONTRACTOR's intention to cover the same and ENGINEER has not acted with reasonable promptness in response to such notice.

13.7. Neither observations by ENGINEER nor inspections, tests or approvals by others shall relieve CONTRACTOR from CONTRACTOR's obligations to perform the Work in accordance with the Contract Documents.

Uncovering Work:

13.8. If any Work is covered contrary to the written request of ENGINEER, it must, if requested by ENGINEER, be uncovered for ENGINEER's observation and replaced at CONTRACTOR's expense.

13.9. If ENGINEER considers it necessary or advisable that covered Work be observed by ENGINEER or inspected or tested by others, CONTRACTOR, at ENGINEER's request, shall uncover, expose or otherwise make available for observation, inspection or testing as ENGINEER may require, that portion of the Work in question, furnishing all necessary labor, material and equipment. If it is found that such Work is *defective,* CONTRACTOR shall bear all direct, indirect and consequential costs of such uncovering, exposure, observation, inspection and testing and of satisfactory reconstruction, (including but not limited to fees and charges of engineers, architects, attorneys and other professionals), and OWNER shall be entitled to an appropriate decrease in the Contract Price, and, if the parties are unable to agree as to the amount thereof, may make a claim therefor as provided in Article 11. If, however, such Work is not found to be *defective,* CONTRACTOR shall be allowed an increase in the Contract Price or an extension of the Contract Time, or both, directly attributable to such uncovering, exposure, observation, inspection, testing and reconstruction; and, if the parties are unable to agree as to the amount or extent thereof, CONTRACTOR may make a claim therefor as provided in Articles 11 and 12.

Owner May Stop the Work:

13.10. If the Work is *defective,* or CONTRACTOR fails to supply sufficient skilled workers or suitable materials or equipment, or fails to furnish or perform the Work in such a way that the completed Work will conform to the Contract Documents, OWNER may order CONTRACTOR to stop the Work, or any portion thereof, until the cause for such order has been eliminated; however, this right of OWNER to stop the Work shall not give rise to any duty on the part of OWNER to exercise this right for the benefit of CONTRACTOR or any other party.

Correction or Removal of Defective Work:

13.11. If required by ENGINEER, CONTRACTOR shall promptly, as directed, either correct all *defective* Work, whether or not fabricated, installed or completed, or, if the Work has been rejected by ENGINEER, remove it from the site and replace it with *nondefective* Work. CONTRACTOR shall bear all direct, indirect and consequential costs of such correction or removal (including but not limited to fees and charges of engineers, architects, attorneys and other professionals) made necessary thereby.

One Year Correction Period:

13.12. If within one year after the date of Substantial Completion or such longer period of time as may be prescribed by Laws or Regulations or by the terms of any applicable special guarantee required by the Contract Documents or by any specific provision of the Contract Documents, any Work is found to be *defective,* CONTRACTOR shall promptly, without cost to OWNER and in accordance with OWNER's written instructions, either correct such *defective* Work, or, if it has been rejected by OWNER, remove it from the site and replace it with *nondefective* Work. If CONTRACTOR does not promptly comply with the terms of such instructions, or in an emergency where delay would cause serious risk of loss or damage, OWNER may have the *defective* Work corrected or the rejected Work removed and replaced, and all direct, indirect and consequential costs of such removal and replacement (including but not limited to fees and charges of engineers, architects, attorneys and other professionals) will be paid by CONTRACTOR. In special circumstances where a particular item of equipment is placed in continuous service before Substantial Completion of all the Work, the correction period for that item may start to run from an earlier date if so provided in the Specifications or by Written Amendment.

Acceptance of Defective Work:

13.13. If, instead of requiring correction or removal and replacement of *defective* Work, OWNER (and, prior to ENGINEER's recommendation of final payment, also ENGINEER) prefers to accept it, OWNER may do so. CONTRACTOR shall bear all direct, indirect and consequential costs attributable to OWNER's evaluation of and determination to accept such *defective* Work (such costs to be approved by ENGINEER as to reasonableness and to include but not be limited to fees and charges of engineers, architects, attorneys and other professionals). If any such acceptance occurs prior to ENGINEER's recommendation of final payment, a Change Order will be issued incorporating the necessary revisions in the Contract Documents with respect to the Work; and OWNER shall be entitled to an appropriate decrease in the Contract Price, and, if the parties are unable to agree as to the amount thereof, OWNER may make a claim therefor as provided in Article 11. If the acceptance occurs after such recommendation, an appropriate amount will be paid by CONTRACTOR to OWNER.

OWNER May Correct Defective Work:

13.14. If CONTRACTOR fails within a reasonable time after written notice of ENGINEER to proceed to correct and to correct *defective* Work or to remove and replace rejected Work as required by ENGINEER in accordance with paragraph 13.11, or if CONTRACTOR fails to perform the Work in accordance with the Contract Documents, or if CONTRACTOR fails to comply with any other provision of the Contract Documents, OWNER may, after seven days' written notice to CONTRACTOR, correct and remedy any such deficiency. In exercising the rights and remedies under this paragraph OWNER shall proceed expeditiously. To the extent necessary to complete corrective and remedial action, OWNER may exclude CONTRACTOR from all or part of the site, take possession of all or part of the Work, and suspend CONTRACTOR's services related thereto, take possession of CONTRACTOR's tools, appliances, construction equipment and machinery at the site and incorporate in the Work all materials and equipment stored at the site or for which OWNER has paid CONTRACTOR but which are stored elsewhere. CONTRACTOR shall allow OWNER, OWNER's representatives, agents and employees such access to the site as may be necessary to enable OWNER to exercise the rights and remedies under this paragraph. All direct, indirect and consequential costs of OWNER in exercising such rights and remedies will be charged against CONTRACTOR in an amount approved as to reasonableness by ENGINEER, and a Change Order will be issued incorporating the necessary revisions in the Contract Documents with respect to the Work; and OWNER shall be entitled to an appropriate decrease in the Contract Price, and, if the parties are unable to agree as to the amount thereof, OWNER may make a claim therefor as provided in Article 11. Such direct, indirect and consequential costs will include but not be limited to fees and charges of engineers, architects, attorneys and other professionals, all court and arbitration costs and all costs of repair and replacement of work of others destroyed or damaged by correction, removal or replacement of CONTRACTOR's *defective* Work. CONTRACTOR shall not be allowed an extension of the Contract Time because of any delay in performance of the Work attributable to the exercise by OWNER of OWNER's rights and remedies hereunder.

ARTICLE 14—PAYMENTS TO CONTRACTOR AND COMPLETION

Schedule of Values:

14.1. The schedule of values established as provided in paragraph 2.9 will serve as the basis for progress payments and will be incorporated into a form of Application for Payment acceptable to ENGINEER. Progress payments on account of Unit Price Work will be based on the number of units completed.

Application for Progress Payment:

14.2. At least twenty days before each progress payment is scheduled (but not more often than once a month), CONTRACTOR shall submit to ENGINEER for review an Application for Payment filled out and signed by CONTRACTOR covering the Work completed as of the date of the Application and accompanied by such supporting documentation as is required by the Contract Documents. If payment is requested on the basis of materials and equipment not incorporated in the Work but delivered and suitably stored at the site or at another location agreed to in writing, the Application for Payment shall also be accompanied by a bill of sale, invoice or other documentation warranting that OWNER has received the materials and equipment free and clear of all liens, charges, security interests and encumbrances (which are hereinafter in these General Conditions referred to as "Liens") and evidence that the materials and equipment are covered by appropriate property insurance and other arrangements to protect OWNER's interest therein, all of which will be satisfactory to OWNER. The amount of retainage with respect to progress payments will be as stipulated in the Agreement.

CONTRACTOR's Warranty of Title:

14.3. CONTRACTOR warrants and guarantees that title to all Work, materials and equipment covered by any Application for Payment, whether incorporated in the Project or not, will pass to OWNER no later than the time of payment free and clear of all Liens.

Review of Applications for Progress Payment:

14.4. ENGINEER will, within ten days after receipt of each Application for Payment, either indicate in writing a recommendation of payment and present the Application to OWNER, or return the Application to CONTRACTOR indicating in writing ENGINEER's reasons for refusing to recommend payment. In the latter case, CONTRACTOR may make the necessary corrections and resubmit the Application. Ten days after presentation of the Application for Payment with ENGINEER's recommendation, the amount recommended will (subject to the provisions of the last sentence of paragraph 14.7) become due and when due will be paid by OWNER to CONTRACTOR.

14.5. ENGINEER's recommendation of any payment requested in an Application for Payment will constitute a representation by ENGINEER to OWNER, based on ENGINEER's on-site observations of the Work in progress as an experienced and qualified design professional and on ENGINEER's review of the Application for Payment and the accompanying data and schedules that the Work has progressed to the point indicated; that, to the best of ENGINEER's knowledge, information and belief, the quality of the Work is in accordance with the Contract Documents (subject to an evaluation of the Work as a functioning whole prior to or upon Substantial Completion, to the results of any subsequent tests called for in the Contract Documents, to a final determination of quantities and classifications for Unit Price Work under paragraph 9.10, and to any other qualifications stated in the recommendation); and that CONTRACTOR is entitled to payment of the amount recommended. However, by recommending any such payment ENGINEER will not thereby be deemed to have represented that exhaustive or continuous on-site inspections have been made to check the quality or the quantity of the Work beyond the responsibilities specifically assigned to ENGINEER in the Contract Documents or that there may not be other matters or issues between the parties that might entitle CONTRACTOR to be paid additionally by OWNER or OWNER to withhold payment to CONTRACTOR.

14.6. ENGINEER's recommendation of final payment will constitute an additional representation by ENGINEER to OWNER that the conditions precedent to CONTRACTOR's being entitled to final payment as set forth in paragraph 14.13 have been fulfilled.

14.7. ENGINEER may refuse to recommend the whole or any part of any payment if, in ENGINEER's opinion, it would be incorrect to make such representations to OWNER. ENGINEER may also refuse to recommend any such payment, or, because of subsequently discovered evidence or the results of subsequent inspections or tests, nullify any such payment previously recommended, to such extent as may be necessary in ENGINEER's opinion to protect OWNER from loss because:

14.7.1. the Work is *defective,* or completed Work has been damaged requiring correction or replacement,

14.7.2. the Contract Price has been reduced by Written Amendment or Change Order,

14.7.3. OWNER has been required to correct *defective* Work or complete Work in accordance with paragraph 13.14, or

14.7.4. of ENGINEER's actual knowledge of the occurrence of any of the events enumerated in paragraphs 15.2.1 through 15.2.9 inclusive.

OWNER may refuse to make payment of the full amount recommended by ENGINEER because claims have been made against OWNER on account of CONTRACTOR's performance or furnishing of the Work or Liens have been filed in connection with the Work or there are other items entitling OWNER to a set-off against the amount recommended, but OWNER must give CONTRACTOR immediate written notice (with a copy to ENGINEER) stating the reasons for such action.

Substantial Completion:

14.8. When CONTRACTOR considers the entire Work ready for its intended use CONTRACTOR shall notify OWNER and ENGINEER in writing that the entire Work is substantially complete (except for items specifically listed by CONTRACTOR as incomplete) and request that ENGINEER issue a certificate of Substantial Completion. Within a reasonable time thereafter, OWNER, CONTRACTOR and ENGINEER shall make an inspection of the Work to determine the status of completion. If ENGINEER does not consider the Work substantially complete, ENGINEER will notify CONTRACTOR in writing giving the reasons therefor. If ENGINEER considers the Work substantially complete, ENGINEER will prepare and deliver to OWNER a tentative certificate of Substantial Completion which shall fix the date of Substantial Completion. There shall be attached to the certificate a tentative list of items to be completed or corrected before final payment. OWNER shall have seven days after receipt of the tentative certificate during which to make written objection to ENGINEER as to any provisions of the certificate or attached list. If, after considering such objections, ENGINEER concludes that the Work is not substantially complete, ENGINEER will within fourteen days after submission of the tentative certificate to OWNER notify CONTRACTOR in writing, stating the reasons therefor. If, after consideration of OWNER's objections, ENGINEER considers the Work substantially complete, ENGINEER will within said fourteen days execute and deliver to OWNER and CONTRACTOR a definitive certificate of Substantial Completion (with a revised tentative list of items to be completed or corrected) reflecting such changes from the tentative certificate as ENGINEER believes justified after consideration of any objections from OWNER. At the time of delivery of the tentative certificate of Substantial Completion ENGINEER will deliver to OWNER and CONTRACTOR a written recommendation as to division of responsibilities pending final payment between OWNER and CONTRACTOR with respect to security, operation, safety, maintenance, heat, utilities, insurance and warranties. Unless OWNER and CONTRACTOR agree otherwise in writing and so inform ENGINEER prior to ENGINEER's issuing the definitive certificate of Substantial Completion, ENGINEER's aforesaid recommendation will be binding on OWNER and CONTRACTOR until final payment.

14.9. OWNER shall have the right to exclude CONTRACTOR from the Work after the date of Substantial Completion, but OWNER shall allow CONTRACTOR reasonable access to complete or correct items on the tentative list.

Partial Utilization:

14.10. Use by OWNER of any finished part of the Work, which has specifically been identified in the Contract Documents, or which OWNER, ENGINEER and CONTRACTOR agree constitutes a separately functioning and useable part of the Work that can be used by OWNER without significant interference with CONTRACTOR's performance of the remainder of the Work, may be accomplished prior to Substantial Completion of all the Work subject to the following:

14.10.1. OWNER at any time may request CONTRACTOR in writing to permit OWNER to use any such part of the Work which OWNER believes to be ready for its intended use and substantially complete. If CONTRACTOR agrees, CONTRACTOR will certify to OWNER and ENGINEER that said part of the Work is substantially complete and request ENGINEER to issue a certificate of Substantial Completion for that part of the Work. CONTRACTOR at any time may notify OWNER and ENGINEER in writing that CONTRACTOR considers any such part of the Work ready for its intended use and substantially complete and request ENGINEER to issue a certificate of Substantial Completion for that part of the Work. Within a reasonable time after either such request, OWNER, CONTRACTOR and ENGINEER shall make an inspection of that part of the Work to determine its status of completion. If ENGINEER does not consider that part of the Work to be substantially complete, ENGINEER will notify OWNER and CONTRACTOR in writing giving the reasons therefor. If ENGINEER considers that part of the Work to be substantially complete, the provisions of paragraphs 14.8 and 14.9 will apply with respect to certification of Substantial Completion of that part of the Work and the division of responsibility in respect thereof and access thereto.

14.10.2. OWNER may at any time request CONTRACTOR in writing to permit OWNER to take over operation of any such part of the Work although it is not substantially complete. A copy of such request will be sent to ENGINEER and within a reasonable time thereafter OWNER, CONTRACTOR and ENGINEER shall make an inspection of that part of the Work to determine its status of completion and will prepare a list of the items remaining to be completed or corrected thereon before final payment. If CONTRACTOR does not object in writing to OWNER and ENGINEER that such part of the Work is not ready for separate operation by OWNER, ENGINEER will finalize the list of items to be completed or corrected and will deliver such list to OWNER and CONTRACTOR together with a written recommendation as to the division of responsibilities pending final payment between OWNER and CONTRACTOR with respect to security, operation, safety, maintenance, utilities, insurance, warranties and guarantees for that part of the Work which will become binding upon OWNER and CONTRACTOR at the time when OWNER takes over such operation (unless they shall have otherwise agreed in writing and so informed ENGINEER). During such operation and prior to Substantial Completion of such part of the Work, OWNER shall allow CONTRACTOR reasonable access to complete or correct items on said list and to complete other related Work.

14.10.3. No occupancy or separate operation of part of the Work will be accomplished prior to compliance with the requirements of paragraph 5.15 in respect of property insurance.

Final Inspection:

14.11. Upon written notice from CONTRACTOR that the entire Work or an agreed portion thereof is complete, ENGINEER will make a final inspection with OWNER and CONTRACTOR and will notify CONTRACTOR in writing of all particulars in which this inspection reveals that the Work is incomplete or *defective*. CONTRACTOR shall immediately take such measures as are necessary to remedy such deficiencies.

Final Application for Payment:

14.12. After CONTRACTOR has completed all such corrections to the satisfaction of ENGINEER and delivered all maintenance and operating instructions, schedules, guarantees, Bonds, certificates of inspection, marked-up record documents (as provided in paragraph 6.19) and other documents—all as required by the Contract Documents, and after ENGINEER has indicated that the Work is acceptable (subject to the provisions of paragraph 14.16), CONTRACTOR may make application for final payment following the procedure for progress payments. The final Application for Payment shall be accompanied by all documentation called for in the Contract Documents, together with complete and legally effective releases or waivers (satisfactory to OWNER) of all Liens arising out of or filed in connection with the Work. In lieu thereof and as approved by OWNER, CONTRACTOR may furnish receipts or releases in full; an affidavit of CONTRACTOR that the releases and receipts include all labor, services, material and equipment for which a Lien could be filed, and that all payrolls, material and equipment bills, and other indebtedness connected with the Work for which OWNER or OWNER's property might in any way be responsible, have been paid or otherwise satisfied; and consent of the surety, if any, to final payment. If any Subcontractor or Supplier fails to furnish a release or receipt in full, CONTRACTOR may furnish a Bond or other collateral satisfactory to OWNER to indemnify OWNER against any Lien.

Final Payment and Acceptance:

14.13. If, on the basis of ENGINEER's observation of the Work during construction and final inspection, and ENGINEER's review of the final Application for Payment and accompanying documentation—all as required by the Contract Documents, ENGINEER is satisfied that the Work has been completed and CONTRACTOR's other obligations under the Contract Documents have been fulfilled, ENGINEER will, within ten days after receipt of the final Application for Payment, indicate in writing ENGINEER's recommendation of payment and present the Application to OWNER for payment. Thereupon ENGINEER will give written notice to OWNER and CONTRACTOR that the Work is acceptable subject to the provisions of paragraph 14.16.

Otherwise, ENGINEER will return the Application to CONTRACTOR, indicating in writing the reasons for refusing to recommend final payment, in which case CONTRACTOR shall make the necessary corrections and resubmit the Application. Thirty days after presentation to OWNER of the Application and accompanying documentation, in appropriate form and substance, and with ENGINEER's recommendation and notice of acceptability, the amount recommended by ENGINEER will become due and will be paid by OWNER to CONTRACTOR.

14.14. If, through no fault of CONTRACTOR, final completion of the Work is significantly delayed and if ENGINEER so confirms, OWNER shall, upon receipt of CONTRACTOR's final Application for Payment and recommendation of ENGINEER, and without terminating the Agreement, make payment of the balance due for that portion of the Work fully completed and accepted. If the remaining balance to be held by OWNER for Work not fully completed or corrected is less than the retainage stipulated in the Agreement, and if Bonds have been furnished as required in paragraph 5.1, the written consent of the surety to the payment of the balance due for that portion of the Work fully completed and accepted shall be submitted by CONTRACTOR to ENGINEER with the Application for such payment. Such payment shall be made under the terms and conditions governing final payment, except that it shall not constitute a waiver of claims.

Contractor's Continuing Obligation:

14.15. CONTRACTOR's obligation to perform and complete the Work in accordance with the Contract Documents shall be absolute. Neither recommendation of any progress or final payment by ENGINEER, nor the issuance of a certificate of Substantial Completion, nor any payment by OWNER to CONTRACTOR under the Contract Documents, nor any use or occupancy of the Work or any part thereof by OWNER, nor any act of acceptance by OWNER nor any failure to do so, nor any review and approval of a Shop Drawing or sample submission, nor the issuance of a notice of acceptability by ENGINEER pursuant to paragraph 14.13, nor any correction of *defective* Work by OWNER will constitute an acceptance of Work not in accordance with the Contract Documents or a release of CONTRACTOR's obligation to perform the Work in accordance with the Contract Documents (except as provided in paragraph 14.16).

Waiver of Claims:

14.16. The making and acceptance of final payment will constitute:

14.16.1. a waiver of all claims by OWNER against CONTRACTOR, except claims arising from unsettled Liens, from *defective* Work appearing after final inspection pursuant to paragraph 14.11 or from failure to comply with the Contract Documents or the terms of any special guarantees specified therein; however, it will not constitute a waiver by OWNER of any rights in respect of CONTRACTOR's continuing obligations under the Contract Documents; and

14.16.2. a waiver of all claims by CONTRACTOR against OWNER other than those previously made in writing and still unsettled.

ARTICLE 15—SUSPENSION OF WORK AND TERMINATION

Owner May Suspend Work:

15.1. OWNER may, at any time and without cause, suspend the Work or any portion thereof for a period of not more than ninety days by notice in writing to CONTRACTOR and ENGINEER which will fix the date on which Work will be resumed. CONTRACTOR shall resume the Work on the date so fixed. CONTRACTOR shall be allowed an increase in the Contract Price or an extension of the Contract Time, or both, directly attributable to any suspension if CONTRACTOR makes an approved claim therefor as provided in Articles 11 and 12.

Owner May Terminate:

15.2. Upon the occurrence of any one or more of the following events:

15.2.1. if CONTRACTOR commences a voluntary case under any chapter of the Bankruptcy Code (Title 11, United States Code), as now or hereafter in effect, or if CONTRACTOR takes any equivalent or similar action by filing a petition or otherwise under any other federal or state law in effect at such time relating to the bankruptcy or insolvency;

15.2.2. if a petition is filed against CONTRACTOR under any chapter of the Bankruptcy Code as now or hereafter in effect at the time of filing, or if a petition is filed seeking any such equivalent or similar relief against CONTRACTOR under any other federal or state law in effect at the time relating to bankruptcy or insolvency;

15.2.3. if CONTRACTOR makes a general assignment for the benefit of creditors;

15.2.4. if a trustee, receiver, custodian or agent of CONTRACTOR is appointed under applicable law or under contract, whose appointment or authority to take charge of property of CONTRACTOR is for the purpose of enforcing a Lien against such property or for the purpose of general administration of such property for the benefit of CONTRACTOR's creditors;

15.2.5. if CONTRACTOR admits in writing an inability to pay its debts generally as they become due;

15.2.6. if CONTRACTOR persistently fails to perform the Work in accordance with the Contract Documents (including, but not limited to, failure to supply sufficient skilled workers or suitable materials or equipment or failure to adhere to the progress schedule established under paragraph 2.9 as revised from time to time);

15.2.7. if CONTRACTOR disregards Laws or Regulations of any public body having jurisdiction;

15.2.8. if CONTRACTOR disregards the authority of ENGINEER; or

15.2.9. if CONTRACTOR otherwise violates in any substantial way any provisions of the Contract Documents;

OWNER may, after giving CONTRACTOR (and the surety, if there be one) seven days' written notice and to the extent permitted by Laws and Regulations, terminate the services of CONTRACTOR, exclude CONTRACTOR from the site and take possession of the Work and of all CONTRACTOR's tools, appliances, construction equipment and machinery at the site and use the same to the full extent they could be used by CONTRACTOR (without liability to CONTRACTOR for trespass or conversion), incorporate in the Work all materials and equipment stored at the site or for which OWNER has paid CONTRACTOR but which are stored elsewhere, and finish the Work as OWNER may deem expedient. In such case CONTRACTOR shall not be entitled to receive any further payment until the Work is finished. If the unpaid balance of the Contract Price exceeds the direct, indirect and consequential costs of completing the Work (including but not limited to fees and charges of engineers, architects, attorneys and other professionals and court and arbitration costs) such excess will be paid to CONTRACTOR. If such costs exceed such unpaid balance, CONTRACTOR shall pay the difference to OWNER. Such costs incurred by OWNER will be approved as to reasonableness by ENGINEER and incorporated in a Change Order, but when exercising any rights or remedies under this paragraph OWNER shall not be required to obtain the lowest price for the Work performed.

15.3. Where CONTRACTOR's services have been so terminated by OWNER, the termination will not affect any rights or remedies of OWNER against CONTRACTOR then existing or which may thereafter accrue. Any retention or payment of moneys due CONTRACTOR by OWNER will not release CONTRACTOR from liability.

15.4. Upon seven days' written notice to CONTRACTOR and ENGINEER, OWNER may, without cause and without prejudice to any other right or remedy, elect to abandon the Work and terminate the Agreement. In such case, CONTRACTOR shall be paid for all Work executed and any expense sustained plus reasonable termination expenses, which will include, but not be limited to, direct, indirect and consequential costs (including, but not limited to, fees and charges of engineers, architects, attorneys and other professionals and court and arbitration costs).

Contractor May Stop Work or Terminate:

15.5. If, through no act or fault of CONTRACTOR, the Work is suspended for a period of more than ninety days by OWNER or under an order of court or other public authority, or ENGINEER fails to act on any Application for Payment within thirty days after it is submitted, or OWNER fails for thirty days to pay CONTRACTOR any sum finally determined to be due, then CONTRACTOR may, upon seven days' written notice to OWNER and ENGINEER, terminate the Agreement and recover from OWNER payment for all Work executed and any expense sustained plus reasonable termination expenses. In addition and in lieu of terminating the Agreement, if ENGINEER has failed to act on an Application for Payment or OWNER has failed to make any payment as aforesaid, CONTRACTOR may upon seven days' written notice to OWNER and ENGINEER stop the Work until payment of all amounts then due. The provisions of this paragraph shall not relieve CONTRACTOR of the obligations under paragraph 6.29 to carry on the Work in accordance with the progress schedule and without delay during disputes and disagreements with OWNER.

ARTICLE 16—ARBITRATION

16.1. All claims, disputes and other matters in question between OWNER and CONTRACTOR arising out of, or relating to the Contract Documents or the breach thereof (except for claims which have been waived by the making or acceptance of final payment as provided by paragraph 14.16) will be decided by arbitration in accordance with the Construction Industry Arbitration Rules of the American Arbitration Association then obtaining subject to the limitations of this Article 16. This agreement so to arbitrate and any other agreement or consent to arbitrate entered into in accordance herewith as provided in this Article 16 will be specifically enforceable under the prevailing law of any court having jurisdiction.

16.2. No demand for arbitration of any claim, dispute or other matter that is required to be referred to ENGINEER initially for decision in accordance with paragraph 9.11 will be made until the earlier of (a) the date on which ENGINEER has rendered a decision or (b) the tenth day after the parties have presented their evidence to ENGINEER if a written decision has not been rendered by ENGINEER before that date. No demand for arbitration of any such claim, dispute or other matter will be made later than thirty days after the date on which ENGINEER has rendered a written decision in respect thereof in accordance with paragraph 9.11; and the failure to demand arbitration within said thirty days' period shall result in ENGINEER's decision being final and binding upon OWNER and CONTRACTOR. If ENGINEER renders a decision after arbitration proceedings have been initiated, such decision may be entered as evidence but will not supersede the arbitration proceedings, except where the decision is acceptable to the parties concerned. No demand for arbitration of any written decision of ENGINEER rendered in accordance with paragraph 9.10 will be made later than ten days after the party making such demand has delivered written notice of intention to appeal as provided in paragraph 9.10.

16.3. Notice of the demand for arbitration will be filed in writing with the other party to the Agreement and with the American Arbitration Association, and a copy will be sent to ENGINEER for information. The demand for arbitration will be made within the thirty-day or ten-day period specified in paragraph 16.2 as applicable, and in all other cases within a reasonable time after the claim, dispute or other matter in question has arisen, and in no event shall any such demand be made after the date when institution of legal or equitable proceedings based on such claim, dispute or other matter in question would be barred by the applicable statute of limitations.

16.4. No arbitration arising out of or relating to the Contract Documents shall include by consolidation, joinder or in any other manner any other person or entity (including ENGINEER, ENGINEER's agents, employees or consultants) who is not a party to this contract unless:

16.4.1. the inclusion of such other person or entity is necessary if complete relief is to be afforded among those who are already parties to the arbitration,

16.4.2. such other person or entity is substantially involved in a question of law or fact which is common to those who are already parties to the arbitration and which will arise in such proceedings, and

16.4.3. the written consent of the other person or entity sought to be included and of OWNER and CONTRACTOR has been obtained for such inclusion, which consent shall make specific reference to this paragraph; but no such consent shall constitute consent to arbitration of any dispute not specifically described in such consent or to arbitration with any party not specifically identified in such consent.

16.5. The award rendered by the arbitrators will be final, judgment may be entered upon it in any court having jurisdiction thereof, and will not be subject to modification or appeal except to the extent permitted by Sections 10 and 11 of the Federal Arbitration Act (9 U.S.C. §§10,11).

ARTICLE 17—MISCELLANEOUS

Giving Notice:

17.1. Whenever any provision of the Contract Documents requires the giving of written notice, it will be deemed to have been validly given if delivered in person to the individual or to a member of the firm or to an officer of the corporation for whom it is intended, or if delivered at or sent by registered or certified mail, postage prepaid, to the last business address known to the giver of the notice.

Computation of Time:

17.2.1. When any period of time is referred to in the Contract Documents by days, it will be computed to exclude the first and include the last day of such period. If the last day of any such period falls on a Saturday or Sunday or on a day made a legal holiday by the law of the applicable jurisdiction, such day will be omitted from the computation.

17.2.2. A calendar day of twenty-four hours measured from midnight to the next midnight shall constitute a day.

General:

17.3. Should OWNER or CONTRACTOR suffer injury or damage to person or property because of any error, omission or act of the other party or of any of the other party's employees or agents or others for whose acts the other party is legally liable, claim will be made in writing to the other party within a reasonable time of the first observance of such injury or damage. The provisions of this paragraph 17.3 shall not be construed as a substitute for or a waiver of the provisions of any applicable statute of limitations or repose.

17.4. The duties and obligations imposed by these General Conditions and the rights and remedies available hereunder to the parties hereto, and, in particular but without limitation, the warranties, guarantees and obligations imposed upon CONTRACTOR by paragraphs 6.30, 13.1, 13.12, 13.14, 14.3 and 15.2 and all of the rights and remedies available to OWNER and ENGINEER thereunder, are in addition to, and are not to be construed in any way as a limitation of, any rights and remedies available to any or all of them which are otherwise imposed or available by Laws or Regulations, by special warranty or guarantee or by other provisions of the Contract Documents, and the provisions of this paragraph will be as effective as if repeated specifically in the Contract Documents in connection with each particular duty, obligation, right and remedy to which they apply. All representations, warranties and guarantees made in the Contract Documents will survive final payment and termination or completion of the Agreement.

This document has important legal consequences; consultation with an attorney is encouraged with respect to its completion or modification.

STANDARD FORM OF AGREEMENT BETWEEN OWNER AND CONTRACTOR ON THE BASIS OF A STIPULATED PRICE

Prepared by

ENGINEERS JOINT CONTRACT DOCUMENTS COMMITTEE

and

Issued and Published Jointly By

PROFESSIONAL ENGINEERS IN PRIVATE PRACTICE
A practice division of the
NATIONAL SOCIETY OF PROFESSIONAL ENGINEERS

AMERICAN CONSULTING ENGINEERS COUNCIL

AMERICAN SOCIETY OF CIVIL ENGINEERS

CONSTRUCTION SPECIFICATIONS INSTITUTE

This document has been approved and endorsed by

The Associated General Contractors of America

This Standard Form of Agreement has been prepared for use with the Standard General Conditions of the Construction Contract, No. 1910-8, 1983 edition. Their provisions are interrelated and a change in one may necessitate a change in the others. The suggested language for instructions to bidders contained in the Guide to the Preparation of Instructions to Bidders, No. 1910-12, 1983 edition, is also carefully interrelated with the language of this Agreement. Comments concerning their usage are contained in the Commentary on Agreements for Engineering Services and Contract Documents, No. 1910-9, 1981 edition. See also Guide to the Preparation of Supplementary Conditions, No. 1910-17, 1983 edition.

EJCDC No. 1910-8-A-1 (1983 Edition)

Reprinted 6/88

NOTE TO USER

Certain states and federal agencies require provisions in public contracts which permit Contractors to deposit acceptable securities with Owner or a stakeholder in lieu of retainage. Many Owners will not accept this procedure except where required by Laws or Regulations. In the event such a procedure is required, the provisions of this Agreement and possibly those of the other Contract Documents dealing with retainage should be amended, and an attorney should be consulted to prepare the revised language. Among the issues to be addressed by such language are: initial and subsequent valuations of the securities, right to withdraw excess collateral and obligation to deposit additional collateral as market value changes, who is entitled to interest and dividends on deposited collateral, rights of construction lender in deposited collateral, responsibilities of stakeholder, may collateral be freely sold in the event of Contractor default and method of such sale, and application of Uniform Commercial Code and state and federal security laws to the arrangement.

American Consulting Engineers Council
1015 15th Street, N.W., Washington, DC 20005

American Society of Civil Engineers
345 East 47th Street, New York, NY 10017

Construction Specifications Institute
601 Madison Street, Alexandria, VA 22314

EJCDC
STANDARD FORM OF AGREEMENT BETWEEN OWNER AND CONTRACTOR ON THE BASIS OF A STIPULATED PRICE

THIS AGREEMENT is dated as of the ____________ day of ____________________ in the year 19______ by and between ____

____ (hereinafter called OWNER) and

____

(hereinafter called CONTRACTOR).

OWNER and CONTRACTOR, in consideration of the mutual covenants hereinafter set forth, agree as follows:

Article 1. WORK.

CONTRACTOR shall complete all Work as specified or indicated in the Contract Documents. The Work is generally described as follows:

The Project for which the Work under the Contract Documents may be the whole or only a part is generally described as follows:

Article 2. ENGINEER.

The Project has been designed by

who is hereinafter called ENGINEER and who is to act as OWNER's representative, assume all duties and responsibilities and have the rights and authority assigned to ENGINEER in the Contract Documents in connection with completion of the Work in accordance with the Contract Documents.

Article 3. CONTRACT TIME.

(Strike Inapplicable Paragraph)

3.1. The Work will be substantially completed on or before ____________________, 19______, and completed and ready for final payment in accordance with paragraph 14.13 of the General Conditions on or before ____________________, 19 ______.

3.1. The Work will be substantially completed within ________days from the date when the Contract Time commences to run as provided in paragraph 2.3 of the General Conditions, and completed and ready for final payment in accordance with paragraph 14.13 of the General Conditions within ________ days from the date when the Contract Time commences to run.

3.2. *Liquidated Damages.* OWNER and CONTRACTOR recognize that time is of the essence of this Agreement and that OWNER will suffer financial loss if the Work is not completed within the times specified in paragraph 3.1 above, plus any extensions thereof allowed in accordance with Article 12 of the General Conditions. They also recognize the delays, expense and difficulties involved in proving in a legal or arbitration proceeding the actual loss suffered by OWNER if the Work is not completed on time. Accordingly, instead of requiring any such proof, OWNER and CONTRACTOR agree that as liquidated damages for delay (but not as a penalty) CONTRACTOR shall pay OWNER ____________ ________________________________ dollars ($____________) for each day that expires after the time specified in paragraph 3.1 for Substantial Completion until the Work is substantially complete. After Substantial Completion if CONTRACTOR shall neglect, refuse or fail to complete the remaining Work within the Contract Time or any proper extension thereof granted by

OWNER, CONTRACTOR shall pay OWNER ______________________________
______________________ dollars ($______________________) for each day that expires after the time specified in paragraph 3.1 for completion and readiness for final payment.

Article 4. CONTRACT PRICE.

4.1. OWNER shall pay CONTRACTOR for completion of the Work in accordance with the Contract Documents in current funds as follows:

[here insert a lump sum, unit prices or both, if necessary attach exhibits and list them in Article 8.]

[CONTRACTOR's Bid may be attached as an exhibit to avoid lengthy retyping of unit price schedules, formulae for escalation of prices, information as to alternatives, etc.]

Article 5. PAYMENT PROCEDURES.

CONTRACTOR shall submit Applications for Payment in accordance with Article 14 of the General Conditions. Applications for Payment will be processed by ENGINEER as provided in the General Conditions.

5.1. *Progress Payments*. OWNER shall make progress payments on account of the Contract Price on the basis of CONTRACTOR's Applications for Payment as recommended by ENGINEER, on or about the ________ day of each month during construction as provided below. All progress payments will be on the basis of the progress of the Work measured by the schedule of values established in paragraph 2.9 of the General Conditions (and in the case of Unit Price Work based on the number of units completed) or, in the event there is no schedule of values, as provided in the General Requirements.

5.1.1. Prior to Substantial Completion, progress payments will be made in an amount equal to the percentage indicated below, but, in each case, less the aggregate of payments previously made and less such amounts as ENGINEER shall determine, or OWNER may withhold, in accordance with paragraph 14.7 of the General Conditions.

_______% of Work completed. If Work has been 50% completed as determined by ENGINEER, and if the character and progress of the Work have been satisfactory to OWNER and ENGINEER, OWNER on recommendation of ENGINEER, may determine that as long as the character and progress of the Work remain satisfactory to them, there will be no additional retainage on account of Work completed in which case the remaining progress payments prior to Substantial Completion will be in an amount equal to 100% of the Work completed.

_______% of materials and equipment not incorporated in the Work (but delivered, suitably stored and accompanied by documentation satisfactory to OWNER as provided in paragraph 14.2 of the General Conditions).

5.1.2. Upon Substantial Completion, in an amount sufficient to increase total payments to CONTRACTOR to _______% of the Contract Price, less such amounts as ENGINEER shall determine, or OWNER may withhold, in accordance with paragraph 14.7 of the General Conditions.

5.2. *Final Payment*. Upon final completion and acceptance of the Work in accordance with paragraph 14.13 of the General Conditions, OWNER shall pay the remainder of the Contract Price as recommended by ENGINEER as provided in said paragraph 14.13.

Article 6. INTEREST.

All moneys not paid when due as provided in Article 14 of the General Conditions shall bear interest at the maximum rate allowed by law at the place of the Project.

Article 7. CONTRACTOR'S REPRESENTATIONS.

In order to induce OWNER to enter into this Agreement CONTRACTOR makes the following representations:

7.1. CONTRACTOR has familiarized itself with the nature and extent of the Contract Documents, Work, site, locality, and all local conditions and Laws and Regulations that in any manner may affect cost, progress, performance or furnishing of the Work.

7.2. CONTRACTOR has studied carefully all reports of explorations and tests of subsurface conditions and drawings of physical conditions which are identified in the Supplementary Conditions as provided in paragraph 4.2 of the General Conditions, and accepts the determination set forth in paragraph SC-4.2 of the Supplementary Conditions of the extent of the technical data contained in such reports and drawings upon which CONTRACTOR is entitled to rely.

7.3. CONTRACTOR has obtained and carefully studied (or assumes responsibility for obtaining and carefully studying) all such examinations, investigations, explorations, tests, reports and studies (in addition to or to supplement those referred to in paragraph 7.2 above) which pertain to the subsurface or physical conditions at or contiguous to the site or otherwise may affect the cost, progress, performance or furnishing of the Work as CONTRACTOR considers necessary for the performance or furnishing of the Work at the Contract Price, within the Contract Time and in accordance with the other terms and conditions of the Contract Documents, including specifically the provisions of paragraph 4.2 of the General Conditions; and no additional examinations, investigations, explorations, tests, reports, studies or similar information or data are or will be required by CONTRACTOR for such purposes.

7.4. CONTRACTOR has reviewed and checked all information and data shown or indicated on the Contract Documents with respect to existing Underground Facilities at or contiguous to the site and assumes responsibility for the accurate location of said Underground Facilities. No additional examinations, investigations, explorations, tests, reports, studies or similar information or data in respect of said Underground Facilities are or will be required by CONTRACTOR in order to perform and furnish the Work at the Contract Price, within the Contract Time and in accordance with the other terms and conditions of the Contract Documents, including specifically the provisions of paragraph 4.3 of the General Conditions.

7.5. CONTRACTOR has correlated the results of all such observations, examinations, investigations, explorations, tests, reports and studies with the terms and conditions of the Contract Documents.

7.6. CONTRACTOR has given ENGINEER written notice of all conflicts, errors or discrepancies that he has discovered in the Contract Documents and the written resolution thereof by ENGINEER is acceptable to CONTRACTOR.

Article 8. CONTRACT DOCUMENTS.

The Contract Documents which comprise the entire agreement between OWNER and CONTRACTOR concerning the Work consist of the following:

8.1. This Agreement (pages 1 to ______, inclusive).

8.2. Exhibits to this Agreement (pages ______ to ______, inclusive).

8.3. Performance and other Bonds, identified as exhibits ______________________ and consisting of ______ pages.

8.4. Notice of Award.

8.5. General Conditions (pages ______ to ______, inclusive).

8.6. Supplementary Conditions (pages ______ to ______, inclusive).

3

8.7. Specifications bearing the title ______________________________

and consisting of _______ divisions and _______ pages, as listed in table of contents thereof.

8.8. Drawings, consisting of a cover sheet and sheets numbered _______ through _______, inclusive with each sheet bearing the following general title:

[Fill in, and, if a set of Drawings is not attached to each signed counterpart of Agreement, so indicate, in which case OWNER and CONTRACTOR should initial or otherwise appropriately identify all Drawings.]

8.9. Addenda numbers _______ to _______, inclusive.

8.10. CONTRACTOR's Bid (pages _______ to _______, inclusive) marked exhibit _______.

[Attach Bid Form only in special circumstances.]

8.11. Documentation submitted by CONTRACTOR prior to Notice of Award (pages ____________ to ____________, inclusive).

8.12. The following which may be delivered or issued after the Effective Date of the Agreement and are not attached hereto: All Written Amendments and other documents amending, modifying, or supplementing the Contract Documents pursuant to paragraphs 3.4 and 3.5 of the General Conditions.

8.13. The documents listed in paragraphs 8.2 et seq. above are attached to this Agreement (except as expressly noted otherwise above).

There are no Contract Documents other than those listed above in this Article 8. The Contract Documents may only be amended, modified or supplemented as provided in paragraphs 3.4 and 3.5 of the General Conditions.

Article 9. MISCELLANEOUS.

9.1. Terms used in this Agreement which are defined in Article 1 of the General Conditions will have the meanings indicated in the General Conditions.

9.2. No assignment by a party hereto of any rights under or interests in the Contract Documents will be binding on another party hereto without the written consent of the party sought to be bound; and specifically but without limitation moneys that may become due and moneys that are due may not be assigned without such consent (except to the extent that the effect of this restriction may be limited by law), and unless specifically stated to the contrary in any written consent to an assignment no assignment will release or discharge the assignor from any duty or responsibility under the Contract Documents.

9.3. OWNER and CONTRACTOR each binds itself, its partners, successors, assigns and legal representatives to the other party hereto, its partners, successors, assigns and legal representatives in respect of all covenants, agreements and obligations contained in the Contract Documents.

Article 10. OTHER PROVISIONS.

IN WITNESS WHEREOF, OWNER and CONTRACTOR have signed this Agreement in triplicate. One counterpart each has been delivered to OWNER, CONTRACTOR and ENGINEER. All portions of the Contract Documents have been signed or identified by OWNER and CONTRACTOR or by ENGINEER on their behalf.

This Agreement will be effective on ______________________________, 19______.

OWNER ______________________________

____

By ______________________________

[CORPORATE SEAL]

Attest ______________________________

Address for giving notices

____

____

(If OWNER is a public body, attach evidence of authority to sign and resolution or other documents authorizing execution of Agreement.)

CONTRACTOR ______________________________

____

By ______________________________

[CORPORATE SEAL]

Attest ______________________________

Address for giving notices

____

____

License No. ______________________________

Agent for service of process: ______________

____

(If CONTRACTOR is a corporation, attach evidence of authority to sign.)

This document has important legal consequences; consultation with an attorney is encouraged with respect to its completion or modification.

STANDARD FORM OF AGREEMENT BETWEEN OWNER AND CONTRACTOR ON THE BASIS OF COST-PLUS

Prepared by

ENGINEERS JOINT CONTRACT DOCUMENTS COMMITTEE

and

Issued and Published Jointly By

PROFESSIONAL ENGINEERS IN PRIVATE PRACTICE
A practice division of the
NATIONAL SOCIETY OF PROFESSIONAL ENGINEERS

AMERICAN CONSULTING ENGINEERS COUNCIL

AMERICAN SOCIETY OF CIVIL ENGINEERS

CONSTRUCTION SPECIFICATIONS INSTITUTE

This document has been approved and endorsed by

The Associated General Contractors of America

This Standard Form of Agreement has been prepared for use with the Standard General Conditions of the Construction Contract, No. 1910-8, 1983 edition. Their provisions are interrelated and a change in one may necessitate a change in the others. The suggested language for instructions of bidders contained in the Guide to the Preparation of Instructions to Bidders, No. 1910-12, 1983 edition, is also carefully interrelated with the language of this Agreement. Comments concerning their usage are contained in the Commentary on Agreements for Engineering Services and Contract Documents, No. 1910-9, 1981 edition. See also Guide to the Preparation of Supplementary Conditions, No. 1910-17, 1983 edition.

EJCDC No. 1910-8-A-2 (1983 Edition)
Reprinted 8/88

Note to User

Certain states and federal agencies require provisions in public contracts which permit Contractors to deposit acceptable securities with Owner or a stakeholder in lieu of retainage. Many Owners will not accept this procedure except where required by Laws or Regulations. In the event such a procedure is required, the provisions of this Agreement and possibly those of the other Contract Documents dealing with retainage should be amended, and an attorney should be consulted to prepare the revised language. Among the issues to be addressed by such language are: initial and subsequent valuations of the securities, right to withdraw excess collateral and obligation to deposit additional collateral as market value changes, who is entitled to interest and dividends on deposited collateral, responsibilities of stakeholder, may collateral be freely sold in the event of Contractor default and method of such sale, and application of Uniform Commercial Code and state and federal security laws to the arrangement.

1420 King Street, Alexandria, VA 22314

American Consulting Engineers Council
1015 15th Street, N.W., Washington, D.C. 20005

American Society of Civil Engineers
345 East 47th Street, New York, NY 10017

Construction Specifications Institute
601 Madison Street, Alexandria, VA 22314

STANDARD FORM OF AGREEMENT BETWEEN OWNER AND CONTRACTOR ON THE BASIS OF COST-PLUS

THIS AGREEMENT is dated as of the __________ day of ____________________ in the year 19____

by and between ____

____ (hereinafter called OWNER) and

____

(hereinafter called CONTRACTOR).

OWNER and CONTRACTOR, in consideration of the mutual covenants hereinafter set forth, agree as follows:

Article 1. WORK.

CONTRACTOR shall complete all Work as specified or indicated in the Contract Documents. The Work is generally described as follows:

The Project for which the Work under the Contract Documents may be the whole or only a part is generally described as follows:

Article 2. ENGINEER.

The Project has been designed by

who is hereinafter called ENGINEER and who is to act as OWNER's representative, assume all duties and responsibilities and have the rights and authority assigned to ENGINEER in the Contract Documents in connection with completion of the Work in accordance with the Contract Documents.

Article 3. CONTRACT TIME.

(Strike Inapplicable Paragraph)

3.1. The Work will be substantially completed on or before ____________________, 19______, and completed and ready for final payment in accordance with paragraph 14.13 of the General Conditions on or before ____________________, 19______.

3.1. The Work will be substantially completed within __________days from the date when the Contract Time commences to run as provided in paragraph 2.3 of the General Conditions, and completed and ready for final payment in accordance with paragraph 14.13 of the General Conditions within ________ days from the date when the Contract Time commences to run.

3.2. *Liquidated Damages*. OWNER and CONTRACTOR recognize that time is of the essence of this Agreement and that OWNER will suffer financial loss if the Work is not completed within the times specified in paragraph 3.1 above, plus any extensions thereof allowed in accordance with Article 12 of the General Conditions. They also recognize the delays, expense and difficulties involved in proving in a legal or arbitration proceeding the actual loss suffered by OWNER if the Work is not completed on time. Accordingly, instead of requiring any such proof, OWNER and CONTRACTOR agree that as liquidated damages for delay (but not as a penalty) CONTRACTOR shall pay OWNER __________ ____ dollars ($____________________) for each day that expires after the time specified in paragraph 3.1 for Substantial Completion until the Work is substantially complete. After Substantial Completion if CONTRACTOR shall neglect, refuse or fail to complete the remaining Work within the Contract Time or any proper extension thereof granted by OWNER, CONTRACTOR shall pay OWNER ______________________________ dollars ($______________) for each day that expires after the time specified in paragraph 3.1 for completion and readiness for final payment.

Article 4. CONTRACT PRICE.

OWNER shall pay CONTRACTOR for completion of the Work in accordance with the Contract Documents an amount in current funds equal to the Cost of the Work plus a Contractor's Fee for overhead and profit, both of which shall be determined as provided below.

Article 5. COST OF WORK.

Cost of the Work shall be determined as provided in paragraphs 11.4 and 11.5 of the General Conditions, but, in addition to any limitations therein set forth, it shall not include costs in excess to any Guaranteed Maximum Contract Price as set forth in Article 7 hereof. Whenever any portion of the Work is to be performed for CONTRACTOR by Subcontractors, CONTRACTOR shall have identified in writing, given to OWNER prior to the Notice of Award, those portions of the Work that he proposes to subcontract, and after the Notice of Award may only subcontract other portions of the Work with OWNER's written consent.

Article 6. CONTRACTOR'S FEE.

The Contractor's Fee shall be determined as follows:

(Strike Inapplicable Paragraph)

6.1. A fixed fee of ______________________________ dollars ($______________) which shall be subject to increases or decreases for changes in the Work as provided in paragraph 8.1 below.

6.1. A fee based on the following percentages of the various portions of the Cost of the Work:

Payroll costs (see paragraph 11.4.1 of General Conditions) __________ %,

Material and equipment costs (see paragraph 11.4.2 of General Conditions) __________ %,

Amounts paid to Subcontractors (see paragraph 11.4.3 of General Conditions) __________ %,

Amounts paid to special consultants (see paragraph 11.4.4 of General Conditions) __________ %, and

Supplemental costs (see paragraph 11.4.5 of General Conditions) __________ %.

No fee shall be payable on the basis of costs itemized in paragraph 11.5 of the General Conditions.

The provisions in paragraph 11.6 of the General Conditions shall apply only to changes in the Work.

CONTRACTOR guarantees that the maximum amount payable by OWNER in accordance with this paragraph 6.1 as a percentage fee will not exceed ______________________________ dollars ($ __________) subject to increases or decreases for changes in the Work as provided in paragraph 8.3 below.

Article 7. GUARANTEED MAXIMUM CONTRACT PRICE.

CONTRACTOR guarantees that the maximum cost to OWNER of the Cost of the Work and the Contractor's Fee will not exceed ______________________________ dollars ($ ______________) subject to increases or decreases for changes in the Work.

2

Article 8. CHANGES IN THE WORK.

The amount of any increases or decreases in the Contractor's Fee and in any Guaranteed Maximum Contract Price or Fee which results from a Change Order shall be set forth in the applicable Change Order subject to the following:

8.1. If the Contractor's Fee is a fixed fee: any increase or decrease in the Contractor's Fee resulting from net additions or decreases in the Cost of the Work shall be determined in accordance with paragraph 11.6.2 of the General Conditions.

8.2. If the Contractor's Fee is a percentage fee not subject to any guaranteed maximum limitation: Contractor's Fee will adjust automatically as the Cost of the Work changes.

8.3. Wherever there is a Guaranteed Maximum Contract Price or Fee: in the case of net additions in the Work, the amount of any increase in the Guaranteed Maximum shall be determined in accordance with paragraphs 11.3 through 11.6, inclusive, of the General Conditions; and in the case of net deletions in the Work, the amount of any such decrease shall be determined in accordance with paragraph 11.6.2 of the General Conditions and any Guaranteed Maximum shall be reduced by mutual agreement.

Article 9. PAYMENT PROCEDURES.

Contractor shall submit Applications for Payment in accordance with Article 14 of the General Conditions. Applications for Payment will indicate the amount of the Contractor's Fee then payable. Applications for Payment will be processed by ENGINEER as provided in the General Conditions.

9.1. *Progress Payments.* OWNER shall make progress payments on account of the Contract Price on the basis of CONTRACTOR's Applications for Payment as recommended by ENGINEER, on or about the __________ day of each month during construction as provided below. All such payments will be subject to the limitations of any Guaranteed Maximum Contract Price or Fee and will be measured by the schedule of values established in paragraph 2.9 of the General Conditions (and in the case of Unit Price Work based on the number of units completed) or, in the event there is no schedule of values, as provided in the General Requirements.

9.1.1. *For Cost of Work:* Progress payments on account of the Cost of the Work will be made:

9.1.1.1. Prior to Substantial Completion, progress payments will be made in an amount equal to the percentage indicated below, but, in each case, less the aggregate of payments previously made and less such amounts as ENGINEER shall determine, or OWNER may withhold, in accordance with paragraph 14.7 of the General Conditions:

__________ % of Cost of the Work completed. If Work has been 50% completed as determined by ENGINEER, and if the character and progress of the Work have been satisfactory to OWNER and ENGINEER, OWNER on recommendation of ENGINEER, may determine that as long as the character and progress of the Work remain satisfactory to them, there will be no additional retainage on account of Work completed in which case the remaining progress payments prior to Substantial Completion will be in an amount equal to 100% of the Work completed.

__________ % of Cost of the Work applicable to materials and equipment not incorporated in the Work (but delivered, suitably stored and accompanied by documentation satisfactory to OWNER as provided in paragraph 14.2 of the General Conditions).

9.1.1.2. Upon Substantial Completion, in an amount sufficient to increase the total payments to CONTRACTOR to __________ % of the Cost of the Work, less such amounts as ENGINEER shall determine, or OWNER may withhold, in accordance with paragraph 14.7 of the General Conditions.

9.1.2. *For Contractor's Fee:* Progress payments on account of the Contractor's Fee will be made:

9.1.2.1. If the Contractor's Fee is a fixed fee: payments prior to Substantial Completion will be in an amount equal to __________ % of such fee earned to the date of the approved Application for Payment (less in each case payments previously made on account of such fee) based on the progress of the Work measured by the schedule of values established as provided in paragraph 2.9 of the General Conditions, (and in the case of Unit Price Work on the number of units completed) and upon Substantial Completion in an amount sufficient to increase total payments to CONTRACTOR on account of his fee to __________ % of the Contractor's Fee. In the event there is no schedule of values the progress of the Work will be measured as provided in the General Requirements.

9.1.2.2. If the Contractor's Fee is a percentage fee: payments prior to Substantial Completion will be in an amount equal to __________ % of such fee (less in each case payments previously made on account of such fee) based on the Cost of the Work completed, and upon Substantial Completion in an amount sufficient to increase total payments to CONTRACTOR on account of that fee to __________ % of the Contractor's Fee.

9.2. *Final Payment.* Upon final completion and acceptance of the Work in accordance with paragraph 14.13 of the General Conditions, OWNER shall pay the remainder of the Contract Price as recommended by ENGINEER as provided in said paragraph 14.13.

Article 10. INTEREST.

All moneys not paid when due as provided in Article 14 of the General Conditions shall bear interest at the maximum rate allowed by law at the place of the Project.

Article 11. CONTRACTOR'S REPRESENTATIONS.

In order to induce OWNER to enter into this Agreement CONTRACTOR makes the following representations:

11.1. CONTRACTOR has familiarized itself with the nature and extent of the Contract Documents, Work, site, locality, and all local conditions and Laws and Regulations that in any manner may affect cost, progress, performance or furnishing of the Work.

11.2. CONTRACTOR has studied carefully all reports of explorations and tests of subsurface conditions and drawings of physical conditions which are identified in the Supplementary Conditions as provided in paragraph 4.2 of the General Conditions and accepts the determination set forth in paragraph SC-4.2 of the Supplementary Conditions of the extent of the technical data contained in such reports and drawings upon which CONTRACTOR is entitled to rely.

11.3. CONTRACTOR has obtained and carefully studied (or assumes responsibility for obtaining and carefully studying) all such examinations, investigations, explorations, tests, reports and studies (in addition to or to supplement those referred to in paragraph 11.2 above) which pertain to the subsurface or physical conditions at or contiguous to the site or otherwise may affect the cost, progress, performance or furnishing of the Work as CONTRACTOR considers necessary for the performance or furnishing of the Work at the Contract Price, within the Contract Time and in accordance with the other terms and conditions of the Contract Documents, including specifically the provisions of paragraph 4.2 of the General Conditions; and no additional examinations, investigations, explorations, tests, reports, studies or similar information or data are or will be required by CONTRACTOR for such purposes.

11.4. CONTRACTOR has reviewed and checked all information and data shown or indicated on the Contract Documents with respect to existing Underground Facilities at or contiguous to the site and assumes responsibility for the accurate location of said Underground Facilities. No additional examinations, investigations, explorations, tests, reports, studies or similar information or data in respect of said Underground Facilities are or will be required by CONTRACTOR in order to perform and furnish the Work at the Contract Price, within the Contract Time and in accordance with the other terms and conditions of the Contract Documents, including specifically the provisions of paragraph 4.3 of the General Conditions.

11.5. CONTRACTOR has correlated the results of all such observations, examinations, investigations, explorations, tests, reports and studies with the terms and conditions of the Contract Documents.

11.6. CONTRACTOR has given ENGINEER written notice of all conflicts, errors or discrepancies that he has discovered in the Contract Documents and the written resolution thereof by ENGINEER is acceptable to CONTRACTOR.

Article 12. ACCOUNTING RECORDS.

CONTRACTOR shall check all materials, equipment and labor entering into the Work and shall keep such full and detailed accounts as may be necessary for proper financial management under this Agreement, and the accounting methods shall be satisfactory to OWNER. OWNER shall be afforded access to all CONTRACTOR's records, books, correspondence, instructions, drawings, receipts, vouchers, memoranda and similar data relating to the Cost of the Work and Contractor's Fee. CONTRACTOR shall preserve all such documents for a period of three years after the final payment by OWNER.

Article 13. CONTRACT DOCUMENTS.

The Contract Documents which comprise the entire agreement between OWNER and CONTRACTOR concerning the Work consist of the following:

13.1. This Agreement (pages 1 to __________, inclusive).

13.2. Exhibits to this Agreement (pages __________ to __________, inclusive).

13.3. Performance and other Bonds, identified as exhibits __________ and consisting of __________ pages.

13.4. Notice of Award.

13.5. General Conditions (pages __________ to __________, inclusive).

13.6. Supplementary Conditions (pages __________ to __________, inclusive).

13.7. Specifications bearing the title ______________________________ and consisting of __________ divisions and __________ pages, as listed in table of contents thereof.

13.8. Drawings, consisting of a cover sheet and sheets numbered __________ through __________, inclusive with each sheet bearing the following general title:

[Fill in, and, if a set of Drawings is not attached to each signed counterpart of Agreement, so indicate in which case OWNER and CONTRACTOR should initial or otherwise appropriately identify all Drawings.]

13.9. Addenda numbers __________ to __________, inclusive.

13.10. CONTRACTOR's Bid (pages __________ to __________, inclusive) marked exhibit _____.

[Attach Bid Form only in special circumstances.]

13.11. Documentation submitted by CONTRACTOR prior to Notice of Award (pages __________ to __________, inclusive).

13.12. The following which may be delivered or issued after the Effective Date of the Agreement and are not attached hereto: All Written Amendments and other documents amending, modifying or supplementing the Contract Documents pursuant to paragraphs 3.4 and 3.5 of the General Conditions.

13.13. The documents listed in paragraphs 13.2 et seq. above are attached to this Agreement (except as expressly noted otherwise above).

There are no Contract Documents other than those listed above in this Article 13. The Contract Documents may only be amended, modified or supplemented as provided in paragraphs 3.4 and 3.5 of the General Conditions.

Article 14. MISCELLANEOUS.

14.1. Terms used in this Agreement which are defined in Article 1 of the General Conditions will have the meanings indicated in the General Conditions.

14.2. No assignment by a party hereto of any rights under or interests in the Contract Documents will be binding on another party hereto without the written consent of the party sought to be bound; and specifically but without limitation moneys that may become due and moneys that are due may not be assigned without such consent (except to the extent that the effect of this restriction may be limited by law), and unless specifically stated to the contrary in any written consent to an assignment no assignment will release or discharge the assignor from any duty or responsibility under the Contract Documents.

14.3. OWNER and CONTRACTOR each binds itself, its partners, successors, assigns and legal representatives to the other party hereto, its partners, successors, assigns and legal representatives in respect to all covenants, agreements and obligations contained in the Contract Documents.

Article 15. OTHER PROVISIONS.

IN WITNESS WHEREOF, OWNER and CONTRACTOR have signed this Agreement in triplicate. One counterpart each has been delivered to OWNER, CONTRACTOR and ENGINEER. All portions of the Contract Documents have been signed or identified by OWNER and CONTRACTOR or by ENGINEER on their behalf.

This Agreement will be effective on ______________________, 19________.

OWNER ______________________________

____

By ______________________________

[CORPORATE SEAL]

Attest ______________________________

Address for giving notices

____

____

(If OWNER is a public body, attach evidence of authority to sign and resolution or other documents authorizing execution of Agreement.)

CONTRACTOR ______________________________

____

By ______________________________

[CORPORATE SEAL]

Attest ______________________________

Address for giving notices

____

____

License No. ______________________________

Agent for service of process: ______________________

____

(If CONTRACTOR is a corporation, attach evidence of authority to sign).

6

GLOSSARY OF TERMS

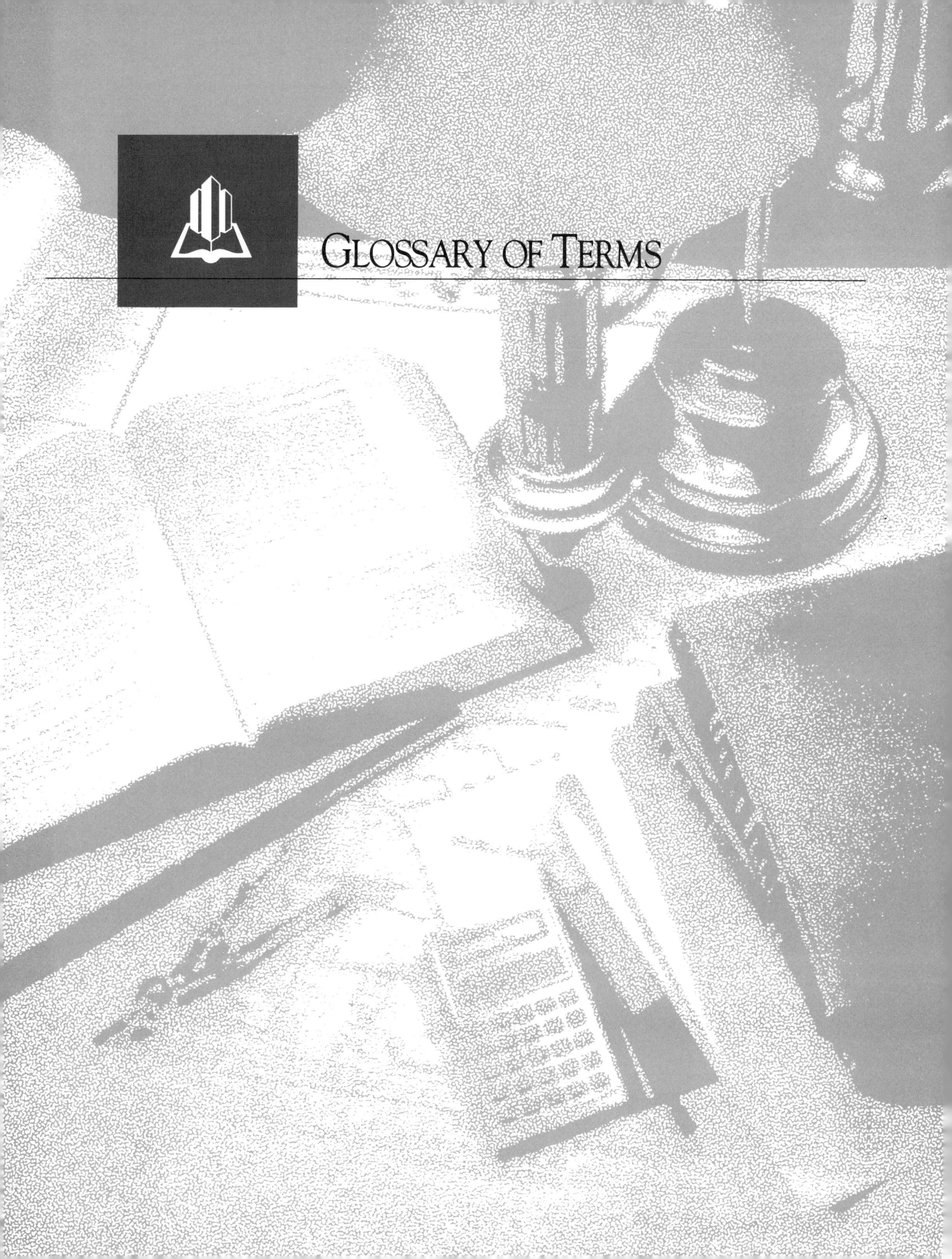

GLOSSARY OF TERMS

Absolute Ownership
The broadest form of estate in real property whereby the owner or owners possess all right, title, and interest in the property. The owner(s) has absolute control over the real property, which includes the right to sell it or give it away, in whole or in part, while alive; or to pass it to heirs or to others upon death.

Actual Authority
Under agency law, the authority that a principal confers on an agent.

Actual Causation
Under tort law, a plaintiff must prove actual causation in a negligence case by showing that the defendant's conduct was connected with the plaintiff's harm.

Adverse Possession
To come to own real property without purchasing it and without doing anything more than openly and physically possessing it for a long period of time in contravention of the rights of the true owner. Although state laws differ, 20 years is the typical length of time that must pass before the person in possession of the land becomes the owner.

Agency Relationship
The relationship between a principal and agent.

Agent
Under agency law, an agent is authorized by the principal to act on the principal's behalf. Generally, an agent's acts bind the principal as though the principal had acted directly. See **Principal**.

Air Rights
The exclusive right of real property owners to possess the air space above their land, as long as they comply with building and zoning laws.

American Institute of Architects (AIA)
A national society of registered architects founded in 1857. It is the single largest representative body of architects in the United States.

Anticipatory Breach of Contract
A term applied to the situation in which a party to a contract indicates to the other party, prior to the expected date of performance, that he or she

does not intend to perform under the contract.

Apparent Authority
Under agency law, the authority that exists when a principal's written or spoken words or conduct lead a third party to reasonably believe that the principal consents to the acts performed on its behalf by the agent.

Arbitration
The process by which parties agree to submit their disputes to the determination of a third, impartial party (referred to as the arbitrator), rather than pursuing their claims before a judge and jury in a court of law.

Architect
An individual who is educated, trained, and experienced in the art and technology of building planning and design, construction administration, and coordination of the specialized skills of structural, mechanical, and electrical design engineers. Use of the term is usually limited by law to those who have been licensed to practice architecture.

Articles of Incorporation (or **Charter)**
A standard form that incorporators are required to file with their state secretary as part of the process of creating a corporation. The articles of incorporation become public record and generally contain the corporation's name and address, its business purpose, the name and address of its managers, and information about the type and amount of shares of stock it will issue.

Assignment
A transfer of rights, frequently involving rights arising under a contract.

Authority
Under agency law, the power that a principal confers on his agent to act on his behalf. See **Actual Authority.**

Berne Convention
An international copyright treaty enacted in 1886.

Bid
A proposal or offer to do specified work for a stipulated sum. See **Offer.**

Board of Directors
A group of individuals, elected by the shareholders, who are ultimately responsible for the management of the corporation.

Breach of Contract
The failure to perform some contractually described obligation in accordance with the terms of the contract.

Bylaws
The internal rules and regulations of a corporation. Bylaws generally provide when and where the stockholders' annual meeting will take place, and rules regarding the election of the directors and officers and their duties and powers.

Capitalization
The total value of a corporation's equity.

Change Order
The client's written order to the contractor, issued after execution of the construction contract, which authorizes a change in the construction work and contract time and/or amount. Typically, the design professional also signs the change order and then issues it to the contractor.

Charter
See **Articles of Incorporation.**

Civil Law
The rights of private individuals and the legal procedures protecting those rights.

Civil Violation
A violation of civil law. See **Civil Law.**

Client
The party who retains the design professional and/or contractor to provide design and/or construction services in connection with a given project. The client is sometimes referred to as the owner.

Closed Corporation
A corporation with a limited number of shareholders.

Closing
The transfer of title to property. A transfer is completed or "closed" when a seller has given a deed to a buyer in return for receipt of the purchase price.

Collective Bargaining
The process by which employers and the representatives of employees (such as unions or other labor organizations) negotiate agreements with respect to wages, hours, and conditions of employment.

Common Law
Principles of conduct based on usage, custom, or commonly held beliefs as interpreted by the courts on a case-by-case basis. Common law is not founded on the will of legislative or rule-making bodies, but on accepted standards of conduct. See **Case Law**.

Comparative Negligence
An allocation of damages according to the relative (comparative) fault of the parties. Comparative negligence is applied in states with comparative negligence statutes.

Compensatory Damages
Damages intended to compensate the plaintiff by granting a monetary amount equal to the loss or injury suffered.

Condominium
A legal system by which individual parts, or units, of real property, such as apartments, stores, or offices, may be owned separately. Each unit owner obtains all the rights incidental to ownership of real property, and shares with the other owners rights to the common areas in the building, facilities, or land of the condominium.

Consequential Damages
Compensation for loss or injury that is not directly attributable to some wrongful act on the part of another, but is the result of one or more of the consequences of such act.

Consideration
The price, motive, or benefit that induces parties to form a contract. A contract is not legally enforceable without consideration.

Contract
A written or oral agreement between two or more parties consisting of one or more legally enforceable promises.

Contract Work Hours and Safety Standards Act
A federal labor law which regulates hours of work and overtime pay for laborers and mechanics on federally funded projects. The Contract Work Hours and Safety Standards Act applies to all contracts on projects valued at more than $2,500.

Contractor
The party who is hired to perform the required construction on a project.

Contribution
A principle in tort law under which a defendant may recover a portion of a judgment that has been rendered against him from other individuals whose negligence contributed to the plaintiff's injury.

Contributory Negligence
A type of negligence in which an injured person's own fault contributed to the harm caused by the defendant.

Cooperative
A form of multiple ownership in which a corporation is formed to own the land and the structures on the land. Unlike a condominium, the individual units of the building are not owned separately. The residents of the units own shares of stock in the corporation in proportion to the size or value of the unit they occupy.

Copeland Anti-Kickback Act
A federal law which makes it a criminal offense for an employer to force, intimidate, or threaten a person employed in constructing or repairing buildings or public works to return or decline to accept any part of his compensation for work performed.

Copyright
A right granted by statute to authors of original literary, artistic, and creative works. A copyright gives the author an exclusive privilege to copy and publish the work. Material that can be protected by a copyright includes literary material, speeches, computer programs, motion pictures, and architects' plans.

Copyright Act of 1976
A federal statute that protects the right to exclusive ownership of original works produced by writers, artists, composers, photographers, architects, and other creators.

Corporation
An entity, created by statute, which exists separate from its owners. A corporation has the capacity to exist indefinitely.

Counteroffer
An offeree's response to an offer which neither accepts it as offered, nor rejects it completely, but proposes modifications to the original offer.

Damages
A measure of monetary compensation that a court or arbitrator awards to a plaintiff for loss or injury suffered by the plaintiff's person, property, or other legally recognizable rights.

Davis-Bacon Act
A federal labor law which requires that laborers and mechanics (those who do physical work) on federally funded construction projects be paid no less than the local prevailing wages. The Davis-Bacon Act applies to all federally funded contracts valued at more than $2,000.

Debt
Money borrowed from others. An obligation or liability to pay or render something to someone else.

Deed
A legal document used to transfer ownership of real property.

Defamation
A statement that exposes a person to ridicule or contempt.

Defendant
An individual against whom a claim or action is brought. See **Plaintiff**.

Defense
An explanation or reason why a person or entity should not be held legally responsible (liable) as a result of claims made against them by another party.

Delegation
The transfer of duties under a contract. When a new party promises to perform the original party's duties under a contract.

Derivative Works
A work based on one or more preexisting works.

Design Professional
A term used generally to refer to architects; civil, structural, mechanical, electrical, plumbing, and heating, ventilating, and air conditioning engineers; interior designers; landscape architects; and others whose services have either traditionally been considered "professional" activities, require licensing or registration by the state, or otherwise require the knowledge and application of design principles appropriate to the problem at hand.

Direct Damages
Damages that can be expected to arise as a matter of course from a breach.

Discrimination in Employment Act
A federal law which makes it illegal for an employer to discriminate on the basis of an employee's age.

Dividends
A share of profits that a corporation pays to its stockholders.

Double Taxation
Taxing a corporation's income twice: once when the corporation pays taxes on its net income, and again when the shareholders pay taxes on their dividends.

Due Process of Law
A person's right to have the opportunity to enforce and protect his rights in a court that has power to hear and determine his case.

Duress
Under contract law, duress occurs when one party is physically forced or verbally threatened to sign a contract. The law will not enforce a contract signed under duress.

Duty
An obligation or standard of conduct established by statute, contract, or common law.

Easement
The limited right of one person to use another person's property in a particular way or for a particular purpose.

Eminent Domain
The power of states to take private property for legitimate public uses, regardless of the wishes of the owner of the real property.

Employer-Independent Contractor Relationship
An employment relationship between an employer and independent contractor. Such a relationship is usually not an agency relationship because an employer does not control the details of an independent contractor's work.

Encumbrances
A liability attached to real property. Mortgages, liens, and certain easements on real property are encumbrances.

Engineer

One who, by education, training, and experience, is skilled in the art and technology of site or building engineering and design. Use of the term is usually limited by law to those who have been duly licensed to practice engineering.

Engineers Joint Contract Documents Committee (EJCDC)

A committee formed in the late 1970s which has drafted standard contracts used on engineering projects.

Equal Pay Act

A federal law which prohibits employers who are subject to the Fair Labor Standards Act from discriminating between employees on the basis of sex.

Equitable Remedies

A remedy other than money damages which is based on a determination of what is fair and just.

Equity

The value of a business or property in excess of any mortgage, liens, or other charges and liabilities.

Escrow

Property (such as money, a deed, stock, or a writing) that is delivered by a person (called a **grantor**) to a third party to be held until certain conditions are fulfilled. When the conditions are satisfied, the third party releases the property to another person (called a **grantee**).

Estate

A legal right, title, or interest in real property. Estates in real property fall into two broad categories: ownership, and rights to possess and use real property owned by another.

Estate for Years

Any ownership of real property which is less than absolute. An estate for years runs for a fixed period of time (this may be any number of days, weeks, months, or years), after which ownership reverts to another.

Executed

Under contract law, a contract that has been signed by both parties.

Executor

A person appointed by the maker of a will to carry out his directions and requests in the will for the disposition of his property.

Express Authority

Actual authority that is given explicitly, either orally or in writing.

Express Contract

A contract that is formed when parties express, through written or spoken words, their agreement to be bound to its terms.

Express Warranty

A promise or a guarantee that a statement about some fact is true and will continue to be true, and is made by one party to another in a manner that is clearly stated, either orally or in writing.

Fair Labor Standards Act (FLSA)

A federal law which establishes minimum wage and hour standards for employers and employees. The Fair Labor Standards Act applies only to employers and employees engaged in commerce or in the production of goods for commerce.

Federal Employer Identification Number

A number assigned by the Internal Revenue Service to identify a business for tax purposes.

Federal Insurance Contribution Act (FICA)

A federal law which imposes a social security tax on employees, employers, and the self-employed.

Federal Unemployment Tax Act (FUTA)

A federal law that imposes an unemployment tax on employers.

Fictitious Name (or **Trade Name)**

A name of a business other than the owner's legal name. A fictitious name must be registered with the city or town in which business is conducted.

Fiduciary

One in whom the confidence and trust of another has been placed and who, as a result, owes a special duty to act in scrupulous good faith for the benefit and in the best interests of the other. See **Agent.**

Fixtures

Tangible goods that are permanently attached to, or embedded in, real property such that they are considered to be part of the real property.

General Conditions

A set of guidelines that define many of the rights, responsibilities, and limitations of authority of the owner and contractor, and include the general procedures governing the performance of the work.

General Liability Insurance

A broad form of liability insurance covering claims for bodily injury and property damage. This type of insurance combines, under one blanket policy, coverage for all liability exposures (except those specifically excluded), and automatically covers new and unknown hazards that may develop.

General Partners

See **General Partnership.**

General Partnership

An association of two or more people to conduct business for profit as co-owners. The participants in a general partnership, called **general partners**, share all aspects of the business including assets, profits/losses, liability, and management responsibility.

Heard Act

A law passed in 1894 which required contractors on federal projects to obtain a penal bond, which was a combined performance and labor and material payment bond. The bond gave subcontractors and suppliers a source of funds should the contractor fail to make payments to them. In 1934 the Heard Act was repealed and replaced with the Miller Act. See **Miller Act**.

Immunity

A defense to a negligence claim.

Implied Authority

Actual authority that is implied from a principal's conduct, and which he or she intends to confer.

Implied Contract

A contract that is formed when parties express, through conduct, their agreement to be bound to its terms. An agreement is inferred from the parties' conduct, rather than their express words.

Implied Warranty

A guarantee created not by the explicit oral or written statements of one party to another, but through a promise that is inferred under the law to exist by virtue of the nature of the transaction or the circumstances between the parties, or by the relative position one party holds with respect to another. A specific warranty assumed to be part of sales of goods in certain situations as provided under the UCC.

Implied Warranty of Fitness for a Particular Purpose

A warranty imposed by the UCC on sellers of goods who, at the time of contracting, have reason to know that the goods are to be used for a particular purpose and that the buyer is relying on his or her skill or judgment to select or furnish suitable goods.

Implied Warranty of Merchantability

Under the UCC, a warranty that is given by all merchants who sell goods.

Impossibility

Physical inability to perform obligations under a contract.

Impracticability

Inability to perform some task or contractual obligation at a commercially reasonable cost.

Incorporated

See **Incorporation.**

Incorporation

The process of forming a corporation. Incorporation procedures are governed by state statutes.

Incorporators

The individuals who form a corporation.

Indemnification

An obligation contractually assumed or legally imposed on one party to protect another against loss or damage from specific liabilities.

Indemnify

See **Indemnification.**

Infringement

Using trademark or copyrighted material without the owner's permission.

Injunctions

A court order instructing a party to stop, or to refrain from stopping, some activity.

Insurance

A contract, typically referred to as an insurance policy, in which the insurer, in return for the premium stated in the policy, agrees to pay the insured up to the limits specified in the policy for losses or damages incurred by the insured.

Integration Clause

A contract clause which states that the written agreement is complete, and all oral and other agreements are integrated, or merged, into it.

Intellectual Property

Property of the intellect, specifically, copyrights, trademarks, service marks, patents, and trade secrets.

Intentional Misrepresentation

An incorrect or false statement that is made deliberately.

Intentional Torts

Those acts performed intentionally by a party which result in harm to another.

Internal Revenue Service

An agency responsible for administering the federal government's tax laws.

Interstate Commerce
The buying, selling, or exchange of goods or property across state lines.

Intestate
To die without having made a will. If a person dies without a will, his or her property is distributed to the heirs under the laws of the state.

Inverse Condemnation
The loss in value of a parcel of real property due to the government's taking of a nearby parcel for public use.

Joint and Several Liability
Where one party is equally responsible to a third party for the acts and omissions of another party or parties.

Joint Tenancy
A form of ownership in which two or more people own equal shares of the same real property. Each co-owner, called a **joint tenant**, has the right to occupy and use all of the property and, therefore, must share with the other owners. The most important distinguishing feature of a joint tenancy is the right of survivorship. This means that when one of the owners dies, his or her share of ownership is split among the surviving owners, thereby increasing the surviving owners' proportionate shares.

Joint Tortfeasors
See **Joint and Several Liability**.

Joint Venture
Sometimes referred to as a contractual joint venture, whereby two or more parties join forces to form an entity with the legal characteristics of a partnership to achieve a specific objective.

Labor and Material Payment Bond
A contract between a contractor and a surety company in which the surety, in return for a premium paid by the contractor, agrees to pay subcontractors, laborers, and material suppliers amounts due to them for their materials and services under the terms of their contractual arrangements with the contractor, should the contractor default in his or her payment obligations to them.

Labor Law
Laws enacted by the federal government which regulate employment.

Land Support
The right of real property owners to expect that land that is in its natural state (without buildings, houses, or structures on it) will be physically supported by adjoining land.

Lanham Act
The primary source of federal trademark law, enacted in 1947.

Lawsuit
A claim brought by a party in court to obtain a judgment against another party for injuries that result from a wrongful act.

Lease
A written agreement between a property owner and a tenant which specifies the length of the tenancy and the rights and responsibilities of each party for a variety of matters. The tenancy ends automatically at the end of the period specified in the lease. A tenancy that is established by a lease is called a **tenancy under a lease.**

Legal Capacity
A person's ability to understand the nature and

effect of his or her conduct. Legal capacity is measured by age and mental competency.

Legal Causation
Under tort law, a plaintiff must prove legal causation in a negligence case by showing that the harm that occurred was a reasonably foreseeable consequence of the wrongful act.

Legal Value
Consideration that is legally sufficient to support a contract. See **Consideration.**

Liability
A situation in which one party is legally obligated to assume responsibility for another party's loss or burden. Liability is created when the law recognizes two elements: the existence of an enforceable legal duty to be performed by one party for the benefit of another; and the failure to perform the duty in accordance with applicable legal standards.

Liable
See **Liability**.

Libel
Defamatory language that is written or printed.

Limited Partners
Partners in a limited partnership who contribute capital and share profits, but have no decision-making power and no control over the day-to-day operations of the limited partnership's business.

Limited Partnership
A form of partnership that is a hybrid between a general partnership and a corporation, and which is composed of both general partners and limited partners.

Liquidated Damages
An amount of damages established in advance by parties to a contract which one party promises to pay to the other in the event that he or she breaches the contract.

Litigation
The process by which parties submit their disputes to the jurisdiction and procedures of federal or state courts for resolution.

Managing Partner
A partner who is responsible for a wide variety of day-to-day decisions on behalf of the partnership.

Marketable Title
Title of sufficient quality that it would not cause subsequent, reasonable, potential buyers to question whether or not they would obtain all rights of ownership or whether or not they could sell the property at a future time without question that they were the rightful owner.

Master
Under agency law, a person who controls or has the right to control another person's performance of work. See **Master-Servant Relationship**.

Master Deed
A deed that creates a condominium and which is used in the conveyancing of a condominium.

Master-Servant Relationship
An employment relationship between employer and employee. Under agency law, the master-servant relationship is an agency relationship in which the servant, as an agent, has been granted certain authority to act on behalf of the master (the principal). The master controls not only the servant's overall conduct, but the details of the servant's work.

Miller Act

A federal labor law which requires general contractors working on federally funded construction projects to obtain performance bonds and labor and material payment bonds to protect the interests of subcontractors and suppliers. The Miller Act applies to all United States government construction contracts valued at more than $25,000.

Misrepresentation

Under contract law, when a party to a contract knowingly states incorrect facts that are important and on which the other party to the contract relies. Misrepresentation is a defense to a breach of contract.

Mistake

Under contract law, a term applied when both parties misunderstood a contract's important terms at the time they entered into it, so as to render the contract unenforceable.

Mitigation of Damages

A duty that the law imposes on an injured party to make a reasonable effort to minimize his or her damages after an injury.

Modifications

A change to a contract which is made after the contract has been signed by both parties.

Money Awards

See **Damages, Money Damages.**

Money Damages

A monetary award that a party who has breached a contract is ordered by a court to pay as compensation to the nonbreaching party.

Multiple Ownership

A form of ownership whereby two or more people or entities own interests in the same real property at the same time. There are three basic forms of multiple ownership of real property. See **Tenancy in Common, Joint Tenancy**, and **Tenancy by the Entirety.**

Mutual Assent

The agreement of two or more parties to be bound to the terms of a contract. A contract is not legally enforceable without mutual assent.

National Labor Relations Act (NLRA)

A federal law which regulates collective bargaining between unions and employers.

National Labor Relations Board (NLRB)

An organization established by the United States government to enforce the National Labor Relations Act.

Negligence

The failure of a party to conform its conduct to the standard of care required by law. The law requires a person to exercise that degree of care which a reasonable person would exercise under the same or similar circumstances. As applied to design professionals, it is the failure to exercise that degree of skill, care, and diligence that other design professionals would exercise under the same or similar circumstances.

Nonconforming Uses

Existing uses of land or structures which do not conform to local zoning ordinances.

Novation

One party's agreement to release another party from a

contract in exchange for a new party as substitute.

Occupational Illnesses
A sickness or disease caused by exposure to toxic substances or other environmental hazards in the workplace.

Occupational Injuries
Injuries that result from work-related accidents.

Occupational Safety and Health Act (OSHA)
A federal law which requires employers to maintain a safe working environment that is free from hazards that could cause death or serious physical illness or injury. Employers who have one or more employees and who are "engaged in a business affecting interstate commerce" (the buying, selling, or exchange of goods or property across state lines) are required to comply with OSHA.

Occupational Safety and Health Administration
A branch of the United States Department of Labor which administers and enforces the Occupational Safety and Health Act.

Offer (or **Bid**)
A proposal or overture made by one person to another to perform a service or to do something that invites the other person's acceptance and the formation of a contract. The person who makes the offer is called an **offeror**, and the person who receives the offer is called an **offeree**. In the construction industry, a bid is an example of an offer.

Offeree
See **Offer.**

Offeror
See **Offer.**

Officers
Employees of the corporation who are elected by the board of directors to manage the daily affairs of the corporation.

Owner
See **Client.**

Parol Evidence
Oral or written evidence of statements made by parties to a contract prior to or at the same time as, the signing of the contract, which vary or contradict the writing in the contract.

Parol Evidence Rule
A rule which requires courts to exclude parol evidence when interpreting a contract. The rule is applied if there is no integration clause in the contract, but the court decides that the writing(s) was intended to be a final expression of the parties' agreement. See **Integration Clause** and **Parol Evidence.**

Partial Performance
Under contract law, a term applied when a party has not fully and completely performed all of his or her obligations of a contract. A party who has only partially performed its contractual obligations generally has no right to recover any part of the contract price from the other party. Partial performance is a breach of contract.

Partners
The members of a partnership.

Patent
A grant made by the federal government to an inventor which gives him or her (or a business entity) the exclusive right to make, use, or sell an invention in the United States for a stated period of time.

Penal Bond
A combined performance and labor and material payment bond.

Performance Bond
A bond issued by a surety company which guarantees the client that if the contractor fails to complete the project in accordance with the terms of the construction agreement, the surety company will either complete the contract itself, or arrange for a client-approved contractor to complete the contract. The surety company will pay the new contractor the amount required to finish the work, minus the unpaid amount under the original contract. However, the surety company is not obligated to pay more than the penal sum or limit of liability stated in the bond.

Personal Property
Generally, all property that is not real property. Tangible goods, such as furniture, cars, books, and equipment; and intangible goods, such as money, notes, bonds, and stocks.

Plain Meaning
A rule of contract interpretation in which a widely known word is given its ordinary meaning.

Plaintiff
The party that initiates a claim or action against another party. See **Defendant**.

Principal
Under agency law, a principal is one who empowers another to act on his or her behalf and to legally bind the principal as though the principal had acted directly. See **Agent**.

Prior Existing Duty
Under contract law, a duty that a party to a contract is already obligated to perform under an original contract, and which cannot be promised again in return for additional consideration (such as money).

Privity of Contract
See **Privity**.

Professional Corporation
A corporation established by a group of professionally licensed individuals to render a specific professional service.

Professional Liability Insurance
Protection afforded the design professional, in the amount stated in the policy, from damages and expenses (such as attorneys' fees and court costs) arising out of the negligent acts, errors, or omissions of the design professional, the design professional's employees, or others for whom the design professional is responsible.

Professional Standard of Care
That degree of learning and skill which the law requires all professionals to exhibit and which is ordinarily possessed by professionals in good standing in that profession in the same locality and under similar circumstances.

Proprietor
See **Sole Proprietorship.**

Proximate Cause
A legally sufficient causal connection between the breach of duty owed under the applicable standard of care and the injury suffered by the party to whom the duty was owed.

Punitive Damages
Also referred to as **exemplary damages**, punitive damages are awarded by a judge to a plaintiff not merely to compensate the plaintiff for losses incurred, but to punish the defendant for wrongful conduct and to use the plight

of the defendant as an example to potential wrongdoers.

Purchase and Sale Agreement
The written contract for the sale of real property. The statute of frauds requires that a contract for the sale of real property must be in writing. See **Statute of Frauds.**

Quantum Meruit
A method of recovering under a contract implied by law.

Quasi-Contract
An obligation imposed by law under which a party who received a benefit must pay the party who gave the benefit, despite the absence of an actual contract between them.

Ratification
Under agency law, a principal's authorization of an agent to perform a certain action *after* the agent has performed that action.

Real Estate
See **Real Property**.

Real Property (or **Real Estate**)
Land and anything growing on the land or constructed on it, such as buildings, stone walls, fences, driveways, garages, underground swimming pools, trees, shrubs, and gardens. Real property also includes **fixtures**, or items that are permanently affixed to real property, such as light fixtures, oil burners, fuse boxes, and plumbing. Real property does *not* include things that are grown on the land for the purpose of being sold, such as crops or trees from a tree farm. Such products that are to be severed from real property and sold are treated as goods.

Rehabilitation Act of 1973
A federal law which requires employers who have contracts with the federal government valued in excess of $2,500, to take affirmative action to hire and promote handicapped people who are qualified for available jobs.

Remedies
The manner in which a breach of contract is cured by a court or arbitrators.

Restitution
A court-ordered money award that is intended to restore the parties to their financial position as it existed before the contract was formed.

Right of Survivorship
A feature of a joint tenancy and a tenancy by the entirety. When one owner dies, the deceased owner's share of ownership is split among the surviving owners, thereby increasing the surviving owners' proportionate shares.

"S" Corporation
A corporation that has elected to be taxed like a partnership, in accordance with the provisions of Subchapter S of the Internal Revenue Code.

Servant
Under agency law, a person whose performance of work is controlled by another person. See **Master-Servant Relationship.**

Service Mark
See **Trademark**.

Share
A fractional part of a corporation's capitalization or stock.

Shareholder
See **Stockholder.**

Slander
Defamatory language that is spoken.

Sole Proprietorship
A business that is owned and controlled by one person (called a **proprietor**). A proprietorship may or may not have employees.

Special Use Permits
Permission granted to a landowner to use his or her property in a way that would not normally be allowed in a zoning district.

Standard Rules of Interpretation
Rules that a court applies to interpret ambiguous or conflicting language in a contract.

Statute
Enactment of the various federal, state, and local rule-making bodies such as Congress, state legislatures, and local planning and zoning boards

Statute of Limitations
A legislative enactment that bars a claim against a party unless the claim is brought within a specified period of time after the event giving rise to the claim has occurred, or the party bringing the claim knew or should have known of the event.

Statute of Repose
A legislative enactment that bars a claim against a party unless the claim is brought within a specified period of time following an event described in the statute, regardless of whether or not the statute of limitations period for that claim has expired.

Statutes of Frauds
Statutes enacted in the various states which require certain contracts to be in writing.

Stock
The total value of the equity in a corporation.

1244 Stock
In the event that a corporation's stock becomes worthless, it may offset up to $50,000 of loss against ordinary income in each taxable year, provided the corporation meets the requirements of Section 1244 of the Internal Revenue Code.

Stockholder
Owners of a specific number of shares of stock in a corporation.

Substantial Performance
A party's performance of most of its contractual obligations, which entitles them to payment of at least a portion of the contract price.

Subsurface Rights
The exclusive right of real property owners to use the ground beneath the surface of their land.

Surety Bond
A legal instrument under which one party agrees to answer to another party for the debt, default, or failure to perform of a third party.

Taft-Hartley Act
A 1947 amendment to the National Labor Relations Act, which imposed restrictions on unions.

Tenancy
A tenant's interest in real property. See **Tenant**.

Tenancy at Will
A tenancy that is established not by written agreement (a lease), but by oral agreement between the parties. Typically, the tenancy is for an unspecified amount of time. Either party may end the tenancy at any time. Rent is

often paid weekly or monthly, and such a pattern may imply a minimum term for the tenancy. If so, the tenancy can normally only be terminated at the end of an established interval (for instance, at the end of the week or month).

Tenancy by the Entirety
A special form of ownership that can only be formed between a husband and wife. When a married couple owns real property as tenants by the entirety, their interests in the property are equal, and each has a right to occupy and use the entire property. Neither spouse, acting alone, may sell its share of the property as long as the other spouse is alive. A tenancy by the entirety carries the right of survivorship; when one spouse dies, the deceased spouse's share of the property passes to the surviving spouse.

Tenancy in Common
A form of ownership among two or more people, each with equal rights to occupy and use all of the real property owned in common. Although the owners all may fully occupy and use the property, they do not necessarily have equal shares of ownership. A major distinguishing feature of a tenancy in common is that when one owner dies, his or her share does not pass to the surviving owner(s) but, instead, passes to those named in the deceased's will or to the deceased's heirs under the law if there is no will.

Tenancy Under a Lease
See **Lease**.

Tenant
In the broadest sense, one who possesses real property, under any kind of right or title, for any period of time. This broad meaning is used with respect to the different forms of multiple ownership. (See **Tenancy in Common**, **Joint Tenancy**, and **Tenancy by the Entirety**.) A more restrictive, but more common, use of the word tenant is one who temporarily possesses and uses another's real property.

Third-Party Beneficiary
A person who benefits from a contract but who is not a party to the contract.

Title
The right of ownership in real property.

Title Search
A search of the history of a title to real property to identify defects and encumbrances, if any.

Title VII of the Civil Rights Act of 1964
A federal law which prohibits employers from discriminating against employees. Under Title VII, discrimination is defined as treating employees differently with respect to compensation, promotions, job conditions, hiring, or discharge on the basis of race, color, religion, sex, or national origin.

Tort
A private or civil wrong or injury, other than a breach of contract, for which the court will provide a remedy in the form of an action for damages. A tort is not a criminal offense. It is a civil offense that arises not merely by the agreement of the parties, but by operation of law.

Trade Name
The name selected for a business or corporation. A trade name is not ordinarily considered a trademark; however, with years of

continuous use, it may develop into a trademark. See **Fictitious Name.**

Trade Secret
Valuable information that a business takes precautions to keep secret. A trade secret could be a formula, pattern, device, or report.

Trademark (or **Service Mark**)
A word, name, symbol, device, or any combination of these which manufacturers and people who provide services use to identify their goods and services and to distinguish them from the goods and services of others.

Trademark Revision Act
An amendment to the Lanham Act.

Trespass
To enter real property owned by others without having a right to do so.

Unconscionability
With respect to contracts, where one party takes advantage of another party's inexperience, or does not give a fair explanation of the contract terms. A contract that is unfair or unconscionable is not enforceable.

"Under Seal"
A procedure for determining whether or not consideration has been provided to form a contract. A seal is a substitute for consideration. In most states, however, there must be consideration even if the contract has a seal on it.

Uniform Commercial Code (UCC)
A model law developed to govern commercial transactions.

Uniform Partnership Act
A model law developed to govern partnerships.

United States Patent and Trademark Office
A federal agency in the Department of Commerce which reviews patent and trademark applications, registers trademarks, and issues patents.

Variance
Permission granted to a landowner to depart from the specific requirements of a zoning ordinance. A variance allows a landowner to use his or her land differently than specified in the ordinance.

Vicarious Liability
An individual or business entity who is held liable as a result of another person's actions or failure to act.

Warranty
A promise that a statement about some fact is true and will continue to be true. With respect to the Uniform Commercial Code, a description of goods, including promises or factual statements (not opinions) relating to the goods, which are intended to induce a buyer to enter a contract.

Water Rights
The right of real property owners to make reasonable changes in the flow of water that is flooding their land.

Works Made for Hire
A rule applied to works produced within the scope of a job whereby the employer is considered the creator and owner of his or her employees' works.

Zoning
A form of public control over land use.

INDEX

INDEX

A

E

J

L